Optimization with PDE Constraints

MATHEMATICAL MODELLING:
Theory and Applications

VOLUME 23

This series is aimed at publishing work dealing with the definition, development and application of fundamental theory and methodology, computational and algorithmic implementations and comprehensive empirical studies in mathematical modelling. Work on new mathematics inspired by the construction of mathematical models, combining theory and experiment and furthering the understanding of the systems being modelled are particularly welcomed.

Manuscripts to be considered for publication lie within the following, non-exhaustive list of areas: mathematical modelling in engineering, industrial mathematics, control theory, operations research, decision theory, economic modelling, mathematical programming, mathematical system theory, geophysical sciences, climate modelling, environmental processes, mathematical modelling in psychology, political science, sociology and behavioural sciences, mathematical biology, mathematical ecology, image processing, computer vision, artificial intelligence, fuzzy systems, and approximate reasoning, genetic algorithms, neural networks, expert systems, pattern recognition, clustering, chaos and fractals.

Original monographs, comprehensive surveys as well as edited collections will be considered for publication.

For other titles published in this series, go to
www.springer.com/series/6299

Optimization with PDE Constraints

M. Hinze
Universität Hamburg
Germany

R. Pinnau
Technische Universität Kaiserslautern
Germany

M. Ulbrich
Technische Universität München
Germany

and

S. Ulbrich
Technische Universität Darmstadt
Germany

Springer

Michael Hinze
Dept. Mathematik
Universität Hamburg
Bundesstr. 55
20146 Hamburg
Germany
michael.hinze@uni-hamburg.de

Rene Pinnau
FB Mathematik
TU Kaiserslautern
Erwin-Schrödinger-Str.
67663 Kaiserslautern
Gebäude 48
Germany
pinnau@mathematik.uni-kl.de

Michael Ulbrich
Technische Universität München
Fakultät für Mathematik
Lehrstuhl für Mathematische Optimierung
Boltzmannstr. 3
85748 Garching
Germany
mulbrich@ma.tum.de

Stefan Ulbrich
Technische Universität Darmstadt
Fachbereich Mathematik, AG10
Schloßgartenstr. 7
64289 Darmstadt
Germany
ulbrich@mathematik.tu-darmstadt.de

ISBN 978-1-4020-8838-4 e-ISBN 978-1-4020-8839-1

Library of Congress Control Number: 2008934394

Printed on acid-free paper.

9 8 7 6 5 4 3 2 1

springer.com

Preface

Solving optimization problems subject to constraints given in terms of partial differential equations (PDEs) with additional constraints on the controls and/or states is one of the most challenging problems in the context of industrial, medical and economical applications, where the transition from model-based numerical simulations to model-based design and optimal control is crucial. For the treatment of such optimization problems the interaction of optimization techniques and numerical simulation plays a central role. After proper discretization, the number of optimization variables varies between 10^3 and 10^{10}. It is only very recently that the enormous advances in computing power have made it possible to attack problems of this size. However, in order to accomplish this task it is crucial to utilize and further explore the specific mathematical structure of optimization problems with PDE constraints, and to develop new mathematical approaches concerning mathematical analysis, structure exploiting algorithms, and discretization, with a special focus on prototype applications.

The present book provides a modern introduction to the rapidly developing mathematical field of optimization with PDE constraints. The first chapter introduces to the analytical background and optimality theory for optimization problems with PDEs. Optimization problems with PDE-constraints are posed in infinite dimensional spaces. Therefore, functional analytic techniques, function space theory, as well as existence- and uniqueness results for the underlying PDE are essential to study the existence of optimal solutions and to derive optimality conditions. These results form the foundation of efficient optimization methods in function space, their adequate numerical realization, mesh independence results and error estimators. The chapter starts with an introduction to the necessary background in functional analysis, Sobolev spaces and the theory of weak solutions for elliptic and parabolic PDEs. These ingredients are then applied to study PDE-constrained optimization problems. Existence results for optimal controls, derivative computations by the sensitivity and adjoint approaches and optimality conditions for problems with control-, state- and general constraints are considered. All concepts are illustrated by elliptic and parabolic optimal control problems. Finally, the optimal control of instationary incompressible Navier-Stokes flow is considered.

The second chapter presents a selection of important algorithms for optimization problems with partial differential equations. The development and analysis of these methods is carried out in a Banach space setting. This chapter starts with introducing a general framework for achieving global convergence. Then, several variants of generalized Newton methods are derived and analyzed. In particular, necessary and sufficient conditions for fast local convergence are derived. Based on this, the concept of semismooth Newton methods for operator equations is introduced. It is shown how complementarity conditions, variational inequalities, and optimality systems can be reformulated as semismooth operator equations. Applications to constrained optimal control problems are discussed, in particular for elliptic partial differential equations and for flow control problems governed by the incompressible instationary Navier-Stokes equations. As a further important concept, the formulation of optimality systems as generalized equations is addressed and the Josephy-Newton method for generalized equations is analyzed. This provides an elegant basis for the motivation and analysis of sequential quadratic programming (SQP) algorithms. The second chapter concludes with a short outline of recent algorithmic advances for state constrained problems and a brief discussion of several further aspects.

The third chapter gives an introduction to discrete concepts for optimization problems with PDE constraints. As models for the state elliptic and parabolic PDEs are considered which are well understood from the analytical point of view. This allows to focus on structural aspects in discretization. The approaches *First discretize, then optimize* and *First optimize, then discretize* are compared and discussed, and a variational discrete concept is introduced which avoids explicit discretization of the controls. Special focus is taken on the treatment of constraints. This includes general constraints on the control, and also pointwise bounds on the state, and on the gradient of the state. The chapter presents the error analysis for the variational discrete concept and accomplishes the analytical findings with numerical examples which confirm the analytical results.

Finally, the fourth chapter is devoted to the study of two industrial applications, in which optimization with partial differential equations plays a crucial role. It provides a survey of the different mathematical settings which can be handled with the general optimal control calculus presented in the previous chapters. The chapter focuses on large scale optimal control problems involving two well-known types of partial differential equations, namely elliptic and parabolic ones. Since real world applications lead generally to mathematically quite involved problems, in particular nonlinear systems of equations are studied. The examples are chosen in such a way that they are up-to-date and modern mathematical tools are used for their specific solution. The industrial fields covered are modern semiconductor design and glass production. Each section starts with a modeling part to introduce the underlying physics and mathematical models, which are then followed by the analytical and numerical study of the related optimal control problems.

Acknowledgements

This Book is based on lecture notes of the autumn school *Modellierung und Optimierung mit Partiellen Differentialgleichungen* which was held in September 2005 at the Universität Hamburg. It was supported by the Collaborative Research Center 609, located at the Technische Universität Dresden, and by the Priority Programme 1253, both sponsored by the Deutsche Forschungsgemeinschaft, as well as by the Schwerpunkt Optimierung und Approximation at the Department Mathematik of the Universität Hamburg. All support is gratefully acknowledged.

Finally we would like to thank a number of colleagues whose collaboration and support influenced the material presented in this book. These include Günter Bärwolff, Martin Burger, Klaus Deckelnick, John Dennis, Michael Ferris, Andreas

Günther, Matthias Heinkenschloss, Michael Herty, Michael Hintermüller, Axel Klar, Karl Kunisch, Günter Leugering, Ulrich Matthes, Christian Meyer, Danny Ralph, Ekkehard Sachs, Anton Schiela, Alexander Schulze, Mohammed Seaid, Norbert Siedow, Guido Thömmes, Philippe Toint, Fredi Tröltzsch, Andreas Unterreiter, Luís Vicente, and Morten Vierling.

Contents

Chapter 1
Analytical Background and Optimality Theory

Stefan Ulbrich

Abstract This chapter provides an introduction to the analytical background and optimality theory for optimization problems with partial differential equations (PDEs). Optimization problems with PDE-constraints are posed in infinite dimensional spaces. Therefore, functional analytic techniques, function space theory, as well as existence- and uniqueness results for the underlying PDE are essential to study the existence of optimal solutions and to derive optimality conditions. These results form the foundation of efficient optimization methods in function space, their adequate numerical realization, mesh independence results and error estimators. The chapter provides first an introduction to the necessary background in functional analysis, Sobolev spaces and the theory of weak solutions for elliptic and parabolic PDEs. These ingredients are then applied to study PDE-constrained optimization problems. Existence results for optimal controls, derivative computations by the sensitivity and adjoint approaches and optimality conditions for problems with control-, state- and general constraints are considered. All concepts are illustrated by elliptic and parabolic optimal control problems. Finally, the optimal control of instationary incompressible Navier-Stokes flow is considered.

1.1 Introduction and Examples

1.1.1 Introduction

The modelling and numerical simulation of complex systems plays an important role in physics, engineering, mechanics, chemistry, medicine, finance, and in other disciplines. Very often, mathematical models of complex systems result in partial differential equations (PDEs). For example heat flow, diffusion, wave propagation, fluid flow, elastic deformation, option prices and many other phenomena can be modelled by using PDEs.

In most applications, the ultimate goal is not only the mathematical modelling and numerical simulation of the complex system, but rather the optimization or optimal control of the considered process. Typical examples are the optimal control of a thermal treatment in cancer therapy and the optimal shape design of an aircraft.

S. Ulbrich (✉)
Fachbereich Mathematik, TU Darmstadt, Darmstadt, Germany
e-mail: ulbrich@mathematik.tu-darmstadt.de

M. Hinze et al., *Optimization with PDE Constraints*,
Mathematical Modelling: Theory and Applications 23,
DOI 10.1007/978-1-4020-8839-1_1, © Springer Science + Business Media B.V. 2009

The resulting optimization problems are very complex and a thorough mathematical analysis is necessary to design efficient solution methods.

There exist many different types of partial differential equations. We will focus on linear and semilinear elliptic and parabolic PDEs. For these PDEs the existence and regularity of solutions is well understood and we will be able to develop a fairly complete theory.

Abstractly speaking, we will consider problems of the following form

$$\min_{w \in W} J(w) \quad \text{subject to} \quad e(w) = 0, \qquad c(w) \in \mathcal{K}, \quad w \in \mathcal{C}, \tag{1.1}$$

where $J : W \to \mathbb{R}$ is the objective function, $e : W \to Z$ and $c : W \to R$ are operators, W, Z, R are real Banach spaces, $\mathcal{K} \subset R$ is a closed convex cone, and $\mathcal{C} \subset W$ is a closed convex set.

In most cases, the spaces W, Z and R are (generalized) function spaces and the operator equation $e(w) = 0$ represents a PDE or a system of coupled PDEs. The constraint

$$c(w) \in \mathcal{K}$$

is considered as an abstract inequality constraint. Sometimes (e.g., in the case of bound constraints), it will be convenient to write these constraints in the form $w \in \mathcal{C}$, where $\mathcal{C} \subset W$ is a closed convex set and to drop the inequality constraints:

$$\min_{w \in W} J(w) \quad \text{s.t.} \quad e(w) = 0, \quad w \in \mathcal{C}. \tag{1.2}$$

Here "s.t." abbreviates "subject to".

To get the connection to finite dimensional optimization, consider the case

$$W = \mathbb{R}^n, \qquad Z = \mathbb{R}^l, \qquad R = \mathbb{R}^m, \qquad \mathcal{K} = (-\infty, 0]^m, \qquad \mathcal{C} = \mathbb{R}^n.$$

Then the problem (1.1) becomes a nonlinear optimization problem

$$\min_{w \in W} J(w) \quad \text{s.t.} \quad e(w) = 0, \qquad c(w) \leq 0. \tag{1.3}$$

Very often, we will have additional structure: The optimization variable w admits a natural splitting into two parts, a state $y \in Y$ and a control (or design) $u \in U$, where Y and U are Banach spaces. Then $W = Y \times U$, $w = (y, u)$, and the problem reads

$$\min_{y \in Y, u \in U} J(y, u) \quad \text{s.t.} \quad e(y, u) = 0, \qquad c(y, u) \in \mathcal{K}. \tag{1.4}$$

Here, $y \in Y$ describes the state (e.g., the velocity field of a fluid) of the considered system, which is described by the equation $e(y, u) = 0$ (in our context usually a PDE). The control (or design, depending on the application) $u \in U$ is a parameter that shall be adapted in an optimal way.

The splitting of the optimization variable $w = (y, u)$ into a state and a control is typical in the optimization with PDE-constraints. Problems with this structure are called *optimal control problems*. In most cases we will consider, the state equation

$e(y,u) = 0$ admits, for every $u \in U$, a unique corresponding solution $y(u)$, because the state equation is a well posed PDE for y in which u appears as a parameter. Several examples will follow below.

We use the finite-dimensional problem (1.3) to give a teaser about important questions we will be concerned with.

1. Existence of solutions

Denote by J^* the optimal objective function value. First, we show, using the properties of the problem at hand, that J^* is achievable and finite. Then, we consider a minimizing sequence (w^k), i.e., $e(w^k) = 0$, $c(w^k) \le 0$, $J(w^k) \to J^*$. Next, we prove that (w^k) is bounded (which has to be verified for the problem at hand). Now we do something that *only works in finite dimensions*: We conclude that, due to boundedness, (w^k) contains a convergent subsequence $(w_k)_K \to \bar{w}$. Assuming the continuity of J, e and c we see that

$$J(\bar{w}) = \lim_{K \ni k \to \infty} J(w^k) = J^*, \qquad e(\bar{w}) = \lim_{K \ni k \to \infty} e(w^k) = 0,$$
$$c(\bar{w}) = \lim_{K \ni k \to \infty} c(w^k) \le 0.$$

Therefore, $\bar{w}$ solves the problem.

We note that for doing the same in Banach space, we need a replacement for the compactness argument, which will lead us to weak convergence and weak compactness. Furthermore, we need the continuity of the function J and of the operators e and c with respect to the norm topology and/or the weak topology.

2. Uniqueness

Uniqueness usually relies on strict convexity of the problem, i.e., J strictly convex, e linear and c_i convex. This approach can be easily transfered to the infinite-dimensional case.

3. Optimality conditions

Assuming continuous differentiability of the functions J, c, and e, and that the constraints satisfy a regularity condition on the constraints, called *constraint qualification* (CQ) at the solution, the following first-order optimality conditions hold true at a solution $\bar{w}$:

Karush-Kuhn-Tucker conditions:

There exist Lagrange multipliers $\bar{p} \in \mathbb{R}^l$ and $\bar{\lambda} \in \mathbb{R}^m$ such that $(\bar{w}, \bar{p}, \bar{\lambda})$ solves the following KKT-system:

$$\begin{aligned} \nabla J(\bar{w}) + c'(\bar{w})^T \bar{\lambda} + e'(\bar{w})^T \bar{p} &= 0, \\ e(\bar{w}) &= 0, \\ c(\bar{w}) \le 0, \quad \bar{\lambda} \ge 0, \quad c(\bar{w})^T \bar{\lambda} &= 0. \end{aligned}$$

Here, the column vector $\nabla J(w) = J'(w)^T \in \mathbb{R}^n$ is the gradient of J (corresponding to the euclidean inner product) and $c'(w) \in \mathbb{R}^{m\times n}$, $e'(w) \in \mathbb{R}^{l\times n}$ are the Jacobian matrices of c and e.

All really efficient optimization algorithms for (1.3) build upon these KKT-conditions. Therefore, it will be very important to derive first order optimality conditions for the infinite-dimensional problem (1.1). Since the KKT-conditions involve derivatives, we have to extend the notion of differentiability to operators between Banach spaces. This will lead us to the concept of Fréchet-differentiability. For concrete problems, the appropriate choice of the underlying function spaces is not always obvious, but it is crucial for being able to prove the Fréchet-differentiability of the function J and the operators c, e and for verifying constraint qualifications.

4. Optimization algorithms

As already said, modern optimization algorithms are based on solving the KKT system. For instance, for problems without inequality constraints, the KKT system reduces to the following $(n+l) \times (n+l)$ system of equations:

$$G(w, p) := \begin{pmatrix} \nabla J(w) + e'(w)^T p \\ e(w) \end{pmatrix} = 0. \tag{1.5}$$

One of the most powerful algorithms for equality constrained optimization, the Lagrange-Newton method, consists in applying Newton's method to the equation (1.5):

Lagrange-Newton method:

For $k = 0, 1, 2, \dots$:

1. STOP if $G(w^k, p^k) = 0$.
2. Compute $s^k = \binom{s_w^k}{s_p^k}$ by solving

$$G'(w^k, p^k)s^k = -G(w^k, p^k)$$

and set $w^{k+1} := w^k + s_w^k$, $p^{k+1} := p^k + s_p^k$.

Since G involves first derivatives, the matrix $G'(w, p)$ involves second derivatives. For the development of Lagrange-Newton methods for the problem class (1.1) we thus need second derivatives of J and e.

There are many more aspects that will be covered, but for the time being we have given sufficient motivation for the material to follow.

1.1.2 Examples for Optimization Problems with PDEs

We give several simple, but illustrative examples for optimization problems with PDEs.

1.1.3 Optimization of a Stationary Heating Process

Consider a solid body occupying the domain $\Omega \subset \mathbb{R}^3$. Let $y(x)$, $x \in \Omega$ denote the temperature of the body at the point x.

We want to heat or cool the body in such a way that the temperature distribution y coincides as good as possible with a desired temperature distribution $y_d : \Omega \to \mathbb{R}$.

1.1.3.1 Boundary Control

If we apply a temperature distribution $u : \partial\Omega \to \mathbb{R}$ to the boundary of Ω then the temperature distribution y in the body is given by the *Laplace equation*

$$-\Delta y(x) = 0, \quad x \in \Omega \tag{1.6}$$

together with the boundary condition of *Robin type*

$$\kappa \frac{\partial y}{\partial \nu}(x) = \beta(x)(u(x) - y(x)), \quad x \in \partial\Omega,$$

where $\kappa > 0$ is the heat conduction coefficient of the material of the body and $\beta : \partial\Omega \to (0, \infty)$ is a positive function modelling the heat transfer coefficient to the exterior.

Here, Δy is the Laplace operator defined by

$$\Delta y(x) = \sum_{i=1}^{n} y_{x_i x_i}(x)$$

with the abbreviation

$$y_{x_i x_i}(x) = \frac{\partial^2 y}{\partial x_i^2}(x)$$

and $\frac{\partial y}{\partial \nu}(x)$ is the derivative in the direction of the outer unit normal $\nu(x)$ of $\partial\Omega$ at x, i.e.,

$$\frac{\partial y}{\partial \nu}(x) = \nabla y(x) \cdot \nu(x), \quad x \in \partial\Omega.$$

As we will see, the Laplace equation (1.6) is an *elliptic* partial differential equation of second order.

In practice, the control u is restricted by additional constraints, for example by upper and lower bounds

$$a(x) \le u(x) \le b(x), \quad x \in \partial\Omega.$$

To minimize the distance of the actual and desired temperature y and y_d, we consider the following optimization problem.

$$\min J(y,u) := \frac{1}{2}\int_\Omega (y(x) - y_d(x))^2\, dx + \frac{\alpha}{2}\int_{\partial\Omega} u(x)^2\, dS(x)$$

$$\begin{aligned} \text{subject to} \quad & -\Delta y = 0 \quad \text{on } \Omega, \\ & \frac{\partial y}{\partial \nu} = \frac{\beta}{\kappa}(u - y) \quad \text{on } \partial\Omega, \end{aligned} \qquad \text{(State equation)}$$

$$a \le u \le b \quad \text{on } \partial\Omega \quad \text{(Control constraints)}.$$

The first term in the objective functional $J(y,u)$ measures the distance of y and y_d, the second term is a regularization term with parameter $\alpha \ge 0$ (typically $\alpha \in [10^{-5}, 10^{-3}]$), which leads to improved smoothness properties of the optimal control for $\alpha > 0$.

If we set

$$e(y,u) := \begin{pmatrix} -\Delta y \\ \frac{\partial y}{\partial \nu} - \frac{\beta}{\kappa}(u - y) \end{pmatrix}, \qquad c(y,u) := \begin{pmatrix} a - u \\ u - b \end{pmatrix},$$

where Y and U are appropriately chosen Banach spaces of functions

$$y : \Omega \to \mathbb{R}, \qquad u : \partial\Omega \to \mathbb{R},$$

$Z = Z_1 \times Z_2$ with appropriately chosen Banach spaces Z_1, Z_2 of functions

$$z_1 : \Omega \to \mathbb{R}, \qquad z_2 : \partial\Omega \to \mathbb{R},$$

$R = U \times U$, and

$$\mathcal{K} = \{(v_1, v_2) \in R : v_i(x) \le 0,\ x \in \partial\Omega,\ i = 1,2\},$$

then the above optimal control problem is of the form (1.1).

One of the crucial points will be to choose the above function spaces in such a way that J, e, and c are continuous and sufficiently often differentiable, to ensure existence of solutions, the availability of optimality conditions, etc.

In many practical situations elliptic PDEs do not possess classical solutions. The theory of weak solutions of elliptic PDEs and an appropriate function space setting will be introduced in Sect. 1.3.1. Optimality conditions for control constraints will be given in Sect. 1.7.2.3 and for state constraints in Sect. 1.7.3.5.

1.1.3.2 Boundary Control with Radiation Boundary

If we take heat radiation at the boundary of the body into account, we obtain a nonlinear Stefan-Boltzmann boundary condition. This leads to the semilinear state equation (i.e., the highest order term is still linear)

$$\begin{aligned} -\Delta y &= 0 \quad \text{on } \Omega, \\ \frac{\partial y}{\partial \nu} &= \frac{\beta}{\kappa}(u^4 - y^4) \quad \text{on } \partial\Omega. \end{aligned}$$

This is a problem of the form (1.1) with

$$e(y,u) := \begin{pmatrix} -\Delta y \\ \frac{\partial y}{\partial \nu} - \frac{\beta}{\kappa}(u^4 - y^4) \end{pmatrix}$$

and the rest as before.

1.1.3.3 Distributed Control

Instead of heating at the boundary it is in some applications also possible to apply a distributed heat source as control. This can for example be achieved by using electro-magnetic induction.

If the boundary temperature is zero then we obtain, similar as above, the problem

$$\min J(y,u) := \frac{1}{2}\int_{\Omega} (y(x) - y_d(x))^2\, dx + \frac{\alpha}{2}\int_{\Omega} u(x)^2\, dx$$

$$\text{subject to} \quad -\Delta y = \gamma u \quad \text{on } \Omega,$$
$$y = 0 \quad \text{on } \partial\Omega,$$
$$a \le u \le b \quad \text{on } \Omega.$$

Here, the coefficient $\gamma : \Omega \to [0,\infty)$ weights the control. The choice $\gamma = 1_{\Omega_c}$ for some control region $\Omega_c \subset \Omega$ restricts the action of the control to the control region Ω_c.

If we assume a surrounding temperature y_a then the state equation changes to

$$-\Delta y = \gamma u \quad \text{on } \partial\Omega,$$
$$\frac{\partial y}{\partial \nu} = \frac{\beta}{\kappa}(y_a - y) \quad \text{on } \partial\Omega.$$

1.1.3.4 Problems with State Constraints

In addition to control constraint also *state constraints*

$$l \le y \le r$$

with functions $l < r$ are of practical interest. They are much harder to handle than control constraints.

1.1.4 Optimization of an Unsteady Heating Processes

In most applications, heating processes are time-dependent. Then the temperature $y : [0,T] \times \Omega \to \mathbb{R}$ depends on space and time. We set

$$\Omega_T := (0,T) \times \Omega, \qquad \Sigma_T := (0,T) \times \partial\Omega.$$

1.1.4.1 Boundary Control

Let y_d be a desired temperature distribution at the end time T and y_0 be the initial temperature of the body. To find a control $u : \Sigma_T \to \mathbb{R}$ that minimizes the distance of the actual temperature $y(T,\cdot)$ at the end time and the desired temperature y_d, we consider similar as above the following optimization problem.

$$\min J(y,u) := \frac{1}{2}\int_{\Omega} (y(T,x) - y_d(x))^2\, dx + \frac{\alpha}{2}\int_0^T \int_{\partial\Omega} u(t,x)^2\, dS(x)\, dt$$

$$\begin{aligned} \text{subject to}\quad & y_t - \Delta y = 0 \quad \text{on } \Omega_T, \\ & \frac{\partial y}{\partial \nu} = \frac{\beta}{\kappa}(u - y) \quad \text{on } \Sigma_T, \\ & y(0,x) = y_0(x) \quad \text{on } \Omega \\ & a \le u \le b \quad \text{on } \Sigma_T. \end{aligned}$$

Here, y_t denotes the partial derivative with respect to time and Δy is the Laplace operator in space. The PDE

$$y_t - \Delta y = 0$$

is called *heat equation* and is the prototype of a *parabolic* partial differential equation.

Similarly, unsteady boundary control with radiation and unsteady distributed control can be derived from the steady counterparts.

The theory of weak solutions for parabolic PDEs and an appropriate functional analytic setting will be introduced in 1.3.2. Optimality conditions for control constraints will be given in Sect. 1.7.2 and for state constraints in Sect. 1.7.3.5.

Optimal control problems with linear state equation and quadratic objective function are called *linear-quadratic*. If the PDE is nonlinear in lower order terms then the PDE is called *semilinear*.

1.1.5 Optimal Design

A very important discipline is optimal design. Here, the objective is to optimize the shape of some object. A typical example is the optimal design of a wing or a whole airplane with respect to certain objective, e.g., minimal drag, maximum lift or a combination of both.

Depending on the quality of the mathematical model employed, the flow around a wing is described by the Euler equations or (better) by the compressible Navier-Stokes equations. Both are systems of PDEs. A change of the wing shape would then result in a change of the spatial flow domain Ω and thus, the design parameter is the domain Ω itself or a description of it (e.g. a surface describing the shape of the wing). Optimization problems of this type are very challenging.

Therefore, we look here at a much simpler example:

Consider a very thin elastic membrane spanned over the domain $\Omega \subset \mathbb{R}^2$. Its thickness $u(x) > 0$, $x \in \Omega$, varies (but is very small). At the boundary of Ω, the membrane is clamped at the level $x_3 = 0$.

Given a vertical force distribution $f : \Omega \to \mathbb{R}$ acting from below, the membrane takes the equilibrium position described by the graph of the function $y : \Omega \to \mathbb{R}$ (we assume that the thickness is negligibly compared to the displacement). For small displacement, the mathematical model for this membrane then is given by the following elliptic PDE:

$$\begin{aligned} -\operatorname{div}(u\nabla y) &= f \quad \text{on } \Omega, \\ y &= 0 \quad \text{on } \partial\Omega. \end{aligned}$$

Here, $\operatorname{div} v = \sum_i (v_i)_{x_i}$ denotes the divergence of $v : \Omega \to \mathbb{R}^2$.

The design goal consists in finding an optimal thickness u subject to the thickness constraints

$$a(x) \le u(x) \le b(x) \quad x \in \Omega$$

and the volume constraint

$$\int_\Omega u(x)\,dx \le V$$

such that the *compliance*

$$J(y) = \int_\Omega f(x)y(x)\,dx$$

of the membrane is as small as possible. The smaller the compliance, the stiffer the membrane with respect to the load f. We obtain the following optimal design problem

$$\begin{aligned} \min J(y) &:= \int_\Omega f(x)y(x)\,dx \\ \text{subject to} \quad -\operatorname{div}(u\nabla y) &= f \quad \text{on } \Omega, \\ y &= 0 \quad \text{on } \partial\Omega, \\ a \le u &\le b \quad \text{on } \Omega, \\ \int_\Omega u(x)\,dx &\le V. \end{aligned}$$

1.2 Linear Functional Analysis and Sobolev Spaces

We have already mentioned that PDEs do in practical relevant situations, e.g. for discontinuous right hand sides, not necessarily have classical solutions. A satisfactory solution theory can be developed by using Sobolev spaces and functional analysis. This will also provide a suitable framework to derive optimality conditions.

We recall first several basics on Banach and Hilbert spaces. Details can be found in any book on linear functional analysis, e.g., [4, 83, 115, 146, 149].

1.2.1 Banach and Hilbert Spaces

1.2.1.1 Basic Definitions

Definition 1.1 (Norm, Banach space) Let X be a real vector space.

(i) A mapping $\|\cdot\| : X \mapsto [0, \infty)$ is a *norm* on X, if
   (1) $\|u\| = 0 \Longleftrightarrow u = 0$,
   (2) $\|\lambda u\| = |\lambda| \|u\| \ \forall u \in X, \lambda \in \mathbb{R}$,
   (3) $\|u + v\| \leq \|u\| + \|v\| \ \forall u, v \in X$.

(ii) A normed real vector space X is called (real) *Banach space* if it is complete, i.e., if any Cauchy sequence (u_n) has a limit $u \in X$, more precisely, if $\lim_{m,n\to\infty} \|u_m - u_n\| = 0$ then there is $u \in X$ with $\lim_{n\to\infty} \|u_n - u\| = 0$.

Example 1.1

1. For $\Omega \subset \mathbb{R}^n$ consider the function space

$$C(\Omega) = \{u : \Omega \to \mathbb{R} : u \text{ continuous}\}.$$

   If Ω is bounded then $C(\bar{\Omega})$ is a Banach space with the sup-norm

$$\|u\|_{C(\bar{\Omega})} = \sup_{x \in \bar{\Omega}} |u(x)|.$$

2. Let $\Omega \subset \mathbb{R}^n$ be open. For a multiindex $\alpha = (\alpha_1, \ldots, \alpha_n) \in \mathbb{N}_0^n$ we define its order by $|\alpha| := \sum_{i=1}^n \alpha_i$ and associate the $|\alpha|$-th order partial derivative at x

$$D^\alpha u(x) := \frac{\partial^{|\alpha|} u}{\partial x_1^{\alpha_1} \cdots \partial x_n^{\alpha_n}}(x).$$

   We define

$$C^k(\Omega) = \left\{u \in C(\Omega) : D^\alpha u \in C(\Omega) \text{ for } |\alpha| \leq k\right\}.$$

   For $\Omega \subset \mathbb{R}^n$ open and bounded let

$$C^k(\bar{\Omega}) = \left\{u \in C^k(\Omega) : D^\alpha u \text{ has a continuous extension to } \bar{\Omega} \text{ for } |\alpha| \leq k\right\}.$$

   Then the spaces $C^k(\bar{\Omega})$ are Banach spaces with the norm

$$\|u\|_{C^k(\bar{\Omega})} := \sum_{|\alpha| \leq k} \|D^\alpha u\|_{C(\bar{\Omega})}.$$

Definition 1.2 (Inner product, Hilbert space) Let H be a real vector space.

(i) A mapping $(\cdot,\cdot) : H \times H \mapsto \mathbb{R}$ is an *inner product* on H, if
  (1) $(u, v) = (v, u)\ \forall u, v \in H$,
  (2) For every $v \in H$ the mapping $u \in H \mapsto (u, v)$ is linear,
  (3) $(u, u) \geq 0\ \forall u \in H$ and $(u, u) = 0 \Longleftrightarrow u = 0$.

(ii) A vector space H with inner product $(\cdot,\cdot)$ and associated norm

$$\|u\| := \sqrt{(u, u)}$$

is called *Pre-Hilbert space*.

(iii) A Pre-Hilbert space $(H, (\cdot,\cdot))$ is called *Hilbert space* if it is complete under its norm $\|u\| := \sqrt{(u, u)}$.

Example 1.2 Let $\emptyset \neq \Omega \subset \mathbb{R}^n$ be open and bounded. Then $(C(\bar{\Omega}), (\cdot,\cdot)_{L^2})$ is a Pre-Hilbert space with the L^2-inner product

$$(u, v)_{L^2} = \int_{\Omega} u(x)\, v(x)\, dx.$$

Note that $(C(\bar{\Omega}), (\cdot,\cdot)_{L^2})$ is not complete (why?).

Theorem 1.1 *Let H be a Pre-Hilbert space. Then the* Cauchy-Schwarz inequality *holds*

$$|(u, v)| \leq \|u\|\,\|v\| \quad \forall u, v \in H.$$

Many spaces arising in applications have the important property that they contain a countable dense subset.

Definition 1.3 A Banach space X is called *separable* if it contains a countable dense subset. I.e., there exists $Y = \{x_i \in X : i \in \mathbb{N}\} \subset X$ such that

$$\forall x \in X,\ \forall \varepsilon > 0 : \exists y \in Y: \quad \|x - y\|_X < \varepsilon.$$

Example 1.3 For bounded Ω the space $C(\bar{\Omega})$ is separable (the polynomials with rational coefficients are dense by Weierstraß's approximation theorem).

1.2.1.2 Linear Operators and Dual Space

Obviously, linear partial differential operators define linear mappings between function spaces. We recall the following definition.

Definition 1.4 (Linear operator) Let X, Y be normed real vector spaces with norms $\|\cdot\|_X$, $\|\cdot\|_Y$.

(i) A mapping $A : X \to Y$ is called *linear operator* if it satisfies

$$A(\lambda u + \mu v) = \lambda Au + \mu Av \quad \forall u, v \in X,\ \lambda, \mu \in \mathbb{R}.$$

The *range* of A is defined by

$$R(A) := \{y \in Y : \exists x \in X : y = Ax\}$$

and the *null space* of A by

$$N(A) := \{x \in X : Ax = 0\}.$$

(ii) By $\mathcal{L}(X, Y)$ we denote the space of all linear operators $A : X \to Y$ that are bounded in the sense that

$$\|A\|_{X,Y} := \sup_{\|u\|_X = 1} \|Au\|_Y < \infty.$$

$\mathcal{L}(X, Y)$ is a normed space with the *operator norm* $\|\cdot\|_{X,Y}$.

Theorem 1.2 *If Y is a Banach space then $\mathcal{L}(X, Y)$ is a Banach space.*

The following theorem tells us, as a corollary, that if Y is a Banach space, any operator $A \in \mathcal{L}(X, Y)$ is determined uniquely by its action on a dense subspace.

Theorem 1.3 *Let X be a normed space, Y be a Banach space and let $U \subset X$ be a dense subspace (carrying the same norm as X). Then for all $A \in \mathcal{L}(U, Y)$, there exists a unique extension $\tilde{A} \in \mathcal{L}(X, Y)$ with $\tilde{A}|_U = A$. For this extension, there holds $\|\tilde{A}\|_{X,Y} = \|A\|_{U,Y}$.*

Definition 1.5 (Linear functionals, dual space)

(i) Let X be a Banach space. A bounded linear operator $u^* : X \to \mathbb{R}$, i.e., $u^* \in \mathcal{L}(X, \mathbb{R})$ is called a *bounded linear functional* on X.
(ii) The space $X^* := \mathcal{L}(X, \mathbb{R})$ of linear functionals on X is called *dual space* of X and is (by Theorem 1.2) a Banach space with the operator norm

$$\|u^*\| := \sup_{\|u\|_X = 1} |u^*(u)|.$$

(iii) We use the notation

$$\langle u^*, u \rangle_{X^*,X} := u^*(u).$$

$\langle \cdot, \cdot \rangle_{X^*,X}$ is called the *dual pairing* of X^* and X.

Of essential importance is the following

Theorem 1.4 (Riesz representation theorem) *The dual space H^* of a Hilbert space H is isometric to H itself. More precisely, for every $v \in H$ the linear functional u^* defined by*

$$\langle u^*, u\rangle_{H^*,H} := (v, u)_H \quad \forall u \in H$$

is in H^* with norm $\|u^*\|_{H^*} = \|v\|_H$. Vice versa, for any $u^* \in H^*$ there exists a unique $v \in H$ such that

$$\langle u^*, u\rangle_{H^*,H} = (v, u)_H \quad \forall u \in H$$

and $\|u^*\|_{H^*} = \|v\|_H$.

In particular, a Hilbert space is reflexive (we will introduce this later in Definition 1.17).

Definition 1.6 Let X, Y be Banach spaces. Then for an operator $A \in \mathcal{L}(X, Y)$ the dual operator $A^* \in \mathcal{L}(Y^*, X^*)$ is defined by

$$\langle A^*u, v\rangle_{X^*,X} = \langle u, Av\rangle_{Y^*,Y} \quad \forall u \in Y^*,\ v \in X.$$

It is easy to check that $\|A^*\|_{Y^*,X^*} = \|A\|_{X,Y}$.

1.2.2 Sobolev Spaces

To develop a satisfactory theory for PDEs, it is necessary to replace the classical function spaces $C^k(\bar{\Omega})$ by *Sobolev spaces* $W^{k,p}(\Omega)$. Roughly speaking, the space $W^{k,p}(\Omega)$ consists of all functions $u \in L^p(\Omega)$ that possess (weak) partial derivatives $D^\alpha u \in L^p(\Omega)$ for $|\alpha| \le k$.

We recall

1.2.2.1 Lebesgue Spaces

Our aim is to characterize the function space $L^p(\Omega)$ that is complete under the L^p-norm, where

$$\|u\|_{L^p(\Omega)} = \left(\int_\Omega |u(x)|^p\,dx\right)^{1/p}, \quad p \in [1, \infty),$$

$$\|u\|_{L^\infty(\Omega)} = \operatorname*{ess\,sup}_{x\in\Omega} |u(x)| \quad \Big(= \sup_{x\in\Omega} |u(x)| \text{ for } u \in C(\bar{\Omega})\Big).$$

1.2.2.2 Lebesgue Measurable Functions and Lebesgue Integral

Definition 1.7 A collection $\mathcal{S} \subset \mathcal{P}(\mathbb{R}^n)$ of subsets of $\mathbb{R}^n$ is called σ-algebra on $\mathbb{R}^n$ if

(i) $\emptyset, \mathbb{R}^n \in \mathcal{S}$,
(ii) $A \in \mathcal{S}$ implies $\mathbb{R}^n \setminus A \in \mathcal{S}$,
(iii) If $(A_k)_{k\in\mathbb{N}} \subset \mathcal{S}$ then $\bigcup_{k=1}^{\infty} A_k \in \mathcal{S}$.

A measure $\mu : \mathcal{S} \to [0, \infty]$ is a mapping with the following properties:

(i) $\mu(\emptyset) = 0$.
(ii) If $(A_k)_{k\in\mathbb{N}} \subset \mathcal{S}$ is a sequence of pairwise disjoint sets then

$$\mu\left(\bigcup_{k=1}^{\infty} A_k\right) = \sum_{k=1}^{\infty} \mu(A_k) \quad (\sigma\text{-additivity}).$$

Of essential importance is the σ-algebra of Lebesgue measurable sets with corresponding Lebesgue measure.

Theorem 1.5 *There exists the σ-algebra $\mathcal{B}_n$ of Lebesgue measurable sets on $\mathbb{R}^n$ and the Lebesgue measure $\mu : \mathcal{B}_n \to [0, \infty]$ with the properties*:

(i) *$\mathcal{B}_n$ contains all open sets (and thus all closed sets).*
(ii) *μ is a measure on $\mathcal{B}_n$.*
(iii) *If B is any ball in $\mathbb{R}^n$ then $\mu(B) = |B|$, where $|B|$ denotes the volume of B.*
(iv) *If $A \subset B$ with $B \in \mathcal{B}_n$ and $\mu(B) = 0$ then $A \in \mathcal{B}_n$ and $\mu(A) = 0$ (i.e., $(\mathbb{R}^n, \mathcal{B}_n, \mu)$ is a complete measure space).*

The sets $A \in \mathcal{B}_n$ are called Lebesgue measurable.

Notation If some property holds for all $x \in \mathbb{R} \setminus N$ with $N \subset \mathcal{B}_n$, $\mu(N) = 0$, then we say that it holds almost everywhere (a.e.).

Definition 1.8 We say that $f : \mathbb{R}^n \to [-\infty, \infty]$ is *Lebesgue measurable* if

$$\left\{x \in \mathbb{R}^n : f(x) > \alpha\right\} \in \mathcal{B}_n \quad \forall \alpha \in \mathbb{R}.$$

If $A \in \mathcal{B}_n$ and $f : A \to [-\infty, \infty]$ then we call f Lebesgue measurable on A if $f 1_A$ is Lebesgue measurable. Here, we use the convention $f 1_A = f$ on A and $f 1_A = 0$ otherwise.

Remark 1.1 For open $\Omega \subset \mathbb{R}^n$ any function $f \in C(\Omega)$ is measurable, since $\{f > \alpha\}$ is relatively open in Ω (and thus open).

We now extend the classical integral to Lebesgue measurable functions.

Definition 1.9 The set of nonnegative elementary functions is defined by

$$E_+(\mathbb{R}^n) := \left\{ f = \sum_{k=1}^{m} \alpha_k 1_{A_k} : (A_k)_{1\le k\le m} \subset \mathcal{B}_n \text{ pairwise disjoint}, \alpha_k \ge 0, m \in \mathbb{N} \right\}.$$

The Lebesgue integral of $f = \sum_{k=1}^{m} \alpha_k 1_{A_k} \in E_+(\mathbb{R}^n)$ is defined by

$$\int_{\mathbb{R}^n} f(x)\, d\mu(x) := \sum_{k=1}^{m} \alpha_k \mu(A_k).$$

An extension to general Lebesgue measurable functions is obtained by the following fact.

Lemma 1.1 *For any sequence (f_k) of Lebesgue measurable functions also*

$$\sup_k f_k, \qquad \inf_k f_k, \qquad \limsup_{k\to\infty} f_k, \qquad \liminf_{k\to\infty} f_k$$

are Lebesgue measurable.

For any Lebesgue measurable function $f \geq 0$ there exists a monotone increasing sequence $(f_k)_{k\in\mathbb{N}} \subset E_+(\mathbb{R}^n)$ with $f = \sup_k f_k$.

This motivates the following definition of the Lebesgue integral.

Definition 1.10 (Lebesgue integral)

(i) For a nonnegative Lebesgue measurable function $f : \mathbb{R}^n \to [0, \infty]$ we define the Lebesgue integral of f by

$$\int_{\mathbb{R}^n} f(x)\, d\mu(x) := \sup_k \int_{\mathbb{R}^n} f_k(x)\, d\mu(x),$$

where $(f_k)_{k\in\mathbb{N}} \subset E_+(\mathbb{R}^n)$ is a monotone increasing sequence with $f = \sup_k f_k$.

(ii) For a Lebesgue measurable function $f : \mathbb{R}^n \to [-\infty, \infty]$ we define the Lebesgue integral by

$$\int_{\mathbb{R}^n} f(x)\, d\mu(x) := \int_{\mathbb{R}^n} f^+(x)\, d\mu(x) - \int_{\mathbb{R}^n} f^-(x)\, d\mu(x)$$

with $f^+ = \max(f, 0)$, $f^- = \max(-f, 0)$ if one of the terms on the right hand side is finite. In this case f is called *integrable*.

(iii) If $A \in \mathcal{B}_n$ and $f : A \to [-\infty, \infty]$ is a function such that $f 1_A$ is integrable then we define

$$\int_A f(x)\, d\mu(x) := \int_{\mathbb{R}^n} f(x) 1_A(x)\, d\mu(x).$$

Notation In the sequel we will write dx instead of $d\mu(x)$.

1.2.2.3 Definition of Lebesgue Spaces

Clearly, we can extend the L^p-norm to Lebesgue measurable functions.

Definition 1.11 Let $\Omega \in \mathcal{B}_n$. We define for $p \in [1, \infty)$ the seminorm

$$\|u\|_{L^p(\Omega)} := \left(\int_\Omega |u(x)|^p dx \right)^{1/p}$$

and

$$\|u\|_{L^\infty(\Omega)} := \operatorname*{ess\,sup}_{x \in \Omega} |u(x)| := \inf\{\alpha \geq 0 : \mu(\{|u| > \alpha\}) = 0\}.$$

Now, for $1 \leq p \leq \infty$ we define the spaces

$$\mathcal{L}^p(\Omega) := \left\{ u : \Omega \to \mathbb{R} \text{ Lebesgue measurable} : \|u\|_{L^p(\Omega)} < \infty \right\}.$$

These are not normed space since there exist measurable functions $u : \Omega \to \mathbb{R}$, $u \neq 0$, with $\|u\|_{L^p} = 0$.

We use the equivalence relation

$$u \sim v \quad \text{in } L^p(\Omega) \quad :\Longleftrightarrow \quad \|u - v\|_{L^p(\Omega)} = 0 \quad \overset{\text{by Lemma 1.2}}{\Longleftrightarrow} \quad u = v \quad \text{a.e.}$$

to define $L^p(\Omega) = \mathcal{L}^p(\Omega)/\sim$ as the space of equivalence classes of a.e. identical functions, equipped with the norm $\|\cdot\|_{L^p}$.

Finally we define

$$\mathcal{L}^p_{\text{loc}}(\Omega) := \left\{ u : \Omega \to \mathbb{R} \text{ Lebesgue measurable} : u \in \mathcal{L}^p(K) \text{ for all } K \subset \Omega \text{ compact} \right\}$$

and set $L^p_{\text{loc}}(\Omega) := \mathcal{L}^p_{\text{loc}}(\Omega)/\sim$.

In the following we will consider elements of L^p and L^p_{loc} as functions that are known up to a set of measure zero.

Remark 1.2 It is easy to see that $L^p(\Omega) \subset L^1_{\text{loc}}(\Omega)$ for all $p \in [1, \infty]$.

We collect several important facts of Lebesgue spaces.

Lemma 1.2 *For all* $u, v \in \mathcal{L}^p(\Omega)$, $p \in [1, \infty]$, *we have*

$$\|u - v\|_{L^p} = 0 \iff u = v \quad \textit{a.e.}$$

Proof The assertion is obvious for $p = \infty$. For $p \in [1, \infty)$ let $w = u - v$.

"$\Longrightarrow$:" We have for all $k \in \mathbb{N}$

$$0 = \|w\|_{L^p} \geq \frac{1}{k} \mu(\{|w| \geq 1/k\})^{1/p}.$$

Hence $\mu(\{w \geq 1/k\}) = 0$ and consequently

$$\mu(w \neq 0) = \mu \left(\bigcup_{k=1}^\infty \{|w| \geq 1/k\} \right) \leq \sum_{k=1}^\infty \mu(\{|w| \geq 1/k\}) = 0.$$

"$\Longleftarrow$:" If $w = 0$ a.e. then $|w|^p = 0$ on $\mathbb{R}^n \setminus N$ for some N with $\mu(N) = 0$. Hence, $|w|^p = \sup_k w_k$ with $(w_k) \subset E_+(\mathbb{R}^n)$, where $w_k = 0$ on $\mathbb{R}^n \setminus N$. Hence $\int_{\mathbb{R}^n} w_k \, dx = 0$ and consequently $\int_{\mathbb{R}^n} |w|^p dx = 0$.

Theorem 1.6 (Fischer-Riesz) *The spaces $L^p(\Omega)$, $p \in [1, \infty]$, are Banach spaces. The space $L^2(\Omega)$ is a Hilbert space with inner product*

$$(u, v)_{L^2} := \int_\Omega uv \, dx.$$

Lemma 1.3 (Hölder inequality) *Let $\Omega \in \mathcal{B}_n$. Then for all $p \in [1, \infty]$ we have with the dual exponent $q \in [1, \infty]$ satisfying $\frac{1}{p} + \frac{1}{q} = 1$ for all $u \in L^p(\Omega)$ and $v \in L^q(\Omega)$ the* Hölder inequality

$$uv \in L^1(\Omega) \quad \textit{and} \quad \|uv\|_{L^1} \le \|u\|_{L^p} \|v\|_{L^q}.$$

Now we can characterize the dual space of L^p-spaces.

Theorem 1.7 *Let $\Omega \in \mathcal{B}_n$, $p \in [1, \infty)$ and $q \in (1, \infty]$ the dual exponent satisfying $\frac{1}{p} + \frac{1}{q} = 1$. Then the dual space $(L^p(\Omega))^*$ can be identified with $L^q(\Omega)$ by means of the isometric isomorphism*

$$v \in L^q(\Omega) \mapsto u^* \in (L^p(\Omega))^*, \quad \textit{where } \langle u^*, u \rangle_{(L^p)^*, L^p} := \int_\Omega u(x)v(x) \, dx.$$

Remark 1.3 Note however that L^1 is only a subspace of $(L^\infty)^*$.

1.2.2.4 Density Results and Convergence Theorems

A fundamental result is the following:

Theorem 1.8 (Dominated convergence theorem) *Let $\Omega \in \mathcal{B}_n$. Assume that $f_k : \Omega \to \mathbb{R}$ are measurable with*

$$f_k \to f \quad \textit{a.e.} \quad \textit{and} \quad |f_k| \le g \quad \textit{a.e.}$$

with a function $g \in L^1(\Omega)$. Then $f_k, f \in L^1(\Omega)$ and

$$\int_\Omega f_k \, dx \to \int_\Omega f \, dx, \qquad f_k \to f \quad \textit{in } L^1(\Omega).$$

Next we state the important fact that the set of "nice" functions

$$C_c^\infty(\Omega) := \left\{ u \in C^\infty(\bar{\Omega}) : \mathrm{supp}(u) \subset \Omega \text{ compact} \right\}$$

is actually dense in $L^p(\Omega)$ for all $p \in [1, \infty)$.

Lemma 1.4 *Let $\Omega \subset \mathbb{R}^n$ be open. Then $C_c^\infty(\Omega)$ is dense in $L^p(\Omega)$ for all $p \in [1, \infty)$.*

A quite immediate consequence is the following useful result.

Lemma 1.5 *Let $\Omega \subset \mathbb{R}^n$ be open and $f \in L^1_{\text{loc}}(\Omega)$ with*

$$\int_\Omega f(x)\varphi(x)\,dx = 0 \quad \forall \varphi \in C_c^\infty(\Omega).$$

Then $f = 0$ a.e.

1.2.2.5 Weak Derivatives

The definition of weak derivatives is motivated by the fact that for any function $u \in C^k(\bar{\Omega})$ and any multiindex $\alpha \in \mathbb{N}_0^n$, $|\alpha| \le k$, the identity holds (integrate $|\alpha|$-times by parts)

$$\int_\Omega D^\alpha u \varphi\,dx = (-1)^{|\alpha|} \int_\Omega u D^\alpha \varphi\,dx, \quad \forall \varphi \in C_c^\infty(\Omega). \tag{1.7}$$

This motivates the definition

Definition 1.12 Let $\Omega \subset \mathbb{R}^n$ be open and let $u \in L^1_{\text{loc}}(\Omega)$. If there exists a function $w \in L^1_{\text{loc}}(\Omega)$ such that

$$\int_\Omega w\varphi\,dx = (-1)^{|\alpha|} \int_\Omega u D^\alpha \varphi\,dx, \quad \forall \varphi \in C_c^\infty(\Omega) \tag{1.8}$$

then $D^\alpha u := w$ is called the α-th weak partial derivative of u.

Remark 1.4

1. By Lemma 1.5, (1.8) determines the weak derivative $D^\alpha u \in L^1_{\text{loc}}(\Omega)$ uniquely.
2. For $u \in C^k(\bar{\Omega})$ and $\alpha \in \mathbb{N}_0^n$, $|\alpha| \le k$, the classical derivative $w = D^\alpha u$ satisfies (1.7) and thus (1.8). Hence, the weak derivative is consistent with the classical derivative.

1.2.2.6 Regular Domains and Integration by Parts

Let $\Omega \subset \mathbb{R}^n$ be open. For $k \in \mathbb{N}_0$ and $\beta \in (0, 1]$ let

$$C^{k,\beta}(\bar{\Omega}) = \left\{u \in C^k(\bar{\Omega}) : D^\alpha u \ \beta\text{-Hölder continuous for } |\alpha| = k\right\}.$$

Here, $f : \bar{\Omega} \to \mathbb{R}$ is β-Hölder continuous if there exists a constant $C > 0$ such that

$$|f(x) - f(y)| \leq C\|x - y\|_2^\beta \quad \forall x, y \in \bar{\Omega},$$

where $\|\cdot\|_2$ denotes the euclidean norm on $\mathbb{R}^n$. Of course, 1-Hölder continuity is Lipschitz continuity.

We set $C^{k,0}(\bar{\Omega}) := C^k(\bar{\Omega})$. If Ω is bounded then $C^{k,\beta}(\bar{\Omega})$ is a Banach space with the norm

$$\|u\|_{C^{k,\beta}(\bar{\Omega})} := \|u\|_{C^k(\bar{\Omega})} + \sum_{|\alpha|=k} \sup_{x,y\in\bar{\Omega},x\neq y} \frac{|D^\alpha u(x) - D^\alpha u(y)|}{\|x - y\|_2^\beta}.$$

Definition 1.13 ($C^{k,\beta}$-boundary, unit normal field) Let $\Omega \subset \mathbb{R}^n$ be open and bounded.

(a) We say that Ω has a $C^{k,\beta}$*-boundary*, $k \in \mathbb{N}_0 \cup \{\infty\}$, $0 \leq \beta \leq 1$, if for any $x \in \partial\Omega$ there exists $r > 0$, $l \in \{1, \ldots, n\}$, $\sigma \in \{-1, +1\}$, and a function $\gamma \in C^{k,\beta}(\mathbb{R}^{n-1})$ such that

$$\Omega \cap B(x; r) = \{y \in B(x; r) : \sigma y_l < \gamma(y_1, \ldots y_{l-1}, y_{l+1}, \ldots, y_n)\},$$

where $B(x; r)$ denotes the open ball around x with radius r. Instead of $C^{0,1}$-boundary we say also *Lipschitz-boundary*.

(b) If $\partial\Omega$ is $C^{0,1}$ then we can define a.e. the *unit outer normal field* $\nu : \partial\Omega \to \mathbb{R}^n$, where $\nu(x)$, $\|\nu(x)\|_2 = 1$, is the outward pointing unit normal of $\partial\Omega$ at x.

(c) Let $\partial\Omega$ be $C^{0,1}$. We call the directional derivative

$$\frac{\partial u}{\partial \nu}(x) := \nu(x) \cdot \nabla u(x), \quad x \in \partial\Omega,$$

the *normal derivative* of u.

We recall the Gauß-Green theorem (integration by parts formula).

Theorem 1.9 *Let $\Omega \subset \mathbb{R}^n$ be open and bounded with $C^{0,1}$-boundary. Then for all $u, v \in C^1(\bar{\Omega})$*

$$\int_\Omega u_{x_i}(x)v(x)\,dx = -\int_\Omega u(x)v_{x_i}(x)\,dx + \int_{\partial\Omega} u(x)v(x)\nu_i(x)\,dS(x).$$

1.2.2.7 Sobolev Spaces

We will now introduce subspaces $W^{k,p}(\Omega)$ of functions $u \in L^p(\Omega)$, for which the weak derivatives $D^\alpha u$, $|\alpha| \leq k$, are in $L^p(\Omega)$.

Definition 1.14 Let $\Omega \subset \mathbb{R}^n$ be open. For $k \in \mathbb{N}_0$, $p \in [1, \infty]$, we define the *Sobolev space* $W^{k,p}(\Omega)$ by

$$W^{k,p}(\Omega) = \{u \in L^p(\Omega) : u \text{ has weak derivatives } D^\alpha u \in L^p(\Omega) \text{ for all } |\alpha| \le k\} \tag{1.9}$$

equipped with the norm

$$\|u\|_{W^{k,p}(\Omega)} := \Big(\sum_{|\alpha|\le k} \|D^\alpha u\|_{L^p}^p\Big)^{1/p}, \quad p \in [1, \infty),$$

$$\|u\|_{W^{k,\infty}(\Omega)} := \sum_{|\alpha|\le k} \|D^\alpha u\|_{L^\infty(\Omega)}.$$

Notations

1. In the case $p = 2$ one writes $H^k(\Omega) := W^{k,2}(\Omega)$. We note that $W^{0,p}(\Omega) = L^p(\Omega)$ for $p \in [1, \infty]$.
2. For weak partial derivatives we use also the notation $u_{x_i}, u_{x_i x_j}, u_{x_i x_j x_k}, \ldots$.
3. For $u \in H^1(\Omega)$ we set

$$\nabla u(x) = \begin{pmatrix} u_{x_1}(x) \\ \vdots \\ u_{x_n}(x) \end{pmatrix}.$$

One can show that the following density results hold.

Theorem 1.10 *Let $\Omega \subset \mathbb{R}^n$ be open. Then the following holds.*

(i) *The set $C^\infty(\Omega) \cap W^{k,p}(\Omega)$, $k \in \mathbb{N}_0$, $1 \le p < \infty$, is dense in $W^{k,p}(\Omega)$. Hence, $W^{k,p}(\Omega)$ is the completion of $\{u \in C^\infty(\Omega) : \|u\|_{W^{k,p}} < \infty\}$ with respect to the norm $\|\cdot\|_{W^{k,p}}$.*
(ii) *If Ω is a bounded domain with $C^{0,1}$-boundary then $C^\infty(\bar{\Omega})$ is dense in $W^{k,p}(\Omega)$, $k \in \mathbb{N}_0$, $1 \le p < \infty$.*

Remark 1.5 Simple examples show that weak differentiability does not necessarily ensure continuity. We have for example with $\Omega := B(0; 1)$ and $u(x) := \|x\|^{-\beta}$ that

$$u \in W^{1,p}(\Omega) \iff \beta < \frac{n-p}{p}.$$

Theorem 1.11 *Let $\Omega \subset \mathbb{R}^n$ be open, $k \in \mathbb{N}_0$, and $p \in [1, \infty]$. Then $W^{k,p}(\Omega)$ is a Banach space.*

Moreover, the space $H^k(\Omega) = W^{k,2}(\Omega)$ is a Hilbert space with inner product

$$(u, v)_{H^k(\Omega)} = \sum_{|\alpha|\le k} (D^\alpha u, D^\alpha v)_{L^2(\Omega)}.$$

To incorporate homogeneous boundary conditions already in the function space we define the following subspace.

Definition 1.15 Let $\Omega \subset \mathbb{R}^n$ be open. For $k \in \mathbb{N}_0$, $p \in [1, \infty]$, we denote by

$$W_0^{k,p}(\Omega)$$

the closure of $C_c^\infty(\Omega)$ in $W^{k,p}(\Omega)$ (i.e., for any $u \in W_0^{k,p}(\Omega)$ there exists a sequence $(\varphi_i) \subset C_c^\infty(\Omega)$ with $\lim_{i\to\infty} \|u - \varphi_i\|_{W^{k,p}(\Omega)} = 0$). The space is equipped with the same norm as $W^{k,p}(\Omega)$ and is a Banach space. The space $H_0^k(\Omega) = W_0^{k,2}(\Omega)$ is a Hilbert space.

Remark 1.6 $W_0^{k,p}(\Omega)$ contains exactly all $u \in W^{1,p}(\Omega)$ such that $D^\alpha u = 0$ for $|\alpha| \le k-1$ on $\partial\Omega$ with an appropriate interpretation of the *traces* $D^\alpha u|_{\partial\Omega}$.

We consider next the appropriate assignment of boundary values (so called *boundary traces*) for functions $u \in W^{k,p}(\Omega)$ if Ω has Lipschitz-boundary.

If $u \in W^{k,p}(\Omega) \cap C(\bar{\Omega})$ then the boundary values can be defined in the classical sense by using the continuous extension. However, since $\partial\Omega$ is a set of measure zero and functions $u \in W^{k,p}(\Omega)$ are only determinded up to a set of measure zero, the definition of boundary values requires care. We resolve the problem by defining a *trace operator.*

Theorem 1.12 *Assume that $\Omega \subset \mathbb{R}^n$ is open and bounded with Lipschitz-boundary. Then for all $p \in [1, \infty]$ there exists a unique bounded linear operator*

$$T : W^{1,p}(\Omega) \to L^p(\partial\Omega)$$

such that

$$Tu = u|_{\partial\Omega} \quad \forall u \in W^{1,p}(\Omega) \cap C(\bar{\Omega}).$$

Here, $\|T\|_{W^{1,p}(\Omega),L^p(\partial\Omega)}$ depends only on Ω and p. Tu is called the trace *of u on $\partial\Omega$.*

1.2.2.8 Poincaré's Inequality

We have seen that the trace of functions in $H_0^k(\Omega)$, $k \ge 0$, vanishes. For the treatment of boundary value problems it will be useful that the semi-norm

$$|u|_{H^k(\Omega)} := \left(\sum_{|\alpha|=k} \|D^\alpha u\|_{L^2}^2 \right)^{1/2} \tag{1.10}$$

defines an equivalent norm on the Hilbert space $H_0^k(\Omega)$. It is obvious that

$$|u|_{H^k(\Omega)} \le \|u\|_{H^k(\Omega)}.$$

We will now see that also

$$\|u\|_{H^k(\Omega)} \leq C\,|u|_{H^k(\Omega)} \quad \forall u \in H_0^k(\Omega). \tag{1.11}$$

Theorem 1.13 (Poincaré's inequality) *Let $\Omega \subset \mathbb{R}^n$ be open and bounded. Then there exists a constant $C > 0$ with*

$$|u|_{H^k(\Omega)} \leq \|u\|_{H^k(\Omega)} \leq C\,|u|_{H^k(\Omega)} \quad \forall u \in H_0^k(\Omega). \tag{1.11}$$

1.2.2.9 Sobolev Imbedding Theorem

Sobolev spaces are embedded in classical spaces:

Theorem 1.14 *Let $\Omega \subset \mathbb{R}^n$ be open, bounded with Lipschitz-boundary. Let $m \in \mathbb{N}$, $1 \leq p < \infty$.*

(i) *For all $k \in \mathbb{N}_0$, $0 < \beta < 1$ with*

$$m - \frac{n}{p} \geq k + \beta$$

one has the continuous embedding

$$W^{m,p}(\Omega) \hookrightarrow C^{k,\beta}(\bar{\Omega}).$$

More precisely, there exists a constant $C > 0$ such that for all $u \in W^{m,p}(\Omega)$ possibly after modification on a set of measure zero $u \in C^{k,\beta}(\bar{\Omega})$ and

$$\|u\|_{C^{k,\beta}(\bar{\Omega})} \leq C\|u\|_{W^{m,p}(\Omega)}.$$

(ii) *For all $k \in \mathbb{N}_0$, $0 \leq \beta \leq 1$ with*

$$m - \frac{n}{p} > k + \beta$$

one has the compact embedding

$$W^{m,p}(\Omega) \hookrightarrow\hookrightarrow C^{k,\beta}(\bar{\Omega}),$$

i.e., closed balls in $W^{m,p}(\Omega)$ are relatively compact in $C^{k,\beta}(\bar{\Omega})$.

(iii) *For $q \geq 1$ and $l \in \mathbb{N}_0$ with $m - n/p \geq l - n/q$ one has the continuous embedding*

$$W^{m,p}(\Omega) \hookrightarrow W^{l,q}(\Omega).$$

The embedding is compact if $m > l$ and $m - n/p > l - n/q$.

For $l = 0$ we have $W^{0,q}(\Omega) = L^q(\Omega)$.

For arbitrary open bounded $\Omega \subset \mathbb{R}^n$ (i), (ii), (iii) *hold for $W_0^{m,p}(\Omega)$ instead of $W^{m,p}(\Omega)$.*

Proof See for example [1, 4, 47].

Example 1.4 For $n \le 3$ we have the continuous embedding $H^1(\Omega) \hookrightarrow L^6(\Omega)$ and the compact embedding $H^2(\Omega) \hookrightarrow\hookrightarrow C(\bar{\Omega})$.

1.2.2.10 The Dual Space H^{-1} of H_0^1

The dual space of the Hilbert space $H_0^1(\Omega)$ is denoted by $H^{-1}(\Omega)$. This space can be characterized as follows:

Theorem 1.15 *For the space $H^{-1}(\Omega)$, $\Omega \subset \mathbb{R}^n$ open, the following holds:*

$$H^{-1}(\Omega) = \left\{ v \in H_0^1(\Omega) \mapsto (f^0, v)_{L^2} + \sum_{j=1}^{n} (f^j, v_{x_j})_{L^2} : f^j \in L^2(\Omega) \right\}.$$

Furthermore,

$$\|f\|_{H^{-1}} = \min\left\{ \left(\sum_{j=0}^{n} \|f^j\|_{L^2}^2 \right)^{1/2} : \langle f, v\rangle_{H^{-1},H_0^1} = (f^0, v)_{L^2} + \sum_{j=1}^{n} (f^j, v_{x_j})_{L^2},\ f^j \in L^2(\Omega) \right\}.$$

Proof "$\subset$": Let $f \in H^{-1}(\Omega)$. By the Riesz representation theorem, there exists a unique $u \in H_0^1(\Omega)$ with

$$(u, v)_{H^1} = \langle f, v\rangle_{H^{-1},H_0^1} \quad \forall v \in H_0^1(\Omega).$$

Set $f^0 = u$, $f^j = u_{x_j}$, $j \ge 1$.
Then

$$(f^0, v)_{L^2} + \sum_{j=1}^{n} (f^j, v_{x_j})_{L^2} = (u, v)_{L^2} + \sum_{j=1}^{n} (u_{x_j}, v_{x_j})_{L^2}$$
$$= (u, v)_{H^1} = \langle f, v\rangle_{H^{-1},H_0^1} \quad \forall v \in H_0^1(\Omega).$$

"$\supset$": For $g_0, \ldots, g_n \in L^2(\Omega)$, consider

$$g : v \in H_0^1(\Omega) \mapsto (g^0, v)_{L^2} + \sum_{j=1}^{n} (g^j, v_{x_j})_{L^2}.$$

Obviously, g is linear. Furthermore, for all $v \in H_0^1(\Omega)$, there holds

$$\begin{aligned}
&\left|(g^0, v)_{L^2} + \sum_{j=1}^n (g^j, v_{x_j})_{L^2}\right| \\
&\quad \le \|g^0\|_{L^2}\|v\|_{L^2} + \sum_{j=1}^n \|g^j\|_{L^2}\|v_{x_j}\|_{L^2} \\
&\quad \le \left(\sum_{j=0}^n \|g^j\|_{L^2}^2\right)^{1/2} \left(\|v\|_{L^2}^2 + \sum_{j=1}^n \|v_{x_j}\|_{L^2}\right)^{1/2} = \left(\sum_{j=0}^n \|g^j\|_{L^2}^2\right)^{1/2} \|v\|_{H^1}.
\end{aligned}$$

This shows $g \in H^{-1}(\Omega)$ and

$$\|g\|_{H^{-1}} \le \left(\sum_{j=0}^n \|g^j\|_{L^2}^2\right)^{1/2}. \tag{1.12}$$

To show the formula for $\|g\|_{H^{-1}}$ let $g^0, \ldots, g^n \in L^2(\Omega)$ be an arbitrary representation of g. Moreover let u be the Riesz representation of g and choose

$$(\bar{g}^0, \ldots, \bar{g}^n) := (u, u_{x_1}, \ldots, u_{x_n})$$

as above. Then by the Riesz representation theorem

$$\|g\|_{H^{-1}}^2 = \|u\|_{H^1}^2 = \|u\|_{L^2}^2 + \sum_{j=1}^n \|u_{x_j}\|_{L^2}^2 = \sum_{j=0}^n \|\bar{g}^j\|_{L^2}^2 \le \sum_{j=0}^n \|g^j\|_{L^2}^2,$$

where the last inequality follows from (1.12). This shows that $\bar{g}^0, \ldots, \bar{g}^n$ is the representation with minimum norm and yields $\|g\|_{H^{-1}}$.

1.2.3 Weak Convergence

In infinite dimensional spaces bounded, closed sets are no longer compact. In order to obtain compactness results, one has to use the concept of weak convergence.

Definition 1.16 Let X be a Banach space. We say that a sequence $(x_k) \subset X$ *converges weakly* to $x \in X$, written

$$x_k \rightharpoonup x,$$

if

$$\langle x^*, x_k \rangle_{X^*,X} \to \langle x^*, x \rangle_{X^*,X} \quad \text{as } k \to \infty \quad \forall x^* \in X^*.$$

It is easy to check that strong convergence $x_k \to x$ implies weak convergence $x_k \rightharpoonup x$. Moreover, one can show:

Theorem 1.16

(i) *Let X be a normed space and let $(x_k) \subset X$ be weakly convergent to $x \in X$. Then (x_k) is bounded.*
(ii) *Let $C \subset X$ be a closed convex subset of the normed space X. Then C is weakly sequentially closed.*

Definition 1.17 A Banach space X is called *reflexive* if the mapping $x \in X \mapsto \langle \cdot, x\rangle_{X^*,X} \in (X^*)^*$ is surjective, i.e., if for any $x^{**} \in (X^*)^*$ there exists $x \in X$ with

$$\langle x^{**}, x^*\rangle_{(X^*)^*,X^*} = \langle x^*, x\rangle_{X^*,X} \quad \forall x^* \in X^*.$$

Remark 1.7 Note that for any $x \in X$ the mapping $x^{**} := \langle \cdot, x\rangle_{X^*,X}$ is in $(X^*)^*$ with $\|x^{**}\|_{(X^*)^*} \le \|x\|_X$, since

$$|\langle x^*, x\rangle_{X^*,X}| \le \|x^*\|_{X^*}\|x\|_X.$$

One can show that actually $\|x^{**}\|_{(X^*)^*} = \|x\|_X$.

Remark 1.8 L^p is for $1 < p < \infty$ reflexive, since we have the isometric isomorphisms $(L^p)^* = L^q$, $1/p + 1/q = 1$, and thus $((L^p)^*)^* = (L^q)^* = L^p$. Moreover, any Hilbert space is reflexive by the Riesz representation theorem.

The following result is important.

Theorem 1.17 (Weak sequential compactness) *Let X be a reflexive Banach space. Then the following holds*

(i) *Every bounded sequence $(x_k) \subset X$ contains a weakly convergent subsequence, i.e., there are $(x_{k_i}) \subset (x_k)$ and $x \in X$ with $x_{k_i} \rightharpoonup x$.*
(ii) *Every bounded, closed and convex subset $C \subset X$ is weakly sequentially compact, i.e., every sequence $(x_k) \subset C$ contains a weakly convergent subsequence $(x_{k_i}) \subset (x_k)$ with $x_{k_i} \rightharpoonup x$, where $x \in C$.*

For a proof see for example [4, 149].

Theorem 1.18 (Lower semicontinuity) *Let X be a Banach space. Then any continuous, convex functional $F : X \to \mathbb{R}$ is weakly lower semicontinuous, i.e.*

$$x_k \rightharpoonup x \quad \Longrightarrow \quad \liminf_{k\to\infty} F(x_k) \ge F(x).$$

Finally, it is valuable to have mappings that map weakly convergent sequences to strongly convergent ones.

Definition 1.18 (Compact operator) A linear operator $A : X \to Y$ between normed spaces is called *compact* if it maps bounded sets to relatively compact sets, i.e.,

$$M \subset X \text{ bounded} \quad \Longrightarrow \quad \overline{AM} \subset Y \text{ compact}.$$

Remark 1.9 Since compact sets are bounded (why?), compact operators are automatically bounded and thus continuous.

For a compact embedding $X \hookrightarrow\hookrightarrow Y$ the imbedding operator $I_{X,Y} : x \in X \mapsto x \in Y$ is compact.

The connection to weak/strong convergence is as follows.

Lemma 1.6 *Let $A : X \to Y$ be a compact operator between normed spaces. Then, for all $(x_k) \subset X$, $x_k \rightharpoonup x$, there holds*

$$Ax_k \to Ax \quad \text{in } Y.$$

Proof From $x_k \rightharpoonup x$ and $A \in \mathcal{L}(X,Y)$ we see that $Ax_k \rightharpoonup Ax$. Since (x_k) is bounded (Theorem 1.16), there exists a bounded set $M \subset X$ with $x \in M$ and $(x_k) \subset M$. Now assume $Ax_k \not\to Ax$. Then there exist $\varepsilon > 0$ and a subsequence $(Ax_k)_K$ with $\|Ax_k - Ax\|_Y \geq \varepsilon$ for all $k \in K$. Since $\overline{AM}$ is compact, the sequence $(Ax_k)_K$ possesses a convergent subsequence $(Ax_k)_{K'} \to y$. The continuity of the norm implies

$$\|y - Ax\|_Y \geq \varepsilon.$$

But since $(Ax_k)_{K'} \rightharpoonup Ax$ and $(Ax_k)_{K'} \to y$ we must have $y = Ax$, which is a contradiction.

1.3 Weak Solutions of Elliptic and Parabolic PDEs

1.3.1 Weak Solutions of Elliptic PDEs

In this section we sketch the theory of weak solutions for elliptic second order partial differential equations. For more details we refer, e.g., to [4, 47, 90, 115, 133, 146].

1.3.1.1 Weak solutions of the Poisson equation

Dirichlet Boundary Conditions

We start with the elliptic boundary value problem

$$-\Delta y = f \quad \text{on } \Omega, \tag{1.13}$$

$$y = 0 \quad \text{on } \partial\Omega \quad \text{(Dirichlet condition)}, \tag{1.14}$$

where $\Omega \subset \mathbb{R}^n$ is an open, bounded set and $f \in L^2(\Omega)$. This admits discontinuous right hand sides f, e.g. source terms f that act only on a subset of Ω. Since a classical solution $y \in C^2(\Omega) \cap C^1(\bar{\Omega})$ exists at best for continuous right hand sides, we need a generalized solution concept. It is based on a *variational formulation* of (1.13)–(1.14).

To this end let us assume that $y \in C^2(\Omega) \cap C^1(\bar{\Omega})$ is a classical solution of (1.13)–(1.14). Then we have $y \in H_0^1(\Omega)$ by Remark 1.6. Multiplying by $v \in C_c^\infty(\Omega)$ and integrating over Ω yields

$$-\int_\Omega \Delta y v \, dx = \int_\Omega f v \, dx \quad \forall v \in C_c^\infty(\Omega). \tag{1.15}$$

It is easy to see that (1.13) and (1.15) are equivalent for classical solutions. Now integration by parts gives

$$-\int_\Omega y_{x_i x_i} v \, dx = \int_\Omega y_{x_i} v_{x_i} \, dx - \int_{\partial\Omega} y_{x_i} v \nu_i \, dS(x) = \int_\Omega y_{x_i} v_{x_i} \, dx. \tag{1.16}$$

Note that the boundary integral vanishes, since $v|_{\partial\Omega} = 0$. Thus, (1.15) is equivalent to

$$\int_\Omega \nabla y \cdot \nabla v \, dx = \int_\Omega f v \, dx \quad \forall v \in C_c^\infty(\Omega). \tag{1.17}$$

We note that this variational equation makes already perfect sense in a larger space:

Lemma 1.7 *The mapping*

$$(y, v) \in H_0^1(\Omega)^2 \mapsto a(y, v) := \int_\Omega \nabla y \cdot \nabla v \, dx \in \mathbb{R}$$

is bilinear and bounded:

$$|a(y, v)| \leq \|y\|_{H^1} \|v\|_{H^1}. \tag{1.18}$$

For $f \in L^2(\Omega)$, the mapping

$$v \in H_0^1(\Omega) \mapsto \int_\Omega f v \, dx \in \mathbb{R}$$

is linear and bounded:

$$\left| \int_\Omega f v \, dx \right| = (f, v)_{L^2} \leq \|f\|_{L^2} \|v\|_{L^2} \leq \|f\|_{L^2} \|v\|_{H_0^1}. \tag{1.19}$$

Proof Clearly, $a(y, v)$ is bilinear. The boundedness follows from

$$|a(y, v)| \leq \int_\Omega |\nabla y(x)^T \nabla v(x)| \, dx \leq \int_\Omega \|\nabla y(x)\|_2 \|\nabla v(x)\|_2 \, dx$$

$$\le \| \|\nabla y\|_2 \|_{L^2} \| \|\nabla v\|_2 \|_{L^2} = |y|_{H^1} |v|_{H^1} \le \|y\|_{H^1} \|v\|_{H^1},$$

where we have applied the Cauchy-Schwarz inequality.

The second assertion is trivial.

By density and continuity, we can extend (1.17) to $y \in H_0^1(\Omega)$ and $v \in H_0^1(\Omega)$. We arrive at the *variational formulation*

$$\int_\Omega \nabla y \cdot \nabla v \, dx = \int_\Omega f v \, dx \quad \forall v \in H_0^1(\Omega). \tag{1.20}$$

We summarize: (1.13) and (1.20) are equivalent for a classical solution $y \in C^2(\Omega) \cap C^1(\bar{\Omega})$. But the variational formulation (1.20) makes already perfectly sense for $y \in H_0^1(\Omega)$ and $f \in L^2(\Omega)$. This motivates the following definition.

Definition 1.19 A function $y \in H_0^1(\Omega)$ is called *weak solution* of the boundary value problem (1.13)–(1.14) if it satisfies the *variational formulation* or *weak formulation*

$$\int_\Omega \nabla y \cdot \nabla v \, dx = \int_\Omega f v \, dx \quad \forall v \in H_0^1(\Omega). \tag{1.20}$$

In order to allow a uniform treatment of more general equations than (1.13)–(1.14), we introduce the following abstract notation. Let

$$V = H_0^1(\Omega),$$

$$a(y, v) = \int_\Omega \nabla y \cdot \nabla v \, dx, \quad y, v \in V, \tag{1.21}$$

$$F(v) = (f, v)_{L^2(\Omega)}, \quad v \in V. \tag{1.22}$$

Then $a : V \times V \to \mathbb{R}$ is a bilinear form, $F \in V^*$ is a linear functional on V and (1.20) can be written as

$$\text{Find } y \in V\text{:} \quad a(y, v) = F(v) \quad \forall v \in V. \tag{1.23}$$

Remark 1.10 Since $a(y, \cdot) \in V^*$ for all $y \in V$ and $y \in V \mapsto a(y, \cdot) \in V^*$ is continuous and linear, there exists a bounded linear operator $A : V \to V^*$ with

$$a(y, v) = \langle Ay, v \rangle_{V^*,V} \quad \forall y, v \in V. \tag{1.24}$$

Then (1.23) can be written in the form

$$\text{Find } y \in V\text{:} \quad Ay = F. \tag{1.25}$$

We have the following important existence and uniqueness result for solutions of (1.23).

Lemma 1.8 (Lax-Milgram lemma) *Let V be a real Hilbert space with inner product $(\cdot,\cdot)_V$ and let $a : V \times V \to \mathbb{R}$ be a bilinear form that satisfies with constants $\alpha_0, \beta_0 > 0$*

$$|a(y,v)| \le \alpha_0 \|y\|_V \|v\|_V \quad \forall y, v \in V \quad (\textit{boundedness}), \tag{1.26}$$

$$a(y,y) \ge \beta_0 \|y\|_V^2 \quad \forall y \in V \quad (V\textit{-coercivity}). \tag{1.27}$$

Then for any bounded linear functional $F \in V^*$ the variational equation (1.23) *has a unique solution $y \in V$. Moreover, u satisfies*

$$\|y\|_V \le \frac{1}{\beta_0} \|F\|_{V^*}. \tag{1.28}$$

In particular the operator A defined in (1.24) *satisfies*

$$A \in \mathcal{L}(V, V^*), \qquad A^{-1} \in \mathcal{L}(V^*, V), \qquad \|A^{-1}\|_{V^*,V} \le \frac{1}{\beta_0}.$$

Proof See for example [47, 115].

Remark 1.11 If $a(\cdot,\cdot)$ is symmetric, i.e., if $a(y,v) = a(v,y)$ for all $y, v \in V$, then the Lax-Milgram lemma is an immediate consequence of the Riesz representation theorem. In fact, in this case $(u,v) := a(u,v)$ defines a new inner product on V and the existence of a unique solution of (1.23) follows directly from the Riesz representation theorem.

Application of the Lax-Milgram lemma to (1.20) yields

Theorem 1.19 (Existence and uniqueness for the Dirichlet problem) *Let $\Omega \subset \mathbb{R}^n$ be open and bounded. Then the bilinear form a in* (1.21) *is bounded and V-coercive for $V = H_0^1(\Omega)$ and the associated operator $A \in \mathcal{L}(V, V^*)$ in* (1.24) *has a bounded inverse. In particular,* (1.13)–(1.14) *has for all $f \in L^2(\Omega)$ a unique weak solution $y \in H_0^1(\Omega)$ given by* (1.20) *and satisfies*

$$\|y\|_{H^1(\Omega)} \le C_D \|f\|_{L^2(\Omega)},$$

where C_D depends on Ω but not on f.

Proof We verify the hypotheses of Lemma 1.8. Clearly, $a(y,u)$ in (1.21) is bilinear. The boundedness (1.26) follows from (1.18). Using Poincaré's inequality (1.11) we obtain

$$a(y,y) = \int_\Omega \nabla y \cdot \nabla y \, dx = |y|_{H_0^1(\Omega)}^2 \ge \frac{1}{C^2} \|y\|_{H_0^1(\Omega)}^2 = \frac{1}{C^2} \|y\|_V^2$$

which shows the V-coercivity (1.27).

Finally, the definition of F in (1.22) yields

$$\|F\|_{V^*} = \sup_{\|v\|_V=1} F(v) = \sup_{\|v\|_V=1} (f, v)_{L^2(\Omega)} \leq \sup_{\|v\|_V=1} \|f\|_{L^2(\Omega)}\|v\|_{L^2(\Omega)} \leq \|f\|_{L^2(\Omega)}.$$

Thus, the assertion holds with $C_D = C^2$ by the Lax-Milgram lemma.

A refined analysis shows that the weak solution y is bounded and (after a modification on a set of measure zero) continuous if Ω has Lipschitz-boundary.

Theorem 1.20 (Boundedness and continuity for the Dirichlet problem) *Let in addition to the assumptions of the previous theorem $\Omega \subset \mathbb{R}^n$ be open and bounded with Lipschitz-boundary and let $r > n/2$, $n \geq 2$. Then for any $f \in L^r(\Omega)$ there exists a unique weak solution $y \in H_0^1(\Omega) \cap C(\bar{\Omega})$ of* (1.13)–(1.14) *and there exists a constant $C_\infty > 0$ with*

$$\|y\|_{H^1(\Omega)} + \|y\|_{C(\bar{\Omega})} \leq C_\infty \|f\|_{L^r(\Omega)},$$

where C_∞ depends on Ω but not on f.

Proof See [86, Thm. B.4].

Boundary Conditions of Robin Type

We have seen that for example in heating applications the boundary condition is sometimes of Robin type. We consider now problems of the form

$$-\Delta y + c_0 y = f \quad \text{on } \Omega, \tag{1.29}$$

$$\frac{\partial y}{\partial \nu} + \alpha y = g \quad \text{on } \partial\Omega \quad \text{(Robin condition)}, \tag{1.30}$$

where $f \in L^2(\Omega)$ and $g \in L^2(\partial\Omega)$ are given and $c_0 \in L^\infty(\Omega)$, $\alpha \in L^\infty(\partial\Omega)$ are nonnegative coefficients.

Weak solutions can be defined similarly as above. If y is a classical solution of (1.29)–(1.30) then for any test function $v \in C^1(\bar{\Omega})$ integration by parts, see (1.16), yields as above

$$\begin{aligned}\int_\Omega (-\Delta y + c_0 y) v \, dx &= \int_\Omega \nabla y \cdot \nabla v \, dx + (c_0 y, v)_{L^2(\Omega)} - \int_{\partial\Omega} \frac{\partial y}{\partial \nu} v \, dS(x) \\ &= \int_\Omega f v \, dx \quad \forall v \in C^1(\bar{\Omega}).\end{aligned}$$

Inserting the boundary condition $\frac{\partial y}{\partial \nu} = -\alpha y + g$ we arrive at

$$\int_\Omega \nabla y \cdot \nabla v \, dx + (c_0 y, v)_{L^2(\Omega)} + (\alpha y, v)_{L^2(\partial\Omega)}$$

$$= (f, v)_{L^2(\Omega)} + (g, v)_{L^2(\partial\Omega)} \quad \forall v \in H^1(\Omega). \tag{1.31}$$

The extension to $v \in H^1(\Omega)$ is possible, since for $y \in H^1(\Omega)$ both sides are continuous with respect to $v \in H^1(\Omega)$ and since $C^1(\bar{\Omega})$ is dense in $H^1(\Omega)$.

Definition 1.20 A function $y \in H^1(\Omega)$ is called *weak solution* of the boundary value problem (1.29)–(1.30) if it satisfies the *variational formulation* or *weak formulation* (1.31).

To apply the general theory, we set

$$\begin{aligned} V &= H^1(\Omega), \\ a(y, v) &= \int_\Omega \nabla y \cdot \nabla v \, dx + (c_0 y, v)_{L^2(\Omega)} + (\alpha y, v)_{L^2(\partial\Omega)}, \quad y, v \in V, \\ F(v) &= (f, v)_{L^2(\Omega)} + (g, v)_{L^2(\partial\Omega)}, \quad v \in V. \end{aligned} \tag{1.32}$$

The Lax-Milgram lemma yields similarly as above

Theorem 1.21 (Existence and uniqueness for Robin boundary conditions) *Let $\Omega \subset \mathbb{R}^n$ be open and bounded with Lipschitz-boundary and let $c_0 \in L^\infty(\Omega)$, $\alpha \in L^\infty(\partial\Omega)$ be nonnegative with $\|c_0\|_{L^2(\Omega)} + \|\alpha\|_{L^2(\partial\Omega)} > 0$. Then the bilinear form a in* (1.32) *is bounded and V-coercive for $V = H^1(\Omega)$ and the associated operator $A \in \mathcal{L}(V, V^*)$ in* (1.24) *has a bounded inverse. In particular,* (1.29)–(1.30) *has for all $f \in L^2(\Omega)$ and $g \in L^2(\partial\Omega)$ a unique weak solution $y \in H^1(\Omega)$ given by* (1.31) *and satisfies*

$$\|y\|_{H^1(\Omega)} \le C_R(\|f\|_{L^2(\Omega)} + \|g\|_{L^2(\partial\Omega)}),$$

where C_R depends on Ω, α, c_0 but not on f, g.

Proof The proof is an application of the Lax-Milgram lemma. The boundedness of $a(y, v)$ and of $F(v)$ follows by the trace theorem. The V-coercivity is a consequence of a generalized Poincaré inequality.

A refined analysis yields the following result [108], [133].

Theorem 1.22 (Boundedness and continuity for Robin boundary conditions) *Let the assumptions of the previous theorem hold and let $r > n/2$, $s > n - 1$, $n \ge 2$. Then for any $f \in L^r(\Omega)$ and $g \in L^s(\partial\Omega)$ there exists a unique weak solution $y \in H^1(\Omega) \cap C(\bar{\Omega})$ of* (1.29)–(1.30). *There exists a constant $C_\infty > 0$ with*

$$\|y\|_{H^1(\Omega)} + \|y\|_{C(\bar{\Omega})} \le C_\infty(\|f\|_{L^r(\Omega)} + \|g\|_{L^s(\partial\Omega)}),$$

where C_∞ depends on Ω, α, c_0 but not on f, g.

1.3.1.2 Weak Solutions of Uniformly Elliptic Equations

The results can be extended to general second order elliptic PDEs of the form

$$Ly = f \quad \text{on } \Omega \tag{1.33}$$

with

$$Ly := -\sum_{i,j=1}^{n} (a_{ij} y_{x_i})_{x_j} + c_0 y, \quad a_{ij}, c_0 \in L^\infty,\ c_0 \geq 0,\ a_{ij} = a_{ji} \tag{1.34}$$

and L is assumed to be *uniformly elliptic* in the sense that there is a constant $\theta > 0$ such that

$$\sum_{i,j=1}^{n} a_{ij}(x)\xi_i\xi_j \geq \theta \|\xi\|^2 \quad \text{for almost all } x \in \Omega \text{ and all } \xi \in \mathbb{R}^n. \tag{1.35}$$

For example in the case of Dirichlet boundary conditions

$$y|_{\partial\Omega} = 0 \tag{1.36}$$

the weak formulation of (1.33) is given by

$$\text{Find } y \in V := H_0^1(\Omega):\quad a(y, v) = (f, v)_{L^2(\Omega)} \quad \forall v \in V \tag{1.37}$$

with the bilinear form

$$a(y, v) = \int_\Omega \left(\sum_{i,j=1}^{n} a_{ij} y_{x_i} v_{x_j} + c_0 y v \right) dx. \tag{1.38}$$

Theorem 1.23 (Existence, uniqueness and continuity for the general Dirichlet problem) *Let L be a uniformly elliptic second order operator according to* (1.34), (1.35). *Then the statements of Theorem* 1.19 *and Theorem* 1.20 *hold also for the weak solution of* (1.33), (1.36) *defined by* (1.37), (1.38). *The constants C_D and C_∞ depend only on a_{ij}, c_0, Ω.*

Proof It is easy to check that the Lax-Milgram lemma is applicable. The uniform boundedness and continuity of weak solutions is shown in [86, Thm. B.4].

In the case of the Robin boundary condition the normal derivative has to be replaced by the conormal derivative

$$\frac{\partial y}{\partial \nu_A}(x) := \nabla y(x) \cdot A(x)\nu(x), \quad A(x) = (a_{ij}(x)). \tag{1.39}$$

The weak formulation for Robin boundary conditions

$$\frac{\partial y}{\partial \nu_A} + \alpha y = g \quad \text{on } \partial\Omega \tag{1.40}$$

is consequently

$$\text{Find } y \in V := H^1(\Omega): \quad a(y, v) = (f, v)_{L^2(\Omega)} + (g, v)_{L^2(\partial\Omega)} \quad \forall v \in V \tag{1.41}$$

with the bilinear form

$$a(y, v) = \int_\Omega \left(\sum_{i,j=1}^n a_{ij} y_{x_i} v_{x_j} + c_0 y v \right) dx + (\alpha y, v)_{L^2(\partial\Omega)}. \tag{1.42}$$

Theorem 1.24 (Existence, uniqueness and continuity for Robin boundary conditions) *Let L be a uniformly elliptic second order operator according to* (1.34), (1.35). *Then the statements of Theorem* 1.21 *and Theorem* 1.22 *hold also for the weak solution of* (1.33), (1.40) *defined by* (1.41), (1.42). *The constants C_R and C_∞ depend only on $a_{ij}, c_0, \alpha, \Omega$.*

Proof Again the Lax-Milgram lemma is applicable. The uniform boundedness and continuity of weak solutions is shown in [108, 133].

1.3.1.3 An Existence and Uniqueness Result for Semilinear Elliptic Equations

We finally state an existence and uniqueness result for a uniformly elliptic semilinear equation

$$\begin{aligned} Ly + d(x, y) &= f \quad \text{on } \Omega \\ \frac{\partial y}{\partial \nu_A} + \alpha y + b(x, y) &= g \quad \text{on } \partial\Omega \end{aligned} \tag{1.43}$$

where the operator L is given by

$$Ly := -\sum_{i,j=1}^n (a_{ij} y_{x_i})_{x_j} + c_0 y, \quad a_{ij}, c_0 \in L^\infty, \ c_0 \geq 0, \ a_{ij} = a_{ji} \tag{1.34}$$

and L is assumed to be uniformly elliptic in the sense that there is a constant $\theta > 0$ such that

$$\sum_{i,j=1}^n a_{ij}(x) \xi_i \xi_j \geq \theta \|\xi\|^2 \quad \text{for almost all } x \in \Omega \text{ and all } \xi \in \mathbb{R}^n. \tag{1.35}$$

Moreover, we assume that $0 \le \alpha \in L^\infty(\partial\Omega)$ and that the functions $d : \Omega \times \mathbb{R} \to \mathbb{R}$, and $b : \partial\Omega \times \mathbb{R} \to \mathbb{R}$ satisfy

$$\begin{aligned} &d(x,\cdot) \text{ is continuous and monotone increasing for a.a. } x \in \Omega, \\ &b(x,\cdot) \text{ is continuous and monotone increasing for a.a. } x \in \partial\Omega, \\ &d(\cdot,y) \in L^\infty(\Omega), b(\cdot,y) \in L^\infty(\partial\Omega) \quad \text{for all } y \in \mathbb{R}. \end{aligned} \tag{1.44}$$

Analogous to (1.41), a weak solution of (1.43) is given by

$$\begin{aligned} &\text{Find } y \in V := H^1(\Omega): \\ &\quad a(y,v) + (d(\cdot,y),v)_{L^2(\Omega)} + (b(\cdot,y),v)_{L^2(\partial\Omega)} \\ &\qquad = (f,v)_{L^2(\Omega)} + (g,v)_{L^2(\partial\Omega)} \quad \forall v \in V \end{aligned} \tag{1.45}$$

with the bilinear form (1.42).

Under the assumptions in (1.44) the theory of maximal monotone operators and a technique of Stampacchia can be applied to extend Theorem 1.22 to the semilinear elliptic equation (1.43), see for example [133]. The proof of continuity can be obtained by the techniques in [108].

Theorem 1.25 *Let $\Omega \subset \mathbb{R}^n$ be open and bounded with Lipschitz-boundary, let $c_0 \in L^\infty(\Omega)$, $\alpha \in L^\infty(\partial\Omega)$ be nonnegative with $\|c_0\|_{L^2(\Omega)} + \|\alpha\|_{L^2(\partial\Omega)} > 0$ and let* (1.35), (1.44) *be satisfied. Moreover, let $r > n/2$, $s > n-1$, $n \ge 2$. Then* (1.43), (1.34) *has for any $f \in L^r(\Omega)$ and $g \in L^s(\partial\Omega)$ a unique weak solution $y \in H^1(\Omega) \cap C(\bar{\Omega})$. There exists a constant $C_\infty > 0$ with*

$$\|y\|_{H^1(\Omega)} + \|y\|_{C(\bar{\Omega})} \le C_\infty(\|f - d(\cdot,0)\|_{L^r(\Omega)} + \|g - b(\cdot,0)\|_{L^s(\partial\Omega)}),$$

where C_∞ depends on Ω, α, c_0 but not on f, g, b, d.

Remark 1.12 An analogous result holds also in the case of homogeneous Dirichlet boundary conditions. In fact, it is easy to check that the proof in [133] applies also in this case. The continuity of the solution follows from [86, Thm. B.4].

1.3.1.4 Regularity Results

We have already seen that for sufficiently regular data weak solutions are in $C(\bar{\Omega})$. Under certain conditions one can also show that weak solutions y of (1.33) live actually in a higher Sobolev space if $f \in L^2$ and $a_{ij} \in C^1$.

Interior Regularity

For coefficients $a_{ij} \in C^1(\Omega)$ or $a_{ij} \in C^{0,1}(\bar{\Omega})$ the weak solution of (1.33) satisfies actually $u \in H^2(\Omega')$ for all $\Omega' \subset\subset \Omega$.

Theorem 1.26 *Let L in* (1.34) *be uniformly elliptic with $a_{ij} \in C^1(\Omega)$ or $a_{ij} \in C^{0,1}(\bar{\Omega})$, $c_0 \in L^\infty(\Omega)$. Let $f \in L^2(\Omega)$ and let $y \in H^1(\Omega)$ be a weak solution of $Ly = f$, that is,*

$$a(y, v) = (f, v)_{L^2(\Omega)} \quad \forall v \in H_0^1(\Omega)$$

with a in (1.38) *or* (1.42) *(which coincide for $v \in H_0^1(\Omega)$). Then $y \in H^2(\Omega')$ for all $\Omega' \subset\subset \Omega$ and there is $C > 0$ with*

$$\|y\|_{H^2(\Omega')} \le C(\|y\|_{H^1(\Omega)} + \|f\|_{L^2(\Omega)}), \tag{1.46}$$

where C depends on Ω' but not on f and y.

Proof See for example [47] or [115].

Remark 1.13 Note that the interior regularity result applies to the Dirichlet problem (1.37) as well as to the problem (1.41) with Robin boundary conditions.

The weak solution of the Dirichlet problem (1.37) satisfies in addition

$$\|y\|_{H^1(\Omega)} \le C\|f\|_{L^2(\Omega)}.$$

Inserting in (1.46) gives

$$\|y\|_{H^2(\Omega')} \le C\|f\|_{L^2(\Omega)}.$$

Similarly, the weak solution of (1.41) for Robin boundary condition satisfies

$$\|y\|_{H^2(\Omega')} \le C(\|f\|_{L^2(\Omega)} + \|g\|_{L^2(\partial\Omega)}).$$

If the coefficients are more regular we can iterate this argument to obtain higher interior regularity.

Theorem 1.27 (Higher interior regularity) *Let in addition to the assumptions of Theorem* 1.26 *$a_{ij} \in C^{m+1}(\Omega)$, $c_0 \in C^m(\Omega)$ and $f \in H^m(\Omega)$ hold with some $m \in \mathbb{N}_0$. Then for all $\Omega' \subset\subset \Omega$ the weak solution of $Ly = f$ according to* (1.37) *or* (1.41) *satisfies $y \in H^{m+2}(\Omega')$ and there is $C > 0$ with*

$$\|y\|_{H^{m+2}(\Omega')} \le C(\|y\|_{H^1(\Omega)} + \|f\|_{H^m(\Omega)}). \tag{1.47}$$

Boundary Regularity

If $\partial\Omega$ is sufficiently smooth then in the case of the Dirichlet problem (1.37) the additional regularity of weak solutions holds up to the boundary. We have the following result.

Theorem 1.28 *Let $\Omega \subset \mathbb{R}^n$ be open, bounded with C^2-boundary. Let L in* (1.34) *be uniformly elliptic with $a_{ij} \in C^{0,1}(\bar{\Omega})$, $c_0 \in L^\infty(\Omega)$. Then for any $f \in L^2(\Omega)$ the weak solution $y \in H_0^1(\Omega)$ of the Dirichlet problem* (1.37) *satisfies $y \in H^2(\Omega)$ and*

$$\|y\|_{H^2(\Omega)} \le C(\|y\|_{H^1(\Omega)} + \|f\|_{L^2(\Omega)}), \tag{1.48}$$

where C does not depend on f.

Proof See for example [47] or [115].

Iterating the argument yields

Theorem 1.29 (Higher boundary regularity) *Let in addition to the assumptions of Theorem* 1.28 *$\partial\Omega$ be C^{m+2}, $a_{ij}, c_0 \in C^{m+1}(\bar{\Omega})$ and $f \in H^m(\Omega)$ hold with some $m \in \mathbb{N}_0$. Then the weak solution $y \in H_0^1(\Omega)$ of the Dirichlet problem* (1.37) *satisfies $y \in H^{m+2}(\Omega)$ and*

$$\|y\|_{H^{m+2}(\Omega)} \le C(\|y\|_{H^1(\Omega)} + \|f\|_{H^m(\Omega)}), \tag{1.49}$$

where C does not depend on f.

1.3.2 Weak Solutions of Parabolic PDEs

In this section we describe the basic theory of weak solutions for parabolic second order partial differential equations. For details we refer, e.g., to [47, 90, 115, 133, 146]. Throughout this section let $\Omega \subset \mathbb{R}^n$ be open and bounded and define the cylinder $\Omega_T := (0,T) \times \Omega$ for some $T > 0$. We study the initial-boundary value problem

$$\begin{aligned} y_t + Ly &= f \quad \text{on } \Omega_T, \\ y &= 0 \quad \text{on } [0,T] \times \partial\Omega, \\ y(0,\cdot) &= y_0 \quad \text{on } \Omega, \end{aligned} \tag{1.50}$$

where $f : \Omega_T \to \mathbb{R}$, $y_0 : \Omega \to \mathbb{R}$ are given and $y : \bar{\Omega}_T \to \mathbb{R}$ is the unknown. L denotes for each time t a second order partial differential operator

$$Ly := -\sum_{i,j=1}^{n} (a_{ij}(t,x) y_{x_i})_{x_j} + \sum_{i=1}^{n} b_i(t,x) y_{x_i} + c_0(t,x) y \tag{1.51}$$

in divergence form.

1.3.2.1 Uniformly Parabolic Equations

In analogy to definition (1.35) of uniformly elliptic operators we define a uniformly parabolic operator as follows.

Definition 1.21 The partial differential operator $\frac{\partial}{\partial t} + L$ with L given in (1.51) is called *uniformly parabolic* if there is a constant $\theta > 0$ such that

$$\sum_{i,j=1}^{n} a_{ij}(t,x)\xi_i\xi_j \geq \theta \|\xi\|^2 \quad \text{for almost all } (t,x) \in \Omega_T \text{ and all } \xi \in \mathbb{R}^n. \tag{1.52}$$

It will be convenient to consider a solution y of (1.50) as a Banach space valued function

$$t \in [0,T] \mapsto y(t) \in H_0^1(\Omega).$$

1.3.2.2 Bochner Spaces

Let X be a separable Banach space. We consider mappings

$$t \in [0,T] \mapsto y(t) \in X.$$

We extend the notion of measurability, integrability, and weak differentiability.

Definition 1.22

(i) A function $s : [0,T] \to X$ is called simple if it has the form

$$s(t) = \sum_{i=1}^{m} 1_{E_i}(t) y_i,$$

with Lebesgue measurable sets $E_i \subset [0,T]$ and $y_i \in X$.

(ii) A function $f : t \in [0,T] \mapsto f(t) \in X$ is called *strongly measurable* if there exist simple functions $s_k : [0,T] \to X$ such that

$$s_k(t) \to f(t) \quad \text{for almost all } t \in [0,T].$$

Definition 1.23 (Bochner integral)

(i) For a simple function $s(t) = \sum_{i=1}^{m} 1_{E_i}(t) y_i$ we define the integral

$$\int_0^T s(t)\,dt := \sum_{i=1}^{m} y_i \mu(E_i).$$

(ii) We say that $f : [0, T] \to X$ is Bochner-integrable if there exists a sequence (s_k) of simple functions such that $s_k(t) \to f(t)$ a.e. and

$$\int_0^T \|s_k(t) - f(t)\|_X \, dt \to 0 \quad \text{as } k \to \infty.$$

(iii) If f is Bochner-integrable we define

$$\int_0^T f(t)\, dt := \lim_{k\to\infty} \int_0^T s_k(t)\, dt.$$

Theorem 1.30 *A strongly measurable function $f : [0, T] \to X$ is Bochner-integrable if and only if $t \mapsto \|f(t)\|_X$ is Lebesgue integrable. In this case*

$$\left\| \int_0^T f(t)\, dt \right\|_X \le \int_0^T \|f(t)\|_X \, dt$$

and for all $u^* \in X^*$ the function $t \mapsto \langle u^*, f(t)\rangle_{X^*,X}$ is integrable with

$$\left\langle u^*, \int_0^T f(t)\, dt \right\rangle_{X^*,X} = \int_0^T \langle u^*, f(t)\rangle_{X^*,X} \, dt.$$

Proof See for example Yosida [149].

This motivates the following definition of Banach space valued Lebesgue spaces.

Definition 1.24 Let X be a separable Banach space. We define for $1 \le p < \infty$ the space

$$L^p(0, T; X) := \left\{ y : [0, T] \to X \text{ strongly measurable} : \right.$$
$$\left. \|y\|_{L^p(0,T;X)} := \left(\int_0^T \|y(t)\|_X^p \, dt \right)^{1/p} < \infty \right\}.$$

Moreover, we let

$$L^\infty(0, T; X) := \Big\{ y : [0, T] \to X \text{ strongly measurable} :$$
$$\|y\|_{L^\infty(0,T;X)} := \operatorname*{ess\,sup}_{t\in[0,T]} \|y(t)\|_X < \infty \Big\}.$$

The space $C^k([0, T]; X)$, $k \in \mathbb{N}_0$, is defined as the space of k-times continuously differentiable functions on $[0, T]$ (defined in the usual way).

Definition 1.25 (Weak time derivative) Let $y \in L^1(0, T; X)$. We say that $v \in L^1(0, T; X)$ is the weak derivative of y, written $y_t = v$, if

$$\int_0^T \varphi'(t)y(t)\,dt = -\int_0^T \varphi(t)v(t)\,dt \quad \forall \varphi \in C_c^\infty((0, T)).$$

Lemma 1.9 *For any $y \in L^p(0, T; X)$, $1 \le p < \infty$, there is a sequence (s_k) of simple functions with $s_k \to y$ a.e. and $s_k \to y$ in $L^p(0, T; X)$. Moreover functions of the form*

$$\sum_{i=1}^m \varphi_i(t)y_i, \quad \varphi_i \in C_c^\infty((0, T)),\; y_i \in X$$

are dense in $L^p(0, T; X)$ for $1 \le p < \infty$. In particular, $C_c^\infty((0, T); X)$ as well as $C^k([0, T]; X)$ are dense in $L^p(0, T; X)$ for $1 \le p < \infty$, $k \in \mathbb{N}_0$.

Theorem 1.31 *Let X be a separable Banach space. Then for $1 \le p \le \infty$ the spaces $L^p(0, T; X)$ are Banach spaces.*

For $1 \le p < \infty$ the dual space of $L^p(0, T; X)$ can isometrically be identified with $L^q(0, T; X^*)$, $\frac{1}{p} + \frac{1}{q} = 1$, by means of the pairing

$$\langle v, y\rangle_{L^q(0,T;X^*),L^p(0,T;X)} = \int_0^T \langle v(t), y(t)\rangle_{X^*,X}\,dt.$$

If H is a separable Hilbert space then $L^2(0, T; H)$ is a Hilbert space with inner product

$$(y, v)_{L^2(0,T;H)} := \int_0^T (y(t), v(t))_H\,dt.$$

Proof The proof is similar as for $X = \mathbb{R}$.

We consider now the following setting.

Definition 1.26 (Gelfand triple) Let H, V be separable Hilbert spaces with the continuous and dense imbedding $V \hookrightarrow H$. We identify H with its dual H^*. Then we have the continuous and dense imbeddings

$$V \hookrightarrow H = H^* \hookrightarrow V^*,$$

which is called *Gelfand triple*. Note that the imbedding $H \hookrightarrow V^*$ is given by

$$y \in H \mapsto (y, \cdot)_H \in H^* \subset V^*.$$

Moreover, we introduce the space

$$W(0, T; H, V) := \{y : y \in L^2(0, T; V),\; y_t \in L^2(0, T; V^*)\} \tag{1.53}$$

equipped with the norm

$$\|y\|_{W(0,T;H,V)} = \sqrt{\|y\|^2_{L^2(0,T;V)} + \|y_t\|^2_{L^2(0,T;V^*)}}.$$

Remark 1.14 Given the Gelfand triple $V \hookrightarrow H = H^* \hookrightarrow V^*$ we have for $y \in L^2(0,T;V)$ also $y \in L^2(0,T;V^*)$ and thus $y \in L^1(0,T;V^*)$. Therefore, it makes sense to require that y has a weak derivative $y_t \in L^1(0,T;V^*)$ and to impose the additional condition $y_t \in L^2(0,T;V^*)$.

Theorem 1.32 *Let $V \hookrightarrow H \hookrightarrow V^*$ be a Gelfand triple. Then $W(0,T;H,V)$ is a Hilbert space and we have the continuous imbedding*

$$W(0,T;H,V) \hookrightarrow C([0,T];H).$$

Moreover, for all $y, v \in W(0,T;H,V)$ the integration by parts formula holds

$$(y(t),v(t))_H - (y(s),v(s))_H = \int_s^t (\langle y_t(\tau), v(\tau)\rangle_{V^*,V} + \langle v_t(\tau), y(\tau)\rangle_{V^*,V})\, d\tau. \tag{1.54}$$

Proof See for example [47] or Chap. IV in Gajewski, Gröger, Zacharias [52].

1.3.2.3 Weak Solutions of Uniformly Parabolic Equations

Weak Solutions

We consider the initial-boundary value problem (1.50) for operators L in divergence form (1.51). We will assume that the coefficients satisfy

$$a_{ij}, b_i, c_0 \in L^\infty(\Omega_T), \tag{1.55}$$

and that the source term and initial data satisfy

$$f \in L^2(0,T;H^{-1}(\Omega)), \quad y_0 \in L^2(\Omega), \tag{1.56}$$

where $H^{-1}(\Omega) = H_0^1(\Omega)^*$. We set

$$H := L^2(\Omega), \qquad V := H_0^1(\Omega)$$

and define the Gelfand triple $V \hookrightarrow H \hookrightarrow V^*$, i.e.,

$$H_0^1(\Omega) \hookrightarrow L^2(\Omega) \hookrightarrow H^{-1}(\Omega).$$

To derive a weak formulation of (1.50) consider a function

$$y \in W(0,T;L^2(\Omega),H_0^1(\Omega)) = W(0,T;H,V)$$

where we recall that

$$W(0, T; H, V) = \{y : y \in L^2(0, T; V),\ y_t \in L^2(0, T; V^*)\}.$$

For almost all $t \in [0, T]$ we have

$$a_{ij}(t, \cdot), b_i(t, \cdot), c_0(t, \cdot) \in L^\infty(\Omega), \quad f(t, \cdot) \in H^{-1}(\Omega)$$

and the operator $L(t)$ is a second order operator in divergence form. Now (1.50) yields the boundary value problem

$$L(t)y(t) = f(t) - y_t(t), \quad y(t)|_{\partial\Omega} = 0.$$

Since $f(t) - y_t(t) \in H^{-1}(\Omega) = (H_0^1(\Omega))^*$, the elliptic case motivates to require that for almost all $t \in [0, T]$ the variational equality

$$a(y(t), v; t) = \langle f(t), v\rangle_{H^{-1}, H_0^1} - \langle y_t(t), v\rangle_{H^{-1}, H_0^1} \quad \forall v \in H_0^1(\Omega)$$

holds with the associated bilinear form

$$a(y, v; t) := \int_\Omega \left(\sum_{i,j=1}^n a_{ij}(t) y_{x_i} v_{x_j} + \sum_{i=1}^n b_i(t) y_{x_i} v + c_0(t) y v \right) dx,$$

$$y, v \in H_0^1(\Omega). \tag{1.57}$$

We arrive at the following definition.

Definition 1.27 (Weak solution of parabolic PDE) Let $\Omega \subset \mathbb{R}^n$ be open and bounded. Let the coefficients satisfy (1.55). Consider with

$$H := L^2(\Omega), \qquad V := H_0^1(\Omega)$$

the Gelfand triple $V \hookrightarrow H \hookrightarrow V^*$, i.e.,

$$H_0^1(\Omega) \hookrightarrow L^2(\Omega) \hookrightarrow H^{-1}(\Omega).$$

Then for $f \in L^2(0, T; H^{-1}(\Omega))$, $y_0 \in L^2(\Omega)$ a function

$$y \in W(0, T; L^2, H_0^1)$$

is a *weak solution of the initial-boundary value problem* (1.50) if y satisfies the variational equation

$$\langle y_t(t), v\rangle_{H^{-1}, H_0^1} + a(y(t), v; t) = \langle f(t), v\rangle_{H^{-1}, H_0^1}$$

$$\forall v \in H_0^1(\Omega) \text{ and a.a. } t \in [0, T] \tag{1.58}$$

and the initial condition

$$y(0) = y_0, \tag{1.59}$$

where the bilinear form $a(\cdot, \cdot; t)$ is given in (1.57).

Remark 1.15 Since $W(0,T;L^2,H_0^1)\hookrightarrow C([0,T];L^2(\Omega))$ by Theorem 1.32, the initial condition (1.59) makes sense.

For PDE-constrained optimization the following equivalent weak formulation is more convenient.

Definition 1.28 (Weak solution of parabolic PDE, equivalent formulation) With the same assumptions and notations as in Definition 1.27 the following definition is equivalent. For $f\in L^2(0,T;H^{-1}(\Omega))$, $y_0\in L^2(\Omega)$ a function

$$y\in W(0,T;L^2,H_0^1)$$

is a *weak solution of the initial-boundary value problem* (1.50) if y satisfies the variational equation

$$\begin{aligned}\int_0^T \langle y_t(t),v(t)\rangle_{H^{-1},H_0^1}\,dt+\int_0^T a(y(t),v(t);t)\,dt\\ =\int_0^T \langle f(t),v(t)\rangle_{H^{-1},H_0^1}\,dt\quad \forall v\in L^2(0,T;H_0^1)\end{aligned}\tag{1.60}$$

with $a(\cdot,\cdot;t)$ in (1.57) and the initial condition

$$y(0)=y_0.\tag{1.59}$$

Theorem 1.33 *Definitions* 1.27 *and* 1.28 *are equivalent.*

Proof Let $y\in W(0,T;H,V)$ be a weak solution according to Definiton 1.27. (1.58) implies

$$\begin{aligned}\langle y_t(t),v(t)\rangle_{V^*,V}+a(y(t),v(t);t)=\langle f(t),v(t)\rangle_{V^*,V}\\ \forall v\in L^2(0,T;V)\text{ and a.a. }t\in(0,T).\end{aligned}\tag{1.61}$$

In fact, since $y\in W(0,T;H,V)$ and $f\in L^2(0,T;V^*)$, it is to check that both sides in (1.61) are in $L^1(0,T)$. For a simple function $v(t)=\sum_{i=1}^m 1_{E_i}(t)v_i$, $v_i\in V$, (1.61) is obvious, since then

$$\begin{aligned}&\langle y_t(t),v(t)\rangle_{V^*,V}+a(y(t),v(t);t)-\langle f(t),v(t)\rangle_{V^*,V}\\ &\quad=\sum_{i=1}^m 1_{E_i}(t)(\langle y_t(t),v_i\rangle_{V^*,V}+a(y(t),v_i;t)-\langle f(t),v_i\rangle_{V^*,V})=0\quad\text{for a.a. }t.\end{aligned}$$

For general $v\in L^2(0,T;V)$ choose a sequence v_k of simple functions with $v_k(t)\to v(t)$ almost everywhere. Then we know that (1.61) holds for all v_k outside a set of measure zero (the countable union of the exceptional sets for v_k). Since $v_k(t)\to v(t)$ in V almost everywhere, we conclude that by continuity (1.61) holds also for the limit v. Integrating (1.61) with respect to t shows that (1.60) holds.

Let now $y \in W(0,T;H,V)$ be a weak solution according to Definition 1.28. Then (1.58) must hold. In fact, otherwise we find $w \in V$ and a set E of nonzero measure such that for $v = w$ the difference of the left and right hand side of (1.58) is positive on E. But then (1.60) would not hold for $v(t) = 1_E(t)w$. Hence, (1.60) implies (1.58).

Remark 1.16 By (1.61) the weak formulation (1.58) (or equivalently (1.60)) means that $y_t + Ly = f$ holds in $L^2(0,T;V^*)$.

1.3.2.4 Existence and Uniqueness of Weak Solutions

Let

$$V \hookrightarrow H \hookrightarrow V^*$$

be a Gelfand triple.

Remark 1.17 We recall that the imbedding $H \hookrightarrow V^*$ is given by $\langle v, w\rangle_{V^*,V} = (v,w)_H$ for all $v \in H$, $w \in V$.

In generalization of the weak formulation (1.58), (1.59) we consider the problem to find for $f \in L^2(0,T;V^*)$, $y_0 \in H$ a solution

$$y \in W(0,T;H,V)$$

of the

Abstract Parabolic Evolution Problem

Find $y \in W(0,T;H,V)$ such that

$$\langle y_t(t), v\rangle_{V^*,V} + a(y(t),v;t) = \langle f(t), v\rangle_{V^*,V} \quad \forall v \in V \text{ and a.a. } t \in [0,T] \tag{1.62}$$

with the initial condition

$$y(0) = y_0. \tag{1.63}$$

We will work under the following assumptions:

Assumption 1.34

(i) *$V \hookrightarrow H \hookrightarrow V^*$ is a Gelfand triple, H, V separable Hilbert spaces.*
(ii) *$a(\cdot,\cdot,t) : V \times V \to \mathbb{R}$ is for almost all $t \in (0,T)$ a bilinear form and there are $\alpha, \beta > 0$ and $\gamma \geq 0$ with*

$$|a(v,w;t)| \leq \alpha \|v\|_V \|w\|_V \quad \forall v,w \in V \text{ and a.a. } t \in (0,T), \tag{1.64}$$

$$a(v, v; t) + \gamma \|v\|_H^2 \geq \beta \|v\|_V^2 \quad \forall v \in V \text{ and a.a. } t \in (0, T). \tag{1.65}$$

The mappings $t \mapsto a(v, w; t) \in \mathbb{R}$ *are measurable for all* $v, w \in V$.

(iii) $y_0 \in H$, $f \in L^2(0, T; V^*)$.

Example 1.5 Assumption 1.34 is obviously satisfied for the uniformly parabolic initial boundary value problem (1.50) with $H = L^2(\Omega)$, $V = H_0^1(\Omega)$ and

$$a_{ij}, b_i, c_0 \in L^\infty(\Omega_T), \quad f \in L^2(0, T; H^{-1}(\Omega)).$$

In fact, it is easy to check that the associated bilinear form $a(\cdot, \cdot; t)$ in (1.57) satisfies (1.64) and (1.65).

It is easy to show that under Assumption 1.34 for any $y, v \in L^2(0, T; V)$ the function $t \mapsto a(y(t), v(t); t)$ is in $L^1(0, T)$.

As above (1.62) implies

$$\langle y_t(t), v(t) \rangle_{V^*,V} + a(y(t), v(t); t) = \langle f(t), v(t) \rangle_{V^*,V}$$

$$\forall v \in L^2(0, T; V) \text{ and a.a. } t \in [0, T). \tag{1.61}$$

and (1.62), (1.63) is equivalent to

Abstract Parabolic Evolution Problem, Equivalent Form

Find $y \in W(0, T; H, V)$ such that

$$\int_0^T \langle y_t(t), v(t) \rangle_{V^*,V}\, dt + \int_0^T a(y(t), v(t); t)\, dt = \int_0^T \langle f(t), v(t) \rangle_{V^*,V}\, dt$$

$$\forall v \in L^2(0, T; V) \tag{1.66}$$

with the initial condition

$$y(0) = y_0. \tag{1.63}$$

Energy Estimate and Uniqueness Result

Theorem 1.35 *Let Assumption* 1.34 *hold. Then the abstract parabolic evolution problem has at most one solution* $y \in W(0, T; H, V)$ *and it satisfies the energy estimate*

$$\|y(t)\|_H^2 + \|y\|_{L^2(0,t;V)}^2 + \|y_t\|_{L^2(0,t;V^*)}^2 \leq C(\|y_0\|_H^2 + \|f\|_{L^2(0,t;V^*)}^2) \quad \forall t \in (0, T], \tag{1.67}$$

where $C > 0$ *depends only on* β *and* γ *in Assumption* 1.34.

Proof The proof is obtained by using $v = y$ in (1.61), using Theorem 1.32 and applying the Gronwall lemma. See for example [47].

Existence Result by Galerkin Approximation

One of the milestones in the modern theory of PDEs is the observation that the energy estimate (1.67) can be used as a foundation of an existence proof. To carry out the program, we note that since V is separable there exists a countable set

$$\{v_k : k \in \mathbb{N}\} \subset V$$

of linearly independent elements v_k of V, such that the linear span of $\{v_k : k \in \mathbb{N}\}$ is dense in V (take first a countable dense subset and drop elements that lie in the span of previous elements). Moreover, let

$$V_k := \text{span}\{v_1, \dots, v_k\}.$$

Then $V_k \subset V$ are Hilbert spaces and $\cup V_k$ is dense in V. Since V is dense in H, we find

$$y_{0,k} = \sum_{i=1}^{k} \alpha_{ik} v_i \in V_k \quad \text{with } y_{0,k} \to y_0 \quad \text{in } H.$$

Now fix $k \in \mathbb{N}$. We look for a function

$$y_k(t) := \sum_{i=1}^{k} \varphi_{ik}(t) v_i, \quad \varphi_{ik} \in H^1(0,T), \tag{1.68}$$

satisfying the finite dimensional *Galerkin approximation* of (1.62), (1.63)

$$\langle (y_k)_t(t), v\rangle_{V^*,V} + a(y_k(t), v; t) = \langle f(t), v\rangle_{V^*,V}$$
$$\forall v \in V_k \text{ and a.a. } t \in [0,T], \tag{1.69}$$
$$y(0) = y_{0,k}. \tag{1.70}$$

It is easy to check that functions y_k of the type (1.68) are in $W(0,T;H,V)$ with weak derivative

$$(y_k)_t(t) = \sum_{i=1}^{k} \varphi'_{ik}(t) v_i \in L^2(0,T;V),$$

where $\varphi'_i \in L^2(0,T)$ are the weak derivatives of $\varphi_i \in H^1(0,T)$. Since it is sufficient to test with the basis $\{v_1, \dots, v_k\}$ in (1.69), we conclude that (1.69), (1.70) is equivalent to the system of ODEs for $\varphi_{1k}, \dots, \varphi_{kk}$

$$\sum_{i=1}^{k} (v_i, v_j)_H \varphi'_{ik}(t) + \sum_{i=1}^{k} a(v_i, v_j; t)\varphi_{ik}(t) = \langle f(t), v_j\rangle_{V^*,V},$$
$$1 \le j \le k, \text{ a.a. } t \in [0,T], \tag{1.71}$$
$$\varphi_{ik}(0) = \alpha_{ik}, \quad 1 \le i \le k. \tag{1.72}$$

Here we have used that $V \hookrightarrow H \hookrightarrow V^*$ yields $\langle v_i, v_j \rangle_{V^*,V} = (v_i, v_j)_H$.

We have the following result.

Theorem 1.36 *Let Assumption* 1.34 *hold. Then the Galerkin approximations* (1.69), (1.70) *have unique solutions $y_k \in W(0,T;H,V)$ of the form* (1.68) *and y_k satisfies the energy estimate*

$$\|y_k(t)\|_H^2 + \|y_k\|_{L^2(0,t;V)}^2 + \|(y_k)_t\|_{L^2(0,t;V^*)}^2 \le C(\|y_{0,k}\|_H^2 + \|f\|_{L^2(0,t;V^*)}^2)$$
$$\forall t \in (0,T], \tag{1.73}$$

where $C > 0$ depends only on β, γ in Assumption 1.34.

Proof We know that (1.69), (1.70) is equivalent to (1.71), (1.72). With $A = ((v_i, v_j)_H)_{i,j}$, $M(t) := (a(v_i, v_j; t))_{i,j} \in L^\infty(0,T;\mathbb{R}^{k,k})$, $F(t) := (\langle f(t), v_j \rangle_{V^*,V})_j \in L^2(0,T)$ we can write (1.71), (1.72) as

$$A^T(\varphi'_{ik}(t))_i + M(t)^T(\varphi_{ik}(t))_i = F(t), \quad (\varphi_{ik}(0))_i = (\alpha_{ik})_i.$$

Since v_i are linearly independent, the matrix A is invertible and by standard theory for ODEs with measurable coefficients there exists a unique solution $(\varphi_{ik})_i$ with $\varphi_{ik} \in H^1(0,T)$, see for example [144].

Now Theorem 1.35 applied to (1.69), (1.70) yields the asserted energy estimate (1.73).

Theorem 1.37 *Let Assumption* 1.34 *hold. Then the abstract parabolic evolution problem* (1.62), (1.63) *has a unique solution $y \in W(0,T;H,V)$.*

By Example 1.5 this yields immediately

Corollary 1.1 *Let $\Omega \subset \mathbb{R}^n$ be open and bounded and let $\frac{\partial}{\partial t} + L$ with L in* (1.51) *be uniformly parabolic, where $a_{ij}, b_i, c_0 \in L^\infty(\Omega_T)$. Then for any $f \in L^2(0,T;H^{-1}(\Omega))$ and $y_0 \in L^2(\Omega)$ the initial boundary value problem* (1.50) *has a unique weak solution $y \in W(0,T;L^2,H_0^1)$ and satisfies the energy estimate* (1.67) *with $H = L^2(\Omega)$, $V = H_0^1(\Omega)$, $V^* = H^{-1}(\Omega)$.*

Proof of Theorem 1.37 Since $\|y_{0,k}\|_H \to \|y_0\|_H$, the energy estimate (1.73) yields a constant $C > 0$ such that

$$\|y_k\|_{L^2(0,T;V)} < C, \qquad \|(y_k)_t\|_{L^2(0,T;V^*)} < C.$$

Now $L^2(0,T;V)$, $L^2(0,T;V^*)$ are Hilbert spaces and thus reflexive. Hence, we find by Theorem 1.17 a subsequence (y_{k_i}) with

$$y_{k_i} \rightharpoonup y \quad \text{in } L^2(0,T;V), \qquad (y_{k_i})_t \rightharpoonup w \quad \text{in } L^2(0,T;V^*).$$

It is not difficult to show that this implies $w = y_t$. Now (1.69) implies

$$\int_0^T \big(\langle (y_k)_t(t), v\rangle_{V^*,V} + a(y_k(t), v; t) - \langle f(t), v\rangle_{V^*,V}\big)\,\varphi(t)\,dt = 0$$
$$\forall v \in V_k,\ \varphi \in C_c^\infty((0,T))$$

and the first two terms are bounded linear functionals w.r.t. $(y_k)_t$ and y_k, respectively. Limit transition gives

$$\int_0^T \big(\langle y_t(t), v\rangle_{V^*,V} + a(y(t), v; t) - \langle f(t), v\rangle_{V^*,V}\big)\,\varphi(t)\,dt = 0$$
$$\forall v \in \bigcup_k V_k,\ \varphi \in C_c^\infty((0,T)).$$

This shows (1.62) by Lemma 1.5, where we use that $\bigcup_k V_k$ is dense in V.

Finally, also the initial condition (1.63) holds. In fact, let $\varphi \in C^\infty([0,T])$ with $\varphi(0) = 1$, $\varphi(T) = 0$. Then $t \mapsto w(t) = \varphi(t)v \in W(0,T;H,V)$ for all $v \in V$ and $w(0) = v$, $w(T) = 0$ yields by Theorem 1.32

$$\int_0^T \big(-\langle \varphi'(t)v, y(t)\rangle_{V^*,V} + a(y(t), \varphi(t)v; t) - \langle f(t), \varphi(t)v\rangle_{V^*,V}\big)\,dt = (y(0), v)_H$$
$$\forall v \in V.$$

Similarly, we have by (1.69) and Theorem 1.32

$$\int_0^T \big(-\langle \varphi'(t)v, y_{k_i}(t)\rangle_{V^*,V} + a(y_{k_i}(t), \varphi(t)v; t) - \langle f(t), \varphi(t)v\rangle_{V^*,V}\big)\,dt$$
$$= (y_{0,k}, v)_H \quad \forall v \in V_{k_i}$$

and the left hand side tends to the left hand side of the previous equation by the weak convergence of y_{k_i}. This gives $(y(0), v)_H = \lim_{k\to\infty} (y_{0,k}, v)_H = (y_0, v)_H$ for all $v \in \bigcup_k V_k$ and hence $y(0) = \lim_{k\to\infty} y_{0,k} = y_0$, since $\bigcup_k V_k$ is dense in V.

Operator Formulation

By using the equivalence of (1.58) and (1.61) we can summarize (see Remark 1.16) that for coefficients satisfying (1.55) the weak formulation (1.58), (1.59) (or equivalently (1.60), (1.59)) defines a bounded linear operator

$$A : y \in W(0,T; L^2(\Omega), H_0^1(\Omega)) \mapsto \begin{pmatrix} y_t + Ly \\ y(0,\cdot) \end{pmatrix} \in L^2(0,T; (H_0^1(\Omega))^*) \times L^2(\Omega)$$

in the sense that for all $(f, y_0) \in L^2(0, T; (H_0^1(\Omega))^*) \times L^2(\Omega)$

$$\begin{pmatrix} y_t + Ly \\ y(0, \cdot) \end{pmatrix} = \begin{pmatrix} f \\ y_0 \end{pmatrix} \iff \text{(1.58), (1.59) hold} \iff \text{(1.60), (1.59) hold.}$$

Moreover, A has a bounded inverse by Corollary 1.1.

1.3.2.5 Regularity Results

We assume now in addition that the following assumption holds.

Assumption 1.38 *In addition to Assumption* 1.34 *we assume that*

$$\begin{aligned} & a(v, w; \cdot) \in C^1([0, T]), \qquad a_t(v, w; t) \leq \alpha_1 \|v\|_V \|w\|_V \quad \forall v, w \in V, \\ & y_0 \in \left\{ w \in V : a(w, \cdot; 0) \in H^* \right\}, \\ & f \in W(0, T; H, V). \end{aligned}$$

Theorem 1.39 *Let Assumption* 1.38 *hold. Then the solution of* (1.62) *satisfies in addition* $y_t \in W(0, T; H, V)$ *and satisfies the equation*

$$\begin{aligned} & \langle y_{tt}(t), w \rangle_{V^*, V} + a(y_t(t), w; t) = \langle f_t(t), w \rangle_{V^*, V} - a_t(y(t), w; t), \\ & \langle y_t(0), w \rangle_{V^*, V} = (f(0), w)_H - a(y_0, w; 0) \quad \forall w \in V. \end{aligned} \tag{1.74}$$

Proof See for example [47].

From the temporal regularity we can deduce spatial regularity, if L is for example a uniformly elliptic operator. In fact, we have $y_t, f \in W(0, T; H, V) \hookrightarrow C([0, T]; H)$ and thus

$$\|y_t(t)\|_H + \|f(t)\|_H \leq C \quad \text{for a.a. } t \in [0, T],$$

where $H = L^2(\Omega)$, $V = H_0^1(\Omega)$. This yields

$$\begin{aligned} & a(y(t), w; t) = -\langle y_t(t), w \rangle_{(H_0^1)^*, H_0^1} + (f(t), w)_{L^2} = (-y_t(t) + f(t), w)_{L^2} \\ & \forall w \in H_0^1(\Omega). \end{aligned}$$

Now our regularity results for uniformly elliptic operators imply under the assumptions of Theorem 1.26 or 1.28

$$\|y(t)\|_{H^2(\Omega')} \leq C(\|y_t\|_{L^\infty(0,T;L^2)} + \|f\|_{L^\infty(0,T;L^2)} + \|y\|_{L^\infty(0,T;H^1)}),$$

either for $\Omega' \subset\subset \Omega$ or for $\Omega' = \Omega$ if Ω has C^2-boundary.

1.3.2.6 An Existence and Uniqueness Result for Semilinear Parabolic Equations

We finally state an existence and uniqueness result for a uniformly parabolic semilinear equation

$$\begin{aligned} y_t + Ly + d(t,x,y) &= f \quad \text{on } \Omega_T \\ \frac{\partial y}{\partial \nu_A} + b(t,x,y) &= g \quad \text{on } [0,T] \times \partial\Omega \\ y(0,\cdot) &= y_0, \end{aligned} \tag{1.75}$$

where the operator L is given by

$$Ly := -\sum_{i,j=1}^{n} (a_{ij} y_{x_i})_{x_j}, \quad a_{ij} \in L^\infty(\Omega),\ a_{ij} = a_{ji} \tag{1.34}$$

and L is assumed to be uniformly elliptic in the sense that there is a constant $\theta > 0$ such that

$$\sum_{i,j=1}^{n} a_{ij}(x)\xi_i\xi_j \geq \theta \|\xi\|^2 \quad \text{for almost all } x \in \Omega \text{ and all } \xi \in \mathbb{R}^n. \tag{1.35}$$

Moreover, we assume that the functions $d : \Omega_T \times \mathbb{R} \to \mathbb{R}$, and $b : [0,T] \times \partial\Omega \times \mathbb{R} \to \mathbb{R}$ satisfy

$$\begin{aligned} &d(t,x,\cdot) \text{ is continuous and monotone increasing for a.a. } (t,x) \in \Omega_T, \\ &d(t,x,\cdot) \text{ is locally Lipschitz continuous uniformly for a.a. } (t,x) \in \Omega_T, \\ &b(t,x,\cdot) \text{ is continuous and monotone increasing for a.a. } (t,x) \in [0,T] \times \partial\Omega, \\ &b(t,x,\cdot) \text{ is locally Lipschitz continuous uniformly for a.a. } (t,x) \in [0,T] \times \partial\Omega, \\ &d(\cdot,y) \in L^\infty(\Omega_T), b(\cdot,y) \in L^\infty([0,T] \times \partial\Omega) \quad \text{for all } y \in \mathbb{R}. \end{aligned} \tag{1.76}$$

Under these assumptions one can show the following theorem.

Theorem 1.40 *Let $\Omega \subset \mathbb{R}^n$ be open and bounded with $C^{1,1}$-boundary, let $a_{ij} = a_{ji} \in L^\infty(\Omega)$ and let* (1.35), (1.76) *be satisfied. Moreover, let $r > n/2 + 1$, $s > n + 1$. Then* (1.75), (1.34) *has for any $f \in L^r(\Omega_T)$, $g \in L^s([0,T] \times \partial\Omega)$ and $y_0 \in C(\bar{\Omega})$ a unique weak solution $y \in W(0,T; L^2(\Omega), H^1(\Omega)) \cap C(\bar{\Omega}_T)$. There exists a constant $C_\infty > 0$ with*

$$\begin{aligned} \|y\|_{W(0,T;L^2,H^1)} + \|y\|_{C(\bar{\Omega}_T)} \leq C_\infty(&\|f - d(\cdot,0)\|_{L^r(\Omega_T)} \\ &+ \|g - b(\cdot,0)\|_{L^s([0,T]\times\partial\Omega)} + \|y_0\|_{C(\bar{\Omega})}), \end{aligned}$$

where C_∞ does not depend on f, g, b, d, y_0.

Proof See [23, 114].

1.4 Gâteaux- and Fréchet Differentiability

1.4.1 Basic Definitions

We extend the notion of differentiability to operators between Banach spaces.

Definition 1.29 Let $F : U \subset X \to Y$ be an operator with Banach spaces X, Y and $U \neq \emptyset$ open.

(a) F is called *directionally differentiable* at $x \in U$ if the limit

$$dF(x,h) = \lim_{t \to 0^+} \frac{F(x+th) - F(x)}{t} \in Y$$

exists for all $h \in X$. In this case, $dF(x,h)$ is called directional derivative of F in the direction h.

(b) F is called *Gâteaux differentiable* (G-differentiable) at $x \in U$ if F is directionally differentiable at x and the directional derivative $F'(x) : X \ni h \mapsto dF(x,h) \in Y$ is bounded and linear, i.e., $F'(x) \in \mathcal{L}(X,Y)$.

(c) F is called *Fréchet differentiable* (F-differentiable) at $x \in U$ if F is Gâteaux differentiable at x and if the following approximation condition holds:

$$\|F(x+h) - F(x) - F'(x)h\|_Y = o(\|h\|_X) \quad \text{for } \|h\|_X \to 0.$$

(d) If F is directionally-/G-/F-differentiable at every $x \in V$, $V \subset U$ open, then F is called directionally-/G-/F-differentiable on V.

Higher derivatives can be defined as follows:

If F is G-differentiable in a neighborhood V of x, and $F' : V \to \mathcal{L}(X,Y)$ is itself G-differentiable at x, then F is called twice G-differentiable at x. We write $F''(x) \in \mathcal{L}(X, \mathcal{L}(X,Y))$ for the second G-derivative of F at x. It should be clear now how the kth derivative is defined.

In the same way, we define F-differentiability of order k.

It is easy to see that F-differentiability of F at x implies continuity of F at x. We say that F is k-times continuously F-differentiable if F is k-times F-differentiable and $F^{(k)}$ is continuous.

We collect a couple of facts:

(a) The chain rule holds for F-differentiable operators:

$$H(x) = G(F(x)), \; F, G \text{ F-differentiable at } x \text{ and } F(x) \text{, respectively}$$
$$\Longrightarrow \quad H \text{ F-differentiable at } x \text{ with } H'(x) = G'(F(x))F'(x).$$

Moreover, if F is G-differentiable at x and G is F-differentiable at $F(x)$, then H is G-differentiable and the chain rule holds. As a consequence, also the sum rule holds for F- and G-differentials.

(b) If F is G-differentiable on a neighborhood of x and F' is continuous at x then F is F-differentiable at x.

(c) If $F : X \times Y \to Z$ is F-differentiable at (x, y) then $F(\cdot, y)$ and $F(x, \cdot)$ are F-differentiable at x and y, respectively. These derivatives are called partial derivatives and denoted by $F_x(x, y)$ and $F_y(x, y)$, respectively. There holds (since F is F-differentiable)

$$F'(x, y)(h_x, h_y) = F_x(x, y)h_x + F_y(x, y)h_y.$$

(d) If F is G-differentiable in a neighborhood V of x, then for all $h \in X$ with $\{x + th : t \in [0, 1]\} \subset V$, the following holds:

$$\|F(x + h) - F(x)\|_Y \le \sup_{0<t<1} \|F'(x + th)h\|_Y$$

If $t \in [0, 1] \mapsto F'(x + th)h \in Y$ is continuous, then

$$F(x + h) - F(x) = \int_0^1 F'(x + th)h\,dx,$$

where the Y-valued integral is defined as a Riemann integral.

We only prove the last assertion: As a corollary of the Hahn-Banach theorem, we obtain that for all $y \in Y$ there exists a $y^* \in Y^*$ with $\|y^*\|_{Y^*} = 1$ and

$$\|y\|_Y = \langle y^*, y\rangle_{Y^*,Y}.$$

Hence,

$$\|F(x + h) - F(x)\|_Y = \max_{\|y^*\|_{Y^*}=1} d_{y^*}(1)$$

$$\text{with } d_{y^*}(t) = \langle y^*, F(x + th) - F(x)\rangle_{Y^*,Y}.$$

By the chain rule for G-derivatives, we obtain that d is G-differentiable in a neighborhood of $[0, 1]$ with

$$d'_{y^*}(t) = \langle y^*, F'(x + th)h\rangle_{Y^*,Y}.$$

G-differentiability of $d : (-\varepsilon, 1 + \varepsilon) \to \mathbb{R}$ means that d is differentiable in the classical sense. The mean value theorem yields

$$\langle y^*, F(x + h) - F(x)\rangle_{Y^*,Y} = d_{y^*}(1) = d_{y^*}(1) - d_{y^*}(0) = d'_{y^*}(\tau) \le \sup_{0<t<1} d'_{y^*}(t)$$

for appropriate $\tau \in (0, 1)$. Therefore,

$$\begin{aligned}
\|F(x + h) - F(x)\|_Y = \max_{\|y^*\|_{Y^*}=1} d_{y^*}(1) &\le \sup_{\|y^*\|_{Y^*}=1} \sup_{0<t<1} \langle y^*, F'(x + th)h\rangle_{Y^*,Y} \\
&= \sup_{0<t<1} \sup_{\|y^*\|_{Y^*}=1} \langle y^*, F'(x + th)h\rangle_{Y^*,Y} \\
&= \sup_{0<t<1} \|F'(x + th)h\|_Y.
\end{aligned}$$

1.4.2 Implicit Function Theorem

For optimization problems with PDE-constraints $e(y, u) = 0$ a quite common situation is that $e : Y \times U \to Z$ is continuously F-differentiable and $e_y(y, u) \in \mathcal{L}(Y, Z)$ has a bounded inverse. Then the following well known implicit function theorem shows that $e(y, u) = 0$ defines locally a continuously F-differentiable control-to-state map $u \mapsto y(u)$.

Theorem 1.41 (Implicit Function Theorem) *Let X, Y, Z be Banach spaces and let $F : G \to Z$ be a continuously F-differentiable map from an open set $G \subset X \times Y$ to Z. Let $(\bar{x}, \bar{y}) \in G$ be such that $F(\bar{x}, \bar{y}) = 0$ and that $F_y(\bar{x}, \bar{y}) \in \mathcal{L}(Y, Z)$ has a bounded inverse.*

Then there exists an open neighborhood $U_X(\bar{x}) \times U_Y(\bar{y}) \subset G$ of $(\bar{x}, \bar{y})$ and a unique continuous function $w : U_X(\bar{x}) \to Y$ such that

(i) $w(\bar{x}) = \bar{y}$,
(ii) *For all $x \in U_X(\bar{x})$ there exists exactly one $y \in U_Y(\bar{y})$ with $F(x, y) = 0$, namely $y = w(x)$.*

Moreover, the mapping $w : U_X(\bar{x}) \to Y$ is continuously F-differentiable with derivative

$$w'(x) = F_y(x, w(x))^{-1} F_x(x, w(x)).$$

If $F : G \to Z$ is m-times continuously F-differentiable then also $w : U_X(\bar{x}) \to Y$ is m-times continuously F-differentiable.

Proof See for example [151, Thm. 4.B]

1.5 Existence of Optimal Controls

In the introduction we have discussed several examples of optimal control problems. We will now consider the question whether there exists an optimal solution. In this context the concept of weak convergence will be important.

1.5.1 Existence Result for a General Linear-Quadratic Problem

All linear-quadratic optimization problems in the introduction can be converted to a linear-quadratic optimization problem of the form

$$\begin{aligned} &\min_{(y,u)\in Y\times U} \quad J(y, u) := \frac{1}{2}\|Qy - q_d\|_H^2 + \frac{\alpha}{2}\|u\|_U^2 \\ &\text{subject to} \quad Ay + Bu = g, \quad u \in U_{\text{ad}}, \; y \in Y_{\text{ad}} \end{aligned} \tag{1.77}$$

where H, U are Hilbert spaces, Y, Z are Banach spaces and $q_d \in H$, $g \in Z$, $A \in \mathcal{L}(Y, Z)$, $B \in \mathcal{L}(U, Z)$, $Q \in \mathcal{L}(Y, H)$ and the following assumption holds.

Assumption 1.42

1. $\alpha \geq 0$, $U_{\mathrm{ad}} \subset U$ *is convex, closed and in the case* $\alpha = 0$ *bounded.*
2. $Y_{\mathrm{ad}} \subset Y$ *is convex and closed, such that* (1.77) *has a feasible point.*
3. $A \in \mathcal{L}(Y, Z)$ *has a bounded inverse.*

Definition 1.30 A state-control pair $(\bar{y}, \bar{u}) \in Y_{\mathrm{ad}} \times U_{\mathrm{ad}}$ is called *optimal* for (1.77), if $A\bar{y} + B\bar{u} = g$ and

$$J(\bar{y}, \bar{u}) \leq J(y, u) \quad \forall (y, u) \in Y_{\mathrm{ad}} \times U_{\mathrm{ad}}, \quad Ay + Bu = g.$$

We prove first the following existence result for (1.77).

Theorem 1.43 *Let Assumption* 1.42 *hold. Then problem* (1.77) *has an optimal solution* $(\bar{y}, \bar{u})$. *If* $\alpha > 0$ *then the solution is unique.*

Proof We present first a proof that assumes the reflexivity of Y, since this proof can easily be extended to nonlinear problems. The modification for general Y will be mentioned at the end.

Denote the feasible set by

$$F_{\mathrm{ad}} := \{(y, u) \in Y \times U : (y, u) \in Y_{\mathrm{ad}} \times U_{\mathrm{ad}}, Ay + Bu = g\}.$$

Since $J \geq 0$ and F_{ad} is nonempty, the infimum

$$J^* := \inf_{(y,u) \in F_{\mathrm{ad}}} J(y, u)$$

exists and hence we find a minimizing sequence $(y_k, u_k) \subset F_{\mathrm{ad}}$ with

$$\lim_{k \to \infty} J(y_k, u_k) = J^*.$$

The sequence (u_k) is bounded, since by assumption either U_{ad} is bounded or $\alpha > 0$. In the latter case the boundedness follows from

$$J(y_k, u_k) \geq \frac{\alpha}{2} \|u_k\|_U^2.$$

Since $A \in \mathcal{L}(Y, Z)$, $B \in \mathcal{L}(U, Z)$, and $A^{-1} \in \mathcal{L}(Z, Y)$, this implies that also the state sequence (y_k) given by $y_k = A^{-1}(g - Bu_k)$ is bounded. Since $Y \times U$ is reflexive, Theorem 1.17 yields a weakly convergent subsequence $(y_{k_i}, u_{k_i}) \subset (y_k, u_k)$ and some $(\bar{y}, \bar{u}) \in Y \times U$ with $(y_{k_i}, u_{k_i}) \rightharpoonup (\bar{y}, \bar{u})$ as $i \to \infty$. To show that $(\bar{y}, \bar{u}) \in F_{\mathrm{ad}}$ we note that

$$(y_k, u_k) \subset F_{\mathrm{ad}} \cap (\bar{B}_Y(r) \times \bar{B}_U(r)) =: M$$

for $r > 0$ large enough, where $\bar{B}_Y(r)$, $\bar{B}_U(r)$ denote the closed balls of radius r in Y, U. By assumption $Y_{\text{ad}} \times U_{\text{ad}}$ is closed, convex and thus also F_{ad} is closed and convex. Thus, the set M is bounded, closed and convex and consequently by Theorem 1.17 weakly sequentially compact. Therefore, there exists a weakly convergent subsequence $(y_{k_i}, u_{k_i}) \subset (y_k, u_k)$ and some $(\bar{y}, \bar{u}) \in F_{\text{ad}}$ with $F_{\text{ad}} \ni (y_{k_i}, u_{k_i}) \rightharpoonup (\bar{y}, \bar{u})$ as $i \to \infty$.

Finally, $(y, u) \in Y \times U \to J(y, u)$ is obviously continuous and convex. We conclude by Theorem 1.18 that

$$J^* = \lim_{i \to \infty} J(y_{k_i}, u_{k_i}) \geq J(\bar{y}, \bar{u}) \geq J^*,$$

where the last inequality follows from $(\bar{y}, \bar{u}) \in F_{\text{ad}}$. Therefore, $(\bar{y}, \bar{u})$ is the optimal solution of (1.77). If $\alpha > 0$ then $u \mapsto f(A^{-1}(g - Bu), u)$ is strictly convex, which contradicts the existence of more than one minimizer.

If Y is not reflexive, we can still select a weakly convergent subsequence $(u_{k_i}) \subset (u_k)$ since U is reflexive. But since $y_{k_i} = A^{-1}(g - Bu_{k_i})$ and $A^{-1}B \in \mathcal{L}(U, Y)$, also the subsequence (y_{k_i}) converges weakly in Y and we obtain as above $F_{\text{ad}} \ni (y_{k_i}, u_{k_i}) \rightharpoonup (\bar{y}, \bar{u})$ as $i \to \infty$.

Remark 1.18 Equivalently, one can study the *reduced problem*.

In fact, $Ay + Bu = g$ implies $y = A^{-1}(g - Bu)$ and thus the problem (1.77) is equivalent to

$$\min_{u \in U} \hat{J}(u) \quad \text{s.t.} \quad u \in \hat{U}_{\text{ad}}$$

with

$$\hat{J}(u) = J(A^{-1}(g - Bu), u), \qquad \hat{U}_{\text{ad}} = \left\{u \in U : u \in U_{\text{ad}},\ A^{-1}(g - Bu) \in Y_{\text{ad}}\right\}.$$

It is easy to see that $\hat{J}$ is continuous and convex and $\hat{U}_{\text{ad}}$ is closed and convex. An argumentation as before shows that a minimizing sequence is bounded and thus contains a weakly convergent subsequence convergent to some $\bar{u} \in \hat{U}_{\text{ad}}$. Lower semicontinuity implies the optimality of $\bar{u}$. Setting $\bar{y} = A^{-1}(g - B\bar{u})$, we obtain a solution $(\bar{y}, \bar{u})$ of (1.77).

1.5.2 Existence Results for Nonlinear Problems

The existence result can be extended to nonlinear problems

$$\min_{(y,u) \in Y \times U} J(y, u) \quad \text{subject to} \quad e(y, u) = 0, \quad u \in U_{\text{ad}},\ y \in Y_{\text{ad}}, \tag{1.78}$$

where $J : Y \times U \to \mathbb{R}$, $e : Y \times U \to Z$ are continuous with a Banach space Z and reflexive Banach spaces U, Y.

Similarly as above, existence can be shown under the following assumptions.

Assumption 1.44

1. $U_{\text{ad}} \subset U$ *is convex, bounded and closed.*
2. $Y_{\text{ad}} \subset Y$ *is convex and closed, such that* (1.78) *has a feasible point.*
3. *The state equation* $e(y,u) = 0$ *has a bounded solution operator* $u \in U_{\text{ad}} \mapsto y(u) \in Y$.
4. $(y,u) \in Y \times U \mapsto e(y,u) \in Z$ *is continuous under weak convergence.*
5. J *is sequentially weakly lower semicontinuous.*

Theorem 1.45 *Let Assumption* 1.44 *hold. Then problem* (1.78) *has an optimal solution* $(\bar{y}, \bar{u})$.

Proof We can argue similarly as in the proof of Theorem 1.43. Denote the feasible set of (1.78) by F_{ad}. Assumption 1.44, 1., 3. ensure the existence of a bounded minimizing sequence $(y_k, u_k) \subset F_{\text{ad}}$. Since U, Y are reflexive, we can extract a weakly convergent subsequence $(y_{k_i}, u_{k_i}) \rightharpoonup (\bar{y}, \bar{u})$. By Assumption 1.44, 1., 2., 4. the feasible set F_{ad} of (1.78) is sequentially weakly closed and thus $(\bar{y}, \bar{u}) \in F_{\text{ad}}$. Now Assumption 1.44, 5. can be used to show that $(\bar{y}, \bar{u})$ solves (1.78).

To verify Assumption 1.44, 4. one uses often compact embeddings $Y \hookrightarrow\hookrightarrow \tilde{Y}$ to convert weak convergence in Y to strong convergence in $\tilde{Y}$.

Example 1.6 To show 1.44, 4. for the semilinear state equation

$$y \in Y := H^1(\Omega) \mapsto e(y,u) := -\Delta y + y^3 - u \in Y^* =: Z,$$

one can proceed as follows. Let $\Omega \subset \mathbb{R}^n$ open and bounded with Lipschitz boundary. Then the embedding $Y := H^1(\Omega) \hookrightarrow\hookrightarrow L^5(\Omega)$ is compact for $n = 2, 3$, see Theorem 1.14. Therefore, $y_k \rightharpoonup y$ weakly in Y implies $y_k \to y$ strongly in $L^5(\Omega)$ and thus $y_k^3 \to y^3$ strongly in $L^{5/3}(\Omega) = L^{5/2}(\Omega)^* \hookrightarrow Y^*$ (see below), and thus strongly in Y^*.

To prove $y_k^3 \to y^3$ in $L^{5/3}(\Omega)$, we first observe that $y_k^3, y^3 \in L^{5/3}(\Omega)$ obviously holds, since $y_k, y \in Y \hookrightarrow L^5(\Omega)$. Next, we prove

$$|b^3 - a^3| \le 3(|a|^2 + |b|^2)|b - a|.$$

In fact, for appropriate $t \in [0, 1]$ the mean value theorem yields

$$|b^3 - a^3| = 3|(a + t(b-a))^2(b-a)| \le 3\max(a^2, b^2)|b-a| \le 3(a^2 + b^2)|b-a|.$$

Therefore,

$$\begin{aligned} \|y_k^3 - y^3\|_{L^{5/3}} &\le 3\|(y_k^2 + y^2)|y_k - y|\|_{L^{5/3}} \\ &\le 3\|y_k^2|y_k - y|\|_{L^{5/3}} + 3\|y^2|y_k - y|\|_{L^{5/3}}. \end{aligned}$$

Now the Hölder inequality with $p = 3/2$ and $q = 3$ yields

$$\|v^2 w\|_{L^{5/3}} = \||v|^{10/3}|w|^{5/3}\|_{L^1}^{3/5} \le \||v|^{10/3}\|_{L^{3/2}}^{3/5}\||w|^{5/3}\|_{L^3}^{3/5} = \|v\|_{L^5}^2 \|w\|_{L^5}.$$

This shows

$$\begin{aligned}\|y_k^3 - y^3\|_{L^{5/3}} &\le 3\|y_k^2|y_k - y|\|_{L^{5/3}} + 3\|y^2|y_k - y|\|_{L^{5/3}}\\ &\le 3(\|y_k\|_{L^5}^2 + \|y\|_{L^5}^2)\|y_k - y\|_{L^5} \to 6\|y\|_{L^5}^2 \cdot 0 = 0.\end{aligned}$$

1.5.3 Applications

1.5.3.1 Distributed Control of Elliptic Equations

We apply the result first to the distributed optimal control of a steady temperature distribution with boundary temperature zero.

$$\begin{aligned}\min\ J(y,u) &:= \frac{1}{2}\|y - y_d\|_{L^2(\Omega)}^2 + \frac{\alpha}{2}\|u\|_{L^2(\Omega)}^2\\ \text{subject to}\quad -\Delta y &= \gamma u \quad \text{on } \Omega,\\ y &= 0 \quad \text{on } \partial\Omega,\\ a \le u &\le b \quad \text{on } \Omega,\end{aligned} \tag{1.79}$$

where

$$\gamma \in L^\infty(\Omega) \setminus \{0\}, \quad \gamma \ge 0, a, b \in L^2(\Omega),\ a \le b.$$

The form of J and the assumptions on a, b suggest the choice $U = L^2(\Omega)$ and

$$U_{\text{ad}} = \{u \in U : a \le u \le b\}, \qquad Y_{\text{ad}} = Y.$$

Then $U_{\text{ad}} \subset U$ is bounded, closed and convex.

We know from Theorem 1.19 that the weak formulation of the boundary value problem

$$\begin{aligned}-\Delta y &= \gamma u \quad \text{on } \Omega,\\ y &= 0 \quad \text{on } \partial\Omega,\end{aligned}$$

can be written in the form

$$\text{Find } y \in Y := H_0^1(\Omega):\quad a(y,v) = (\gamma u, v)_{L^2(\Omega)} \quad \forall v \in Y$$

with $a(y,v) = \int_\Omega \nabla y \cdot \nabla v dx$, or short

$$Ay + Bu = 0,$$

where $A \in \mathcal{L}(Y, Y^*)$ is the operator representing a, see (1.24), and $B \in \mathcal{L}(U, Y^*)$ is defined through $Bu = -(\gamma u, \cdot)_{L^2(\Omega)}$. By Theorem 1.19, $A \in \mathcal{L}(Y, Y^*)$ has a bounded inverse. Therefore, Assumption 1.42 is satisfied with the choice $Z = Y^*$. Finally, setting $g = 0$ and $Q = I_{Y,U}$ with the trivial, continuous embedding $I_{Y,U}$: $y \in Y \to y \in U$, (1.79) is equivalent to (1.77).

1.5.3.2 Boundary Control of Semilinear Elliptic Equations

Now consider the following optimal control problem for a semilinear elliptic equation.

$$\begin{aligned} \min\ J(y,u) &:= \frac{1}{2}\|y-y_d\|^2_{L^2(\Omega)} + \frac{\alpha}{2}\|u\|^2_{L^2(\partial\Omega)} \\ \text{subject to}\quad -\Delta y + y^3 &= 0 \quad \text{on } \Omega, \\ \frac{\partial y}{\partial \nu} + y &= u \quad \text{on } \partial\Omega, \\ a \le u &\le b \quad \text{on } \partial\Omega, \end{aligned} \tag{1.80}$$

where $\Omega \subset \mathbb{R}^n$, $n=2$ or $n=3$, is open and bounded with Lipschitz-boundary and

$$a, b \in L^n(\partial\Omega), \quad a \le b.$$

Let $U = L^n(\partial\Omega)$, $Y = H^1(\Omega)$ and

$$U_{\text{ad}} = \{u \in U : a \le u \le b\}, \qquad Y_{\text{ad}} = Y.$$

We verify Assumption 1.44. $U_{\text{ad}} \subset U$ is bounded, closed and convex. If we consider weak solutions according to (1.45) then the PDE-constraint is an operator

$$e : (y,u) \in Y \times U \mapsto e(y,u) := a(y,\cdot) + (y^3,\cdot)_{L^2(\Omega)} + (y-u,\cdot)_{L^2(\partial\Omega)} \in Y^* =: Z,$$

where $a(y,v) = \int_\Omega \nabla y \cdot \nabla v\, dx$ (note that $H^1(\Omega) \hookrightarrow L^6(\Omega)$ for $n \le 3$ and thus $y^3 \in L^2(\Omega)$). We know by Theorem 1.25 that there exists a unique bounded solution operator $u \in U_{\text{ad}} \mapsto y(u) \in Y$. Moreover, $(y,u) \in Y \times U \mapsto e(y,u) \in Z$ is continuous under weak convergence, since the nonlinear term $y \in Y \mapsto y^3 \in Z$ is by Example 1.6 sequentially weakly continuous. Finally, the objective function $J : Y \times U \to \mathbb{R}$ is continuous, convex and thus sequentially lower semicontinuous. Thus, Assumption 1.44 is verified and therefore (1.80) has an optimal solution by Theorem 1.45.

1.6 Reduced Problem, Sensitivities and Adjoints

We consider again optimal control problems of the form

$$\min_{y \in Y, u \in U} J(y,u) \quad \text{subject to} \quad e(y,u) = 0, \quad (y,u) \in W_{\text{ad}}, \tag{1.81}$$

where $J : Y \times U \to \mathbb{R}$ is the objective function, $e : Y \times U \to Z$ is an operator between Banach spaces, and $W_{\text{ad}} \subset W := Y \times U$ is a nonempty closed set.

We assume that J and e are continuously F-differentiable and that the state equation

$$e(y,u) = 0$$

possesses for each $u \in U$ a unique corresponding solution $y(u) \in Y$. Thus, we have a solution operator $u \in U \mapsto y(u) \in Y$. Furthermore, we assume that $e_y(y(u), u) \in \mathcal{L}(Y, Z)$ is continuously invertible. Then the implicit function theorem (Theorem 1.41) ensures that $y(u)$ is continuously differentiable. An equation for the derivative $y'(u)$ is obtained by differentiating the equation $e(y(u), u) = 0$ with respect to u:

$$e_y(y(u), u)y'(u) + e_u(y(u), u) = 0. \tag{1.82}$$

Inserting $y(u)$ in (1.81), we obtain the reduced problem

$$\min_{u \in U} \hat{J}(u) := J(y(u), u) \quad \text{subject to} \quad u \in \hat{U}_{\text{ad}} := \{u \in U : (y(u), u) \in W_{\text{ad}}\}. \tag{1.83}$$

It will be important to investigate the possibilities of computing the derivative of the reduced objective function $\hat{J}$.

Essentially, there are two methods to do this:

- The sensitivity approach,
- The adjoint approach.

1.6.1 Sensitivity Approach

Sensitivities are directional derivatives. For $u \in U$ and a direction $s \in U$, the chain rule yields for the sensitivity of $\hat{J}$:

$$d\hat{J}(u, s) = \langle \hat{J}'(u), s\rangle_{U^*,U} = \langle J_y(y(u), u), y'(u)s\rangle_{Y^*,Y} + \langle J_u(y(u), u), s\rangle_{U^*,U}.$$

In this expression, the sensitivity $dy(u, s) = y'(u)s$ appears. Differentiating $e(y(u), u) = 0$ in the direction s yields

$$e_y(y(u), u)y'(u)s + e_u(y(u), u)s = 0.$$

Hence, the sensitivity $\delta_s y = dy(u, s)$ is given as the solution of the linearized state equation

$$e_y(y(u), u)\delta_s y = -e_u(y(u), u)s.$$

Therefore, to compute the directional derivative $d\hat{J}(u, s) = \langle \hat{J}(u), s\rangle_{U^*,U}$ via the sensitivity approach, the following steps are required:

1. Compute the sensitivity $\delta_s y = dy(u, s)$ by solving
$$e_y(y(u), u)\delta_s y = -e_u(y(u), u)s. \tag{1.84}$$
2. Compute $d\hat{J}(u, s) = \langle \hat{J}'(u), s\rangle_{U^*,U}$ via
$$d\hat{J}(u, s) = \langle J_y(y(u), u), \delta_s y\rangle_{Y^*,Y} + \langle J_u(y(u), u), s\rangle_{U^*,U}.$$

This procedure is expensive if the whole derivative $\hat{J}'(u)$ is required, since this means that for a basis B of U, all the directional derivatives

$$d\hat{J}(u, v), \quad v \in B,$$

have to be computed. Each of them requires the solution of one linearized state equation (1.84) with $s = v$.

This is an effort that grows linearly in the dimension of U.

Actually, computing all sensitivities of $\delta_v y = y'(u)v$, $v \in B$, is equivalent to computing the whole operator $y'(u)$. As we will see now, much less effort is needed to compute the derivative of $\hat{J}$.

1.6.2 Adjoint Approach

We now derive a more efficient way of representing the derivative of $\hat{J}$. From

$$\begin{aligned}\langle \hat{J}'(u), s\rangle_{U^*,U} &= \langle J_y(y(u), u), y'(u)s\rangle_{Y^*,Y} + \langle J_u(y(u), u), s\rangle_{U^*,U} \\ &= \langle y'(u)^* J_y(y(u), u), s\rangle_{U^*,U} + \langle J_u(y(u), u), s\rangle_{U^*,U}\end{aligned}$$

we see that

$$\hat{J}'(u) = y'(u)^* J_y(y(u), u) + J_u(y(u), u).$$

Therefore, not the operator $y'(u) \in \mathcal{L}(U, Y)$, but only the vector $y'(u)^* J_y(y(u), u) \in U^*$ is really required. Since by (1.82)

$$y'(u)^* J_y(y(u), u) = -e_u(y(u), u)^* e_y(y(u), u)^{-*} J_y(y(u), u),$$

it follows that

$$y'(u)^* J_y(y(u), u) = e_u(y(u), u)^* p(u),$$

where the *adjoint state* $p = p(u) \in Z^*$ solves the

Adjoint Equation:

$$e_y(y(u), u)^* p = -J_y(y(u), u). \tag{1.85}$$

We thus have

$$\hat{J}'(u) = e_u(y(u), u)^* p(u) + J_u(y(u), u).$$

The derivative $\hat{J}'(u)$ can thus be computed via the adjoint approach as follows:

1. Compute the adjoint state by solving the adjoint equation

$$e_y(y(u), u)^* p = -J_y(y(u), u).$$

2. Compute $\hat{J}'(u)$ via

$$\hat{J}'(u) = e_u(y(u), u)^* p + J_u(y(u), u).$$

1.6.3 Application to a Linear-Quadratic Optimal Control Problem

We consider the linear-quadratic optimal control problem

$$\begin{aligned}\min_{(y,u)\in Y\times U} \quad & J(y,u) := \frac{1}{2}\|Qy - q_d\|_H^2 + \frac{\alpha}{2}\|u\|_U^2 \\ \text{subject to} \quad & Ay + Bu = g, \quad u \in U_{\text{ad}}, \ y \in Y_{\text{ad}}\end{aligned} \tag{1.86}$$

where H, U are Hilbert spaces, Y, Z are Banach spaces and $q_d \in H$, $g \in Z$, $A \in \mathcal{L}(Y, Z)$, $B \in \mathcal{L}(U, Z)$, $Q \in \mathcal{L}(Y, H)$ and let Assumption 1.42 hold. We obtain the form (1.81) by setting

$$e(y,u) := Ay + Bu - g, \quad W_{\text{ad}} = Y_{\text{ad}} \times U_{\text{ad}}.$$

By assumption, there exists a continuous affine linear solution operator

$$U \ni u \mapsto y(u) = A^{-1}(g - Bu) \in Y.$$

For the derivatives we have

$$\begin{aligned}\langle J_y(y,u), s_y\rangle_{Y^*,Y} &= (Qy - q_d, Qs_y)_H, = \langle Q^*(Qy - q_d), s_y\rangle_{Y^*,Y}, \\ \langle J_u(y,u), s_u\rangle_{U^*,U} &= \alpha(u, s_u)_U, \\ e_y(y,u)s_y &= As_y, \\ e_u(y,u)s_u &= Bs_u.\end{aligned}$$

Therefore,

$$\begin{aligned}J_y(y,u) &= (Qy - q_d, Q\cdot)_H, \\ J_u(y,u) &= \alpha(u, \cdot)_U, \\ e_y(y,u) &= A, \\ e_u(y,u) &= B.\end{aligned}$$

If we choose the Riesz representations $U^* = U$, $H^* = H$, then

$$\begin{aligned}J_y(y,u) &= (Qy - q_d, Q\cdot)_H = \langle Qy - q_d, Q\cdot\rangle_{H^*,H} = \langle Q^*(Qy - q_d), \cdot\rangle_{Y^*,Y} \\ &= Q^*(Qy - q_d), \\ J_u(y,u) &= \alpha(u, \cdot)_U = \alpha u.\end{aligned}$$

The reduced objective function is

$$\hat{J}(u) = J(y(u), u) = \frac{1}{2}\|Q(A^{-1}(g - Bu)) - q_d\|_H^2 + \frac{\alpha}{2}\|u\|_U^2.$$

For evaluation of $\hat{J}$, we first solve the state equation

$$Ay + Bu = g$$

to obtain $y = y(u)$ and then we evaluate $J(y, u)$. In the following, let $y = y(u)$.

Sensitivity Approach:

For $s \in U$, we obtain $d\hat{J}(u, s) = \langle \hat{J}'(u), s\rangle_{U^*,U}$ by first solving the linearized state equation

$$A\delta_s y = -Bs$$

for $\delta_s y$ and then setting

$$d\hat{J}(u, s) = (Qy - q_d, Q\delta_s y)_H + \alpha(u, s)_U.$$

Adjoint Approach:

We obtain $\hat{J}'(u)$ by first solving the adjoint equation

$$A^* p = -(Qy - q_d, Q\cdot)_H \quad (= -Q^*(Qy - q_d) \text{ if } H^* = H)$$

for the adjoint state $p = p(u) \in Z^*$ and then setting

$$\hat{J}'(u) = B^* p + \alpha(u, \cdot)_U \quad (= B^* p + \alpha u \text{ if } U^* = U).$$

Next, let us consider the concrete example of the elliptic control problem

$$\begin{aligned}
\min J(y, u) &:= \frac{1}{2}\int_\Omega (y(x) - y_d(x))^2 dx + \frac{\alpha}{2}\int_\Omega u(x)^2 dx \\
\text{subject to} \quad & -\Delta y = \gamma u \quad \text{on } \Omega, \\
& \frac{\partial y}{\partial \nu} = \frac{\beta}{\kappa}(y_a - y) \quad \text{on } \partial\Omega, \\
& a \le u \le b \quad \text{on } \Omega.
\end{aligned}$$

The appropriate spaces are

$$U = L^2(\Omega), \qquad Y = H^1(\Omega)$$

and we assume

$$\begin{aligned}
&a, b \in U, \qquad y_d \in L^2(\Omega), \qquad \alpha > 0, \\
&y_a \in L^2(\partial\Omega), \qquad \gamma \in L^\infty(\Omega) \setminus \{0\}, \quad \gamma \ge 0.
\end{aligned}$$

The coefficient γ weights the control and y_a can be interpreted as the surrounding temperature in the case of the heat equation. $\beta > 0$ and $\kappa > 0$ are coefficients.

The weak formulation of the state equation is

$$y \in Y, \qquad a(y, v) = (\gamma u, v)_{L^2(\Omega)} + ((\beta/\kappa) y_a, v)_{L^2(\partial\Omega)} \quad \forall v \in Y = H^1(\Omega)$$

with

$$a(y,v)=\int_\Omega \nabla y^T \nabla v dx + ((\beta/\kappa)y, v)_{L^2(\partial\Omega)}.$$

Now let $Z=Y^*$, $H=L^2(\Omega)$ and

- $A\in\mathcal{L}(Y,Y^*)$ the operator induced by a, i.e., $Ay=a(y,\cdot)$,
- $B\in\mathcal{L}(U,Y^*)$, $Bu=-(\gamma u,\cdot)_{L^2(\Omega)}$,
- $g\in Y^*$, $g=((\beta/\kappa)y_a,\cdot)_{L^2(\partial\Omega)}$,
- $U_{\text{ad}}=\{u\in U: a\le u\le b \text{ on } \Omega\}$, $Y_{\text{ad}}=Y$,
- $Q\in\mathcal{L}(Y,H)$, $Qy=y$.

Then, we arrive at a linear quadratic problem of the form (1.86).

We compute the adjoints. Note that all spaces are Hilbert spaces and thus reflexive. In particular, we identify the dual of $U=L^2$ with U by working with $\langle\cdot,\cdot\rangle_{U^*,U}=(\cdot,\cdot)_{L^2(\Omega)}$. We do the same with $H=L^2$. We thus have

$$\begin{aligned}
A^* &\in \mathcal{L}(Z^*,Y^*)=\mathcal{L}(Y^{**},Y^*)=\mathcal{L}(Y,Y^*),\\
B^* &\in \mathcal{L}(Z^*,U^*)=\mathcal{L}(Y^{**},U)=\mathcal{L}(Y,U),\\
Q^* &\in \mathcal{L}(H^*,Y^*)=\mathcal{L}(H,Y^*).
\end{aligned}$$

For A^* we obtain

$$\begin{aligned}
\langle A^*v,w\rangle_{Y^*,Y} &= \langle v,Aw\rangle_{Z^*,Z}=\langle Aw,v\rangle_{Y^*,Y}\\
&= a(w,v)=a(v,w)=\langle Av,w\rangle_{Y^*,Y}\quad \forall v,w\in Y.
\end{aligned}$$

Here, we have used that obviously a is a symmetric bilinear form. Therefore, $A^*=A$.

For B^* we have

$$\begin{aligned}
(B^*v,w)_U &= \langle B^*v,w\rangle_{U^*,U}=\langle v,Bw\rangle_{Z^*,Z}=\langle v,Bw\rangle_{Y,Y^*}=(v,-\gamma w)_{L^2}\\
&= -(\gamma v,w)_U\quad \forall v\in Y,\ w\in U.
\end{aligned}$$

Hence $B^*v=-\gamma v$. Finally, for Q^* we obtain

$$\langle Q^*v,w\rangle_{Y^*,Y}=\langle v,Qw\rangle_{H^*,H}=(v,w)_{L^2(\Omega)}.$$

Therefore, $Q^*v=(v,\cdot)_{L^2(\Omega)}$.

This means that

$$J_y(y,u)=(Q^*(Qy-y_d),\cdot)_{L^2(\Omega)}=(y-y_d,\cdot)_{L^2(\Omega)}.$$

Taking all together, the adjoint equation thus reads

$$Ap=-(y-y_d,\cdot)_{L^2(\Omega)},$$

which is the weak form of

$$-\Delta p = -(y - y_d) \quad \text{on } \Omega,$$
$$\frac{\partial p}{\partial \nu} + \frac{\beta}{\kappa} p = 0 \quad \text{on } \partial\Omega,$$

The adjoint gradient representation then is

$$\hat{J}'(u) = B^* p(u) + J_u(y(u), u) = -\gamma p + \alpha u.$$

1.6.4 A Lagrangian-Based View of the Adjoint Approach

The adjoint gradient representation can also be derived in a different way. Consider (1.81) and define the Lagrange function $L : Y \times U \times Z^* \to \mathbb{R}$,

$$L(y, u, p) = J(y, u) + \langle p, e(y, u)\rangle_{Z^*,Z}.$$

Inserting $y = y(u)$ gives, for arbitrary $p \in Z^*$,

$$\hat{J}(u) = J(y(u), u) = J(y(u), u) + \langle p, e(y(u), u)\rangle_{Z^*,Z} = L(y(u), u, p).$$

Differentiating this, we obtain

$$\langle \hat{J}'(u), s\rangle_{U^*,U} = \langle L_y(y(u), u, p), y'(u)s\rangle_{Y^*,Y} + \langle L_u(y(u), u, p), s\rangle_{U^*,U}. \quad (1.87)$$

Now we choose a special $p = p(u)$, namely such that

$$L_y(y(u), u, p) = 0. \quad (1.88)$$

This is nothing else but the adjoint equation. In fact,

$$\begin{aligned}\langle L_y(y, u, p), d\rangle_{Y^*,Y} &= \langle J_y(y, u), d\rangle_{Y^*,Y} + \langle p, e_y(y, u)d\rangle_{Z^*,Z}\\ &= \langle J_y(y, u) + e_y(y, u)^* p, d\rangle_{Y^*,Y}.\end{aligned}$$

Therefore,

$$L_y(y(u), u, p) = J_y(y(u), u) + e_y(y(u), u)^* p.$$

Now, choosing $p = p(u)$ according to (1.88), we obtain from (1.87) that

$$\hat{J}'(u) = L_u(y(u), u, p(u)) = J_u(y(u), u) + e_u(y(u), u)^* p(u). \quad (1.89)$$

This is exactly the adjoint gradient representation.

1.6.5 Second Derivatives

We can use the Lagrange function based approach to derive the second derivative of $\hat{J}$.

To this end, assume that J and e are twice continuously differentiable. As already noted, for all $p \in Z^*$ we have the identity

$$\hat{J}(u) = J(y(u), u) = L(y(u), u, p).$$

Differentiating this in the direction $s_1 \in U$ yields (see above)

$$\langle \hat{J}'(u), s_1 \rangle_{U^*,U} = \langle L_y(y(u), u, p), y'(u)s_1 \rangle_{Y^*,Y} + \langle L_u(y(u), u, p), s_1 \rangle_{U^*,U}.$$

Differentiating this once again in the direction $s_2 \in U$ gives

$$\begin{aligned}
\langle \hat{J}''(u)s_2, s_1 \rangle_{U^*,U} = {} & \langle L_y(y(u), u, p), y''(u)(s_1, s_2) \rangle_{Y^*,Y} \\
& + \langle L_{yy}(y(u), u, p)y'(u)s_2, y'(u)s_1 \rangle_{Y^*,Y} \\
& + \langle L_{yu}(y(u), u, p)s_2, y'(u)s_1 \rangle_{Y^*,Y} \\
& + \langle L_{uy}(y(u), u, p)y'(u)s_2, s_1 \rangle_{U^*,U} \\
& + \langle L_{uu}(y(u), u, p)s_2, s_1 \rangle_{U^*,U}.
\end{aligned}$$

Now we choose $p = p(u)$, i.e., $L_y(y(u), u, p(u)) = 0$. Then the term containing $y''(u)$ drops out and we arrive at

$$\begin{aligned}
\langle \hat{J}''(u)s_2, s_1 \rangle_{U^*,U} = {} & \langle L_{yy}(y(u), u, p(u))y'(u)s_2, y'(u)s_1 \rangle_{Y^*,Y} \\
& + \langle L_{yu}(y(u), u, p(u))s_2, y'(u)s_1 \rangle_{Y^*,Y} \\
& + \langle L_{uy}(y(u), u, p(u))y'(u)s_2, s_1 \rangle_{U^*,U} \\
& + \langle L_{uu}(y(u), u, p(u))s_2, s_1 \rangle_{U^*,U}.
\end{aligned}$$

This shows

$$\begin{aligned}
\hat{J}''(u) &= y'(u)^* L_{yy}(y(u), u, p(u))y'(u) + y'(u)^* L_{yu}(y(u), u, p(u)) \\
&\quad + L_{uy}(y(u), u, p(u))y'(u) + L_{uu}(y(u), u, p(u)) \\
&= T(u)^* L_{ww}(y(u), u, p(u))T(u)
\end{aligned} \tag{1.90}$$

with

$$T(u) = \begin{pmatrix} y'(u) \\ I_U \end{pmatrix} \in \mathcal{L}(U, Y \times U), \qquad L_{ww} = \begin{pmatrix} L_{yy} & L_{yu} \\ L_{uy} & L_{uu} \end{pmatrix}.$$

Here $I_U \in \mathcal{L}(U, U)$ is the identity.

Note that $y'(u) = -e_y(y(u), u)^{-1} e_u(y(u), u)$ and thus

$$T(u) = \begin{pmatrix} y'(u) \\ I_U \end{pmatrix} = \begin{pmatrix} -e_y(y(u), u)^{-1} e_u(y(u), u) \\ I_U \end{pmatrix}. \tag{1.91}$$

Usually, the Hessian representation (1.90) is not used to compute the whole operator $\hat{J}''(u)$. Rather, it is used to compute operator-vector-products $\hat{J}''(u)s$ as follows:

1. Compute the sensitivity
$$\delta_s y = y'(u)s = -e_y(y(u),u)^{-1}e_u(y(u),u)s.$$
This requircs one linearized state equation solve.
2. Compute
$$\begin{pmatrix} h_1 \\ h_2 \end{pmatrix} = \begin{pmatrix} L_{yy}(y(u),u,p(u))\delta_s y + L_{yu}(y(u),u,p(u))s \\ L_{uy}(y(u),u,p(u))\delta_s y + L_{uu}(y(u),u,p(u))s \end{pmatrix}.$$
3. Compute
$$h_3 = y'(u)^* h_1 = -e_u(y(u),u)^* e_y(y(u),u)^{-*} h_1.$$
This requires an adjoint equation solve.
4. Set $\hat{J}''(u)s = h_2 + h_3$.

This procedure can be used to apply iterative solvers to the Newton equation
$$\hat{J}''(u^k)s^k = -\hat{J}'(u^k).$$

Example For the linear-quadratic optimal control problem (1.86) with $U^* = U$ and $H^* = H$ we have
$$\begin{aligned}
L(y,u,p) &= J(y,u) + \langle p, Ay + Bu\rangle_{Z^*,Z},\\
L_y(y,u,p) &= Q^*(Qy - q_d) + A^* p,\\
L_u(y,u,p) &= \alpha u + B^* p,\\
L_{yy}(y,u,p) &= Q^* Q,\\
L_{yu}(y,u,p) &= 0,\\
L_{yu}(u,y,p) &= 0,\\
L_{uu}(y,u,p) &= \alpha I_U.
\end{aligned}$$

From this, all the steps in the above algorithm can be derived easily.

1.7 Optimality Conditions

1.7.1 Optimality Conditions for Simply Constrained Problems

We consider the problem
$$\min_{w \in W} J(w) \quad \text{s.t.} \quad w \in \mathcal{C}, \tag{1.92}$$

where W is a Banach space, $J : W \to \mathbb{R}$ is Gâteaux-differentiable and $\mathcal{C} \subset W$ is nonempty, closed, and convex.

Theorem 1.46 *Let W be a Banach space and $\mathcal{C} \subset W$ be nonempty and convex. Furthermore, let $J : V \to \mathbb{R}$ be defined on an open neighborhood of $\mathcal{C}$. Let $\bar{w}$ be a local solution of* (1.92) *at which J is Gâteaux-differentiable. Then the following optimality condition holds*:

$$\bar{w} \in \mathcal{C}, \quad \langle J'(\bar{w}), w - \bar{w}\rangle_{W^*,W} \geq 0 \quad \forall w \in \mathcal{C}. \tag{1.93}$$

If J is convex on $\mathcal{C}$, the condition (1.93) *is necessary and sufficient for global optimality.*

If, in addition, J is strictly convex on $\mathcal{C}$, then there exists at most one solution of (1.92), *or, equivalently, of* (1.93).

If W is reflexive, $\mathcal{C}$ is closed and convex, and J is convex and continuous with

$$\lim_{w \in \mathcal{C}, \|w\|_W \to \infty} J(w) = \infty,$$

then there exists a (global = local) solution of (1.92).

Remark 1.19 A condition of the form (1.93) is called variational inequality.

Proof Let $w \in \mathcal{C}$ be arbitrary. By the convexity of $\mathcal{C}$ we have $w(t) = \bar{w} + t(w - \bar{w}) \in \mathcal{C}$ for all $t \in [0, 1]$. Now the optimality of $\bar{w}$ yields

$$J(\bar{w} + t(w - \bar{w})) - J(\bar{w}) \geq 0 \quad \forall t \in [0, 1]$$

and thus

$$\langle J'(\bar{w}), w - \bar{w}\rangle_{W^*,W} = \lim_{t \to 0^+} \frac{J(\bar{w} + t(w - \bar{w})) - J(\bar{w})}{t} \geq 0.$$

Since $w \in \mathcal{C}$ was arbitrary, the proof of (1.93) is complete.

Now let J be convex. Then

$$J(w) - J(\bar{w}) \geq \langle J'(\bar{w}), w - \bar{w}\rangle_{W^*,W} \quad \forall w \in \mathcal{C}. \tag{1.94}$$

In fact, for all $t \in (0, 1]$,

$$J(\bar{w} + t(w - \bar{w})) \leq (1 - t)J(\bar{w}) + tJ(w).$$

Hence,

$$J(w) - J(\bar{w}) \geq \frac{J(\bar{w} + t(w - \bar{w})) - J(\bar{w})}{t} \overset{t \to 0^+}{\longrightarrow} \langle J'(\bar{w}), w - \bar{w}\rangle_{W^*,W}.$$

Now from (1.93) and (1.94) it follows that

$$J(w) - J(\bar{w}) \geq \langle J'(\bar{w}), w - \bar{w}\rangle_{W^*,W} \geq 0 \quad \forall w \in \mathcal{C}.$$

Thus, $\bar{w}$ is optimal.

If J is strictly convex and $\bar{w}_1, \bar{w}_2$ are two global solutions, the point $(\bar{w}_1 + \bar{w}_2)/2 \in \mathcal{C}$ would be a better solution, unless $\bar{w}_1 = \bar{w}_2$.

Finally, let the assumptions of the last assertion hold and let $(w_k) \in \mathcal{C}$ be a minimizing sequence. Then (w_k) is bounded (otherwise $J(w_k) \to \infty$) and thus (w_k) contains a weakly convergent subsequence $(w_k)_K \rightharpoonup \bar{w}$. Since $\mathcal{C}$ is convex and closed, it is weakly closed and thus $\bar{w} \in \mathcal{C}$. From the continuity and convexity of J we conclude that J is weakly sequentially lower semicontinuous and thus

$$J(\bar{w}) \leq \lim_{K \ni k \to \infty} J(w_k) = \inf_{w \in \mathcal{C}} J(w).$$

Thus, $\bar{w}$ solves the minimization problem.

In the case of a closed convex set $\mathcal{C}$ in a *Hilbert space* W, we can rewrite the variational inequality in the form

$$\bar{w} - P(\bar{w} - \gamma \nabla J(w)) = 0$$

where $\gamma > 0$ is a fixed parameter and $\nabla J(w) \in W$ is the Riesz representation of $J'(w) \in W^*$.

To prove this, we need some knowledge about the projection onto closed convex sets.

Lemma 1.10 *Let $\mathcal{C} \subset W$ be a nonempty closed convex subset of the Hilbert space W and denote by $P : W \to \mathcal{C}$ the projection onto $\mathcal{C}$, i.e.,*

$$P(w) \in \mathcal{C}, \quad \|P(w) - w\|_W = \min_{v \in \mathcal{C}} \|v - w\|_W \quad \forall w \in W.$$

Then:

(a) *P is well-defined.*
(b) *For all $w, z \in W$ there holds*:

$$z = P(w) \iff$$
$$z \in \mathcal{C}, \quad (w - z, v - z)_W \leq 0 \quad \forall v \in \mathcal{C}.$$

(c) *P is nonexpansive, i.e.,*

$$\|P(v) - P(w)\|_W \leq \|v - w\|_W \quad \forall v, w \in W.$$

(d) *P is monotone, i.e.,*

$$(P(v) - P(w), v - w)_W \geq 0 \quad \forall v, w \in W.$$

Furthermore, equality holds if and only if $P(v) = P(w)$.

(e) *For all $w \in \mathcal{C}$ and $d \in W$, the function*

$$\phi(t) := \frac{1}{t} \|P(w + td) - w\|_W, \quad t > 0,$$

is nonincreasing.

Proof (a) The function $W \ni w \mapsto \|w\|_W^2$ is strictly convex: For all $w_1, w_2 \in W$, $w_1 \neq w_2$, and all $t \in (0,1)$;

$$\|w_1 + t(w_2 - w_1)\|_W^2 = \|w_1\|_W^2 + 2t(w_1, w_2 - w_1)_W + t^2\|w_2 - w_1\|_W^2 =: p(t).$$

The function on the right is a strictly convex parabola. Hence,

$$\|w_1 + t(w_2 - w_1)\|_W^2 = p(t) < (1-t)p(0) + tp(1) = (1-t)\|w_1\|_2^2 + t\|w_2\|_2^2.$$

Therefore, for all $w \in W$, the function

$$J(v) = \frac{1}{2}\|v - w\|_W^2$$

is strictly convex. Furthermore, it tends to ∞ as $\|v\|_W \to \infty$. Hence, by Theorem 1.46, the problem

$$\min_{v \in \mathcal{C}} J(v)$$

possesses a unique solution $\bar{v}$, and thus $P(w) = \bar{v}$ is uniquely defined.

(b) The function J defined above is obviously F-differentiable with

$$\langle J'(v), s\rangle_{W^*,W} = (v - w, s)_W \quad \forall s \in W.$$

Since $P(w) = \bar{v}$ minimizes J on $\mathcal{C}$, we have by Theorem 1.46 that $z = P(w)$ if and only if $z \in \mathcal{C}$ and

$$z \in \mathcal{C}, \quad \langle J'(z), v - z\rangle_{W^*,W} = (z - w, v - z)_W \geq 0 \quad \forall v \in \mathcal{C}.$$

(c) We use (b)

$$(v - P(v), P(w) - P(v))_W \leq 0,$$
$$(w - P(w), P(v) - P(w))_W \leq 0.$$

Adding these two inequalities gives

$$\begin{aligned}&(w - v + P(v) - P(w), P(v) - P(w))\\ &\quad = (w - v, P(v) - P(w))_W + \|P(v) - P(w)\|_W^2 \leq 0.\end{aligned}$$

Hence, by the Cauchy-Schwarz inequality

$$\|P(v) - P(w)\|_W^2 \leq (v - w, P(v) - P(w))_W \leq \|v - w\|_W \|P(v) - P(w)\|_W. \tag{1.95}$$

(d) The assertion follows immediately from the first inequality in (1.95).

(e) Let $t > s > 0$. If $\|P(w + td) - w\|_W \leq \|P(w + sd) - w\|_W$ then obviously $\phi(s) > \phi(t)$.

Now let $\|P(w + td) - w\|_W > \|P(w + sd) - w\|_W$.

Using the Cauchy-Schwarz inequality, for any $u, v \in W$ we have

$$\begin{aligned}
&\|v\|_W(u, u-v)_W - \|u\|_W(v, u-v)_W \\
&\quad = \|v\|_W\|u\|_W^2 - \|v\|_W(u,v)_W - \|u\|_W(v,u)_W + \|u\|_W\|v\|_W^2 \\
&\quad \ge \|v\|_W\|u\|_W^2 - \|v\|_W\|u\|_W\|v\|_W - \|u\|_W\|v\|_W\|u\|_W + \|u\|_W\|v\|_W^2 = 0.
\end{aligned}$$

Now, set $u := P(w+td) - w$, $v := P(w+sd) - w$, and $w_\tau = w + \tau d$. Then by (b)

$$\begin{aligned}
(u, u-v)_W - (td, P(w_t) - P(w_s))_W &= (P(w_t) - w - td, P(w_t) - P(w_s))_W \\
&= (P(w_t) - w_t, P(w_t) - P(w_s))_W \le 0, \\
(v, u-v)_W - (sd, P(w_t) - P(w_s))_W &= (P(w_s) - w - sd, P(w_t) - P(w_s))_W \\
&= (P(w_s) - w_s, P(w_t) - P(w_s))_W \ge 0.
\end{aligned}$$

Thus,

$$\begin{aligned}
0 &\le \|v\|_W(u, u-v)_W - \|u\|_W(v, u-v)_W \\
&\le \|v\|_W(td, P(w_t) - P(w_s))_W - \|u\|_W(sd, P(w_t) - P(w_s))_W \\
&= (t\|v\|_W - s\|u\|_W)(d, P(w_t) - P(w_s))_W.
\end{aligned}$$

Now, due to the monotonicity of P,

$$(d, P(w_t) - P(w_s))_W = \frac{1}{t-s}(w_t - w_s, P(w_t) - P(w_s))_W > 0,$$

since $P(w_t) \ne P(w_s)$. Therefore,

$$0 \le t\|v\|_W - s\|u\|_W = ts(\phi(s) - \phi(t)).$$

Lemma 1.11 *Let W be a Hilbert space, $\mathcal{C} \subset W$ be nonempty, closed, and convex. Furthermore, let P denote the projection onto $\mathcal{C}$. Then, for all $y \in W$ and all $\gamma > 0$, the following conditions are equivalent*:

$$w \in \mathcal{C}, \quad (y, v-w)_W \ge 0 \quad \forall v \in \mathcal{C}. \tag{1.96}$$

$$w - P(w - \gamma y) = 0. \tag{1.97}$$

Proof Let (1.96) hold. Then with $w_\gamma = w - \gamma y$ we have

$$(w_\gamma - w, v - w)_W = -\gamma(y, v-w)_W \le 0 \quad \forall v \in \mathcal{C}.$$

By Lemma 1.10(b), this implies $w = P(w_\gamma)$ as asserted in (1.97).

Conversely, let (1.97) hold. Then with the same notation as above we obtain $w = P(w_\gamma) \in \mathcal{C}$. Furthermore, Lemma 1.10(b) yields

$$(y, v-w)_W = -\frac{1}{\gamma}(w_\gamma - w, v - w) \ge 0 \quad \forall v \in \mathcal{C}.$$

We obtain the following corollary of Theorem 1.46.

Corollary 1.2 *Let W be a Hilbert space and $\mathcal{C} \subset W$ be nonempty, closed, and convex. Furthermore, let $J : V \to \mathbb{R}$ be defined on an open neighborhood of $\mathcal{C}$. Let $\bar{w}$ be a local solution of* (1.92) *at which J is Gâteaux-differentiable. Then the following optimality condition holds*:

$$\bar{w} = P(\bar{w} - \gamma \nabla J(\bar{w})). \tag{1.98}$$

Here, $\gamma > 0$ is arbitrary but fixed and $\nabla J(w) \in W$ denotes the Riesz-representation of $J'(w) \in W^*$.

Moreover, in the Hilbert space setting (1.98) *is equivalent to* (1.93) *if $\mathcal{C}$ is nonempty, closed, convex and therefore in this case Theorem* 1.46 *holds with* (1.98) *instead of* (1.93).

1.7.2 Optimality Conditions for Control-Constrained Problems

We consider a general possibly nonlinear problem of the form

$$\min_{(y,u)\in Y\times U} J(y,u) \quad \text{subject to} \quad e(y,u) = 0, \quad u \in U_{\text{ad}}. \tag{1.99}$$

We make the

Assumption 1.47

1. *$U_{\text{ad}} \subset U$ is nonempty, convex and closed.*
2. *$J : Y \times U \to \mathbb{R}$ and $e : Y \times U \to Z$ are continuously Fréchet differentiable and U, Y, Z are Banach spaces.*
3. *For all $u \in V$ in a neighborhood $V \subset U$ of U_{ad}, the state equation $e(y,u) = 0$ has a unique solution $y = y(u) \in Y$.*
4. *$e_y(y(u),u) \in \mathcal{L}(Y,Z)$ has a bounded inverse for all $u \in V \supset U_{\text{ad}}$.*

Under these assumptions the mapping $u \in V \mapsto y(u) \in Y$ is continuously F-differentiable by the implicit function theorem.

Obviously, the general linear-quadratic optimization problem

$$\begin{aligned} \min_{(y,u)\in Y\times U} \; & J(y,u) := \frac{1}{2}\|Qy - q_d\|_H^2 + \frac{\alpha}{2}\|u\|_U^2 \\ \text{subject to} \; & Ay + Bu = g, \quad u \in U_{\text{ad}}, \end{aligned} \tag{1.100}$$

is a special case of (1.99), where H, U are Hilbert spaces, Y, Z are Banach spaces and $q_d \in H$, $g \in Z$, $A \in \mathcal{L}(Y,Z)$, $B \in \mathcal{L}(U,Z)$, $Q \in \mathcal{L}(Y,H)$. Moreover, Assumption 1.42 ensures Assumption 1.47, since $e_y(y,u) = A$.

1.7.2.1 A General First Order Optimality Condition

Now consider problem (1.99) and let Assumption 1.47 hold. Then we can formulate the reduced problem

$$\min_{u \in U} \hat{J}(u) \quad \text{s.t.} \quad u \in U_{\text{ad}} \tag{1.101}$$

with the reduced objective functional

$$\hat{J}(u) := J(y(u), u),$$

where $V \ni u \mapsto y(u) \in Y$ is the solution operator of the state equation. We have the following general result.

Theorem 1.48 *Let Assumption* 1.47 *hold. If $\bar{u}$ is a local solution of the reduced problem* (1.101) *then $\bar{u}$ satisfies the variational inequality*

$$\bar{u} \in U_{\text{ad}} \quad \text{and} \quad \langle \hat{J}'(\bar{u}), u - \bar{u} \rangle_{U^*,U} \geq 0 \quad \forall u \in U_{\text{ad}}. \tag{1.102}$$

Proof We can directly apply Theorem 1.46.

Depending on the structure of U_{ad} the variational inequality (1.102) can be expressed in a more convenient form. We show this for the case of box constraints.

Lemma 1.12 *Let $U = L^2(\Omega)$, $a, b \in L^2(\Omega)$, $a \leq b$, and U_{ad} be given by*

$$U_{\text{ad}} = \{u \in L^2(\Omega) : a \leq u \leq b\}.$$

We work with $U^* = U$ write $\nabla \hat{J}(u)$ for the derivative to emphasize that this is the Riesz representation. Then the following conditions are equivalent:

(i) $\bar{u} \in U_{\text{ad}}$,

$$(\nabla \hat{J}(\bar{u}), u - \bar{u})_U \geq 0 \quad \forall u \in U_{\text{ad}}.$$

(ii) $\bar{u} \in U_{\text{ad}}$,

$$\nabla \hat{J}(\bar{u})(x) \quad \begin{cases} = 0, & \text{if } a(x) < \bar{u}(x) < b(x), \\ \geq 0, & \text{if } a(x) = \bar{u}(x) < b(x), \\ \leq 0, & \text{if } a(x) < \bar{u}(x) = b(x), \end{cases} \quad \text{for a.a. } x \in \Omega.$$

(iii) *There are $\bar{\lambda}_a, \bar{\lambda}_b \in U^* = L^2(\Omega)$ with*

$$\nabla \hat{J}(\bar{u}) + \bar{\lambda}_b - \bar{\lambda}_a = 0,$$

$$\bar{u} \geq a, \qquad \bar{\lambda}_a \geq 0, \qquad \bar{\lambda}_a(\bar{u} - a) = 0,$$

$$\bar{u} \leq b, \qquad \bar{\lambda}_b \geq 0, \qquad \bar{\lambda}_b(b - \bar{u}) = 0.$$

(iv) *For any* $\gamma > 0$: $\bar{u} = P_{U_{\text{ad}}}(\bar{u} - \gamma \nabla \hat{J}(\bar{u}))$, *with* $P_{U_{\text{ad}}}(u) = \min(\max(a, u), b)$.

Proof (ii) $\Longrightarrow$ (i): If $\nabla \hat{J}(\bar{u})$ satisfies (ii) then it is obvious that $\nabla \hat{J}(\bar{u})(u - \bar{u}) \geq 0$ a.e. for all $u \in U_{\text{ad}}$ and thus

$$(\nabla \hat{J}(\bar{u}), u - \bar{u})_U = \int_\Omega \nabla \hat{J}(\bar{u})(u - \bar{u})dx \geq 0 \quad \forall u \in U_{\text{ad}}.$$

(i) $\Longrightarrow$ (ii): Clearly, (ii) is the same as

$$\nabla \hat{J}(\bar{u})(x) \begin{cases} \geq 0 & \text{a.e. on } I_a = \{x : a(x) \leq \bar{u}(x) < b(x)\}, \\ \leq 0 & \text{a.e. on } I_b = \{x : a(x) < \bar{u}(x) \leq b(x)\}. \end{cases}$$

Assume this is not true. Then, without loss of generality, there exists a set $M \subset I_a$ of positive measure with $\nabla \hat{J}(\bar{u})(x) < 0$ on M. Now choose $u = \bar{u} + 1_M(b - \bar{u})$. Then $u \in U_{\text{ad}}$, $u - \bar{u} > 0$ on M and $u - \bar{u} = 0$ elsewhere. Hence, we get the contradiction

$$(\nabla \hat{J}(\bar{u}), u - \bar{u})_U = \int_M \underbrace{\nabla \hat{J}(\bar{u})}_{<0} \underbrace{(b - \bar{u})}_{>0} dx < 0.$$

(ii) $\Longrightarrow$ (iii): Let $\bar{\lambda}_a = \max(\nabla \hat{J}(\bar{u}), 0)$, $\bar{\lambda}_b = \max(-\nabla \hat{J}(\bar{u}), 0)$. Then $a \leq \bar{u} \leq b$ and $\bar{\lambda}_a, \bar{\lambda}_b \geq 0$ hold trivially. Furthermore,

$$\bar{u}(x) > a(x) \quad \Longrightarrow \quad \nabla \hat{J}(\bar{u})(x) \leq 0 \quad \Longrightarrow \quad \bar{\lambda}_a(x) = 0,$$

$$\bar{u}(x) < b(x) \quad \Longrightarrow \quad \nabla \hat{J}(\bar{u})(x) \geq 0 \quad \Longrightarrow \quad \bar{\lambda}_b(x) = 0.$$

(iii) $\Longrightarrow$ (ii):

$$a(x) < \bar{u}(x) < b(x) \quad \Longrightarrow \quad \bar{\lambda}_a = \bar{\lambda}_b = 0 \quad \Longrightarrow \quad \nabla \hat{J}(\bar{u}) = 0,$$

$$a(x) = \bar{u}(x) < b(x) \quad \Longrightarrow \quad \bar{\lambda}_b = 0 \quad \Longrightarrow \quad \nabla \hat{J}(\bar{u}) = \bar{\lambda}_a \geq 0,$$

$$a(x) < \bar{u}(x) = b(x) \quad \Longrightarrow \quad \bar{\lambda}_a = 0 \quad \Longrightarrow \quad \nabla \hat{J}(\bar{u}) = -\bar{\lambda}_b \leq 0.$$

(ii) $\Longleftrightarrow$ (iv): This is easily verified.

Alternatively, we can use Lemma 1.11 to prove the equivalence of (i) and (iv).

1.7.2.2 Necessary First Order Optimality Conditions

Next, we use the adjoint representation of the derivative

$$\hat{J}'(u) = e_u(y(u), u)^* p(u) + J_u(y(u), u), \tag{1.103}$$

where the adjoint state $p(u) \in Z^*$ solves the adjoint equation

$$e_y(y(u), u)^* p = -J_y(y(u), u). \tag{1.104}$$

For compact notation, we recall the definition of the Lagrange function associated with (1.99)

$$L : Y \times U \times Z^* \to \mathbb{R}, \qquad L(y,u,p) = J(y,u) + \langle p, e(y,u)\rangle_{Z^*,Z}.$$

The representation (1.103) of $\hat{J}'(\bar{u})$ yields the following corollary of Theorem 1.48.

Corollary 1.3 *Let* $(\bar{y},\bar{u})$ *an optimal solution of the problem* (1.99) *and let Assumption* 1.47 *hold. Then there exists an adjoint state (or Lagrange multiplier)* $\bar{p} \in Z^*$ *such that the following optimality conditions hold*

$$e(\bar{y},\bar{u}) = 0, \tag{1.105}$$

$$e_y(\bar{y},\bar{u})^* \bar{p} = -J_y(\bar{y},\bar{u}), \tag{1.106}$$

$$\bar{u} \in U_{\text{ad}}, \quad \langle J_u(\bar{y},\bar{u}) + e_u(\bar{y},\bar{u})^* \bar{p}, u - \bar{u}\rangle_{U^*,U} \geq 0 \quad \forall u \in U_{\text{ad}}. \tag{1.107}$$

Using the Lagrange function we can write (1.105)–(1.107) *in the compact form*

$$L_p(\bar{y},\bar{u},\bar{p}) = e(\bar{y},\bar{u}) = 0, \tag{1.105}$$

$$L_y(\bar{y},\bar{u},\bar{p}) = 0, \tag{1.106}$$

$$\bar{u} \in U_{\text{ad}}, \quad \langle L_u(\bar{y},\bar{u},\bar{p}), u - \bar{u}\rangle_{U^*,U} \geq 0 \quad \forall u \in U_{\text{ad}}. \tag{1.107}$$

Proof We have only to combine (1.102), (1.104), and (1.103).

To avoid dual operators, one can also use the equivalent variational form

$$\langle e(\bar{y},\bar{u}), p\rangle_{Z,Z^*} = 0 \quad \forall p \in Z^*, \tag{1.108}$$

$$\langle L_y(\bar{y},\bar{u},\bar{p}), v\rangle_{Y^*,Y} = 0 \quad \forall v \in Y \tag{1.109}$$

$$\bar{u} \in U_{\text{ad}}, \quad \langle L_u(\bar{y},\bar{u},\bar{p}), u - \bar{u}\rangle_{U^*,U} \geq 0 \quad \forall u \in U_{\text{ad}}. \tag{1.110}$$

1.7.2.3 Applications

General Linear-Quadratic Problem

We apply the result to the linear-quadratic problem

$$\begin{aligned} &\min_{(y,u)\in Y\times U} \; J(y,u) := \frac{1}{2}\|Qy - q_d\|_H^2 + \frac{\alpha}{2}\|u\|_U^2 \\ &\text{subject to} \quad Ay + Bu = g, \quad u \in U_{\text{ad}} \end{aligned} \tag{1.111}$$

under Assumption 1.42. Then

$$e(y,u) = Ay + Bu - g, \qquad e_y(y,u) = A, \qquad e_u(y,u) = B$$

and Corollary 1.3 is applicable. We only have to compute L_y and L_u for the Lagrange function

$$L(y,u,p) = J(y,u) + \langle p, Ay + Bu - g\rangle_{Z^*,Z}$$
$$= \frac{1}{2}(Qy - q_d, Qy - q_d)_H + \frac{\alpha}{2}(u,u)_U + \langle p, Ay + Bu - g\rangle_{Z^*,Z}.$$

We have with the identification $H^* = H$ and $U^* = U$

$$\begin{aligned} \langle L_y(\bar{y},\bar{u},\bar{p}), v\rangle_{Y^*,Y} &= (Q\bar{y} - q_d, Qv)_H + \langle \bar{p}, Av\rangle_{Z^*,Z} \\ &= \langle Q^*(Q\bar{y} - q_d) + A^*\bar{p}, v\rangle_{Y^*,Y} \quad \forall\, v \in Y \end{aligned} \tag{1.112}$$

and

$$\begin{aligned} (L_u(\bar{y},\bar{u},\bar{p}), w)_U &= \alpha(\bar{u}, w)_U + \langle \bar{p}, Bw\rangle_{Z^*,Z} \\ &= (\alpha\bar{u} + B^*\bar{p}, w)_U \quad \forall w \in U. \end{aligned} \tag{1.113}$$

Thus (1.105)–(1.107) take the form

$$A\bar{y} + B\bar{u} = g, \tag{1.114}$$

$$A^*\bar{p} = -Q^*(Q\bar{y} - q_d), \tag{1.115}$$

$$\bar{u} \in U_{\mathrm{ad}}, \quad (\alpha\bar{u} + B^*\bar{p}, u - \bar{u})_U \geq 0 \quad \forall u \in U_{\mathrm{ad}}. \tag{1.116}$$

Distributed Control of Elliptic Equations

We consider next the distributed optimal control of a steady temperature distribution with boundary temperature zero

$$\begin{aligned} &\min\ J(y,u) := \frac{1}{2}\|y - y_d\|^2_{L^2(\Omega)} + \frac{\alpha}{2}\|u\|^2_{L^2(\Omega)} \\ &\text{subject to} \quad -\Delta y = \gamma u \quad \text{on } \Omega, \\ &\qquad\qquad\quad y = 0 \quad \text{on } \partial\Omega, \\ &\qquad\qquad\quad a \leq u \leq b \quad \text{on } \Omega, \end{aligned} \tag{1.117}$$

where

$$\gamma \in L^\infty(\Omega) \setminus \{0\}, \quad \gamma \geq 0, a, b \in L^2(\Omega), \quad a \leq b.$$

We have already observed in Sect. 1.5.3.1 that (1.117) has the form (1.111) and satisfies Assumption 1.42 with

$$U = H = L^2(\Omega), \qquad Y = H_0^1(\Omega), \qquad Z = Y^*, \qquad g = 0, \qquad Q = I_{Y,H},$$

$U_{\text{ad}} = \{u \in U : a \leq u \leq b\}$ and

$$A \in \mathcal{L}(Y, Y^*), \qquad \langle Ay, v\rangle_{Y^*,Y} = a(y, v) = \int_\Omega \nabla y \cdot \nabla v \, dx,$$

$$B \in \mathcal{L}(U, Y^*), \qquad \langle Bu, v\rangle_{Y^*,Y} = -(\gamma u, v)_{L^2(\Omega)}.$$

Hence, the optimality system is given by (1.114)–(1.116). Moreover, we have $A^* = A$. In fact, as a Hilbert space, Y is reflexive and $Z^* = Y^{**}$ can be identified with Y through

$$\langle p, y^*\rangle_{Y^{**},Y^*} = \langle y^*, p\rangle_{Y^*,Y} \quad \forall y^* \in Y^*, \; p \in Y = Y^{**}.$$

This yields

$$\begin{aligned}\langle A^*v, w\rangle_{Y^*,Y} &= \langle v, Aw\rangle_{Z^*,Z} = \langle Aw, v\rangle_{Y^*,Y} \\ &= a(w, v) = a(v, w) = \langle Av, w\rangle_{Y^*,Y} \quad \forall v, w \in Y.\end{aligned}$$

and thus $A^* = A$.

Instead of interpreting (1.114)–(1.116) for this problem we demonstrate that it is very convenient to work with the form (1.108)–(1.110) of the optimality system. We have

$$\langle p, Ay\rangle_{Z^*,Z} = \langle Ay, p\rangle_{Y^*,Y} = a(y, p) = a(p, y).$$

Let $(\bar{y}, \bar{u}) \in Y \times U$ be an optimal solution. Then by Corollary 1.3 and (1.112), (1.113) the optimality system in the form (1.108)–(1.110) reads

$$a(\bar{y}, v) - (\gamma\bar{u}, v)_{L^2(\Omega)} = 0 \quad \forall v \in Y, \tag{1.118}$$

$$(\bar{y} - y_d, v)_{L^2\Omega} + a(\bar{p}, v) = 0 \quad \forall v \in Y, \tag{1.119}$$

$$a \leq \bar{u} \leq b, \quad (\alpha\bar{u} - \gamma\bar{p}, u - \bar{u})_{L^2(\Omega)} \geq 0, \quad \forall u \in U, \; a \leq u \leq b. \tag{1.120}$$

(1.118)–(1.120) is just an equivalent variational formulation of (1.114)–(1.116) by Corollary 1.3.

Now the adjoint equation (1.119) is just the weak formulation of

$$-\Delta\bar{p} = -(\bar{y} - y_d), \qquad \bar{p}|_{\partial\Omega} = 0.$$

Applying Lemma 1.12 we can summarize

Theorem 1.49 *If* $(\bar{y}, \bar{u})$ *is an optimal solution of* (1.117) *then there exist* $\bar{p} \in H_0^1(\Omega)$, $\bar{\lambda}_a, \bar{\lambda}_b \in L^2(\Omega)$ *such that the following optimality conditions hold in the weak sense.*

$$\begin{aligned}&-\Delta\bar{y} = \gamma\bar{u}, \qquad \bar{y}|_{\partial\Omega} = 0,\\ &-\Delta\bar{p} = -(\bar{y} - y_d), \qquad \bar{p}|_{\partial\Omega} = 0,\\ &\alpha\bar{u} - \gamma\bar{p} + \bar{\lambda}_b - \bar{\lambda}_a = 0,\end{aligned}$$

$$\bar{u} \ge a, \qquad \bar{\lambda}_a \ge 0, \qquad \bar{\lambda}_a(\bar{u} - a) = 0,$$
$$\bar{u} \le b, \qquad \bar{\lambda}_b \ge 0, \qquad \bar{\lambda}_b(b - \bar{u}) = 0.$$

Distributed Control of Semilinear Elliptic Equations

We consider next the distributed optimal control of a semilinear elliptic PDE:

$$\begin{aligned} &\min J(y,u) := \frac{1}{2}\|y - y_d\|^2_{L^2(\Omega)} + \frac{\alpha}{2}\|u\|^2_{L^2(\Omega)} \\ &\text{subject to} \quad -\Delta y + y^3 = \gamma u \quad \text{on } \Omega, \\ &\qquad\qquad y = 0 \quad \text{on } \partial\Omega, \\ &\qquad\qquad a \le u \le b \quad \text{on } \Omega, \end{aligned} \tag{1.121}$$

where

$$\gamma \in L^\infty(\Omega) \setminus \{0\}, \quad \gamma \ge 0, a, b \in L^\infty(\Omega), \quad a \le b.$$

Let $n \le 3$. By the theory of monotone operators one can show, see Theorem 1.25 and Remark 1.12, that there exists a unique bounded solution operator of the state equation

$$u \in U := L^2(\Omega) \to y \in Y := H^1_0(\Omega).$$

Let $A : H^1_0(\Omega) \to H^1_0(\Omega)^*$ be the operator associated with the bilinear form $a(y,v) = \int_\Omega \nabla y \cdot \nabla v \, dx$ for the Laplace operator $-\Delta y$ and let

$$N : y \to y^3.$$

Then the weak formulation of the state equation can be written in the form

$$e(y,u) := Ay + N(y) - \gamma u = 0.$$

By the Sobolev embedding Theorem 1.14 one has for $n \le 3$ the continuous embedding

$$H^1_0(\Omega) \hookrightarrow L^6(\Omega).$$

Moreover, the mapping $N : y \in L^6(\Omega) \to y^3 \in L^2(\Omega)$ is continuously Fréchet differentiable with

$$N'(y)v = 3y^2 v.$$

To show this, it is convenient to prove first the following extension of Hölder's inequality:

Lemma 1.13 *Let $\omega \subset \mathbb{R}^n$ be measurable. Then, for all $p_i, p \in [1,\infty]$ with $1/p_1 + \cdots + 1/p_k = 1/p$ and all $u_i \in L^{p_i}(\Omega)$, there holds $u_1 \cdots u_k \in L^p(\Omega)$ and*

$$\|u_1 \cdots u_k\|_{L^p} \le \|u_1\|_{L^{p_1}} \cdots \|u_k\|_{L^{p_k}}.$$

Proof We use induction. For $k=1$ the assertion is trivial and for $k=2$ we obtain it from Hölder's inequality: From $1/p_1+1/p_2=1/p$ we see that $1/q_1+1/q_2=1$ holds for $q_i=p_i/p$ and thus

$$\begin{aligned}\|u_1u_2\|_{L^p} &= \||u_1|^p|u_2|^p\|_{L^1}^{1/p} \le \||u_1|^p\|_{L^{q_1}}^{1/p}\||u_2|^p\|_{L^{q_2}}^{1/p}\\ &= \||u_1|^{pq_1}\|_{L^1}^{1/p_1}\||u_2|^{pq_2}\|_{L^1}^{1/p_2} = \|u_1\|_{L^{p_1}}\|u_2\|_{L^{p_2}}.\end{aligned}$$

As a consequence, $u_1u_2 \in L^p(\Omega)$ and the assertion is shown for $k=2$.

For $1,\dots,k-1 \to k$, let $q \in [1,\infty]$ be such that

$$\frac{1}{q}+\frac{1}{p_k}=\frac{1}{p}.$$

Then we have $1/p_1+\cdots+1/p_{k-1}=1/q$ and thus (using the assertion for $k-1$), we obtain $u_1\cdots u_{k-1}\in L^q(\Omega)$ and

$$\|u_1\cdots u_{k-1}\|_{L^q}\le \|u_1\|_{L^{p_1}}\cdots\|u_{k-1}\|_{L^{p_{k-1}}}.$$

Therefore, using the assertion for $k=2$,

$$\|u_1\cdots u_k\|_{L^p}\le\|u_1\cdots u_{k-1}\|_{L^q}\|u_k\|_{L^{p_k}}\le\|u_1\|_{L^{p_1}}\cdots\|u_k\|_{L^{p_k}}.$$

We now return to the proof of the F-differentiability of N: We just have to apply the Lemma with $p_1=p_2=p_3=6$ and $p=2$:

$$\begin{aligned}\|(y+h)^3-y^3-3y^2h\|_{L^2} &= \|3yh^2+h^3\|_{L^2}\le 3\|y\|_{L^6}\|h\|_{L^6}^2+\|h\|_{L^6}^3\\ &= O(\|h\|_{L^6}^2)=o(\|h\|_{L^6}).\end{aligned}$$

This shows the F-differentiability of N with derivative N'. Furthermore, to prove the continuity of N', we estimate

$$\begin{aligned}\|(N'(y+h)-N'(y))v\|_{L^2} &= 3\|((y+h)^2-y^2)v\|_{L^2}=3\|(2y+h)hv\|_{L^2}\\ &= 3\|2y+h\|_{L^6}\|h\|_{L^6}\|v\|_{L^6}.\end{aligned}$$

Hence,

$$\|N'(y+h)-N'(y)\|_{L^6,L^2}\le 3\|2y+h\|_{L^6}\|h\|_{L^6}\overset{\|h\|_{L^6}\to 0}{\longrightarrow} 0.$$

Therefore, $e: Y\times U\to Y^* =: Z$ is continuously Fréchet differentiable with

$$e_y(y,u)v = Av+3y^2v, \qquad e_u(y,u)w=-\gamma w.$$

Finally, $e_y(y,u)\in\mathcal{L}(Y,Z)$ has a bounded inverse, since for any $y\in Y$ the equation

$$Av+3y^2v=f$$

has a bounded solution operator $f \in Z \to v \in Y$ by the Lax-Milgram lemma. In fact, $A + 3y^2 I \in \mathcal{L}(Y, Z)$ and corresponds to the bounded and coercive bilinear form $(v, w) \in Y \times Y \mapsto a(v, w) + (3y^2 v, w)_{L^2(\Omega)}$.

Hence, Assumption 1.47 is satisfied. The optimality conditions are now similar to the linear-quadratic problem (1.117): Let $(\bar{y}, \bar{u}) \in Y \times U$ be an optimal solution. Then by Corollary 1.3 the optimality system in the form (1.108)–(1.110) reads

$$a(\bar{y}, v) + (\bar{y}^3, v)_{L^2(\Omega)} - (\gamma \bar{u}, v)_{L^2(\Omega)} = 0 \quad \forall v \in Y, \tag{1.122}$$

$$(\bar{y} - y_d, v)_{L^2\Omega} + a(\bar{p}, v) + (\bar{p}, 3\bar{y}^2 v)_{L^2(\Omega)} = 0 \quad \forall v \in Y, \tag{1.123}$$

$$a \le \bar{u} \le b, \quad (\alpha \bar{u} - \gamma \bar{p}, u - \bar{u})_{L^2(\Omega)} \ge 0, \quad \forall u \in U,\ a \le u \le b. \tag{1.124}$$

Now the adjoint equation (1.123) is just the weak formulation of

$$-\Delta \bar{p} + 3\bar{y}^2 \bar{p} = -(\bar{y} - y_d), \quad \bar{p}|_{\partial\Omega} = 0.$$

Applying Lemma 1.12 we can summarize

Theorem 1.50 *If* $(\bar{y}, \bar{u})$ *is an optimal solution of* (1.121) *then there exist* $\bar{p} \in H_0^1(\Omega)$, $\bar{\lambda}_a, \bar{\lambda}_b \in L^2(\Omega)$ *such that the following optimality system holds in the weak sense.*

$$\begin{aligned}
&-\Delta \bar{y} + \bar{y}^3 = \gamma \bar{u}, && \bar{y}|_{\partial\Omega} = 0,\\
&-\Delta \bar{p} + 3\bar{y}^2 \bar{p} = -(\bar{y} - y_d), && \bar{p}|_{\partial\Omega} = 0,\\
&\alpha \bar{u} - \gamma \bar{p} + \bar{\lambda}_b - \bar{\lambda}_a = 0, &&\\
&\bar{u} \ge a, \quad \bar{\lambda}_a \ge 0, && \bar{\lambda}_a(\bar{u} - a) = 0,\\
&\bar{u} \le b, \quad \bar{\lambda}_b \ge 0, && \bar{\lambda}_b(b - \bar{u}) = 0.
\end{aligned}$$

Boundary Control of Parabolic Equations

We consider finally the optimal boundary control of an unsteady heating process.

$$\begin{aligned}
&\min J(y, u) := \frac{1}{2}\|y(T) - y_d\|^2_{L^2(\Omega)} + \frac{\alpha}{2}\|u\|^2_{L^2((0,T)\times\partial\Omega)}\\
&\text{subject to} \quad y_t - \Delta y = 0 \quad \text{on } \Omega_T,\\
&\qquad\qquad \frac{\partial y}{\partial \nu} = u \quad \text{on } (0, T) \times \partial\Omega,\\
&\qquad y(0, \cdot) = y_0 \quad \text{on } \Omega, \quad a \le u \le b \quad \text{on } (0, T) \times \partial\Omega,
\end{aligned} \tag{1.125}$$

where $a, b \in L^2((0, T) \times \Omega)$, $a < b$, $y_0, y_d \in L^2(\Omega)$. With $V = H^1(\Omega)$, $H = L^2(\Omega)$ and

$$a(y(t), v) := \int_\Omega \nabla y(t) \cdot \nabla v \, dx,$$

$$\langle f(t), v\rangle_{V^*,V} := (u(t), v)_{L^2(\partial\Omega)}, \quad y \in W(0,T;H,V),\ v \in V$$

the weak formulation of the state equation is given by (1.62), (1.63) (or equivalently (1.66), (1.63)). Let

$$U = L^2((0,T)\times\partial\Omega), \qquad Y = W(0,T;L^2(\Omega),H^1(\Omega)),$$
$$Z = L^2(0,T;H^1(\Omega)^*)\times L^2(\Omega).$$

Then it is easy to check that Assumption 1.34 holds. The weak formulation defines a bounded affine linear operator

$$e : (y,u) \in Y\times U \mapsto \begin{pmatrix} Ay+Bu \\ y(0)-y_0 \end{pmatrix} \in Z.$$

By Theorem 1.35 and (1.37) the equation $e(y,u)=0$ has a unique bounded affine linear solution operator $u\in U \mapsto y(u)\in Y$ and $e_y(y,u)\in\mathcal{L}(Y,Z)$, $e_y(y,u)v = \binom{Av}{v(0)}$, has a bounded inverse. Moreover, by using the imbedding $Y \hookrightarrow C([0,T];L^2(\Omega))$, the objective function $J : Y\times U \to \mathbb{R}$ is obviously continuously F-differentiable.

Hence, Assumption 1.47 is satisfied. Let $(\bar{y},\bar{u})\in Y\times U$ be local solution of (1.125), which is a global solution, since the problem is convex. Then Corollary 1.3 yields necessary optimality conditions (1.108)–(1.110), where the Lagrangian is given by

$$\begin{aligned} L(y,u,p,q) &= J(y,u) + \int_0^T \big(c(y(t),p(t)) - (u(t),p(t))_{L^2(\partial\Omega)}\big)\,dt \\ &\quad + (y(0)-y_0,q)_{L^2(\Omega)}, \\ c(y(t),p(t)) &:= \langle y_t(t),p(t)\rangle_{V^*,V} + a(y(t),p(t)) \end{aligned}$$

with $(p,q)\in L^2(0,T;V)\times L^2(\Omega)$. Hence, the optimality system in the form (1.108)–(1.110) reads

$$\int_0^T \big(c(\bar{y}(t),v(t)) - (\bar{u}(t),v(t))_{L^2(\partial\Omega)}\big)\,dt = 0 \quad \forall v\in L^2(0,T;V),$$
$$\int_0^T c(v(t),\bar{p}(t))\,dt + (\bar{y}(T)-y_d, v(T))_{L^2(\Omega)} + (v(0),\bar{q})_{L^2(\Omega)} = 0 \quad \forall v\in Y,$$
$$a\le\bar{u}\le b, \quad (\alpha\bar{u}-\bar{p}, u-\bar{u})_{L^2((0,T)\times\partial\Omega)} \ge 0 \quad \forall u\in U,\ a\le u\le b.$$

Since $e_y(\bar{y},\bar{u})\in\mathcal{L}(Y,Z)$ has a bounded inverse, there exists a unique adjoint state $(\bar{p},\bar{q})\in Z^* = L^2(0,T;V)\times L^2(\Omega)$.

To identify the adjoint equation, assume that $\bar{p}\in W(0,T)$ (which will be justified later). Then integration by parts in the term $\langle v_t(t),\bar{p}(t)\rangle_{V^*,V}$ according to

Theorem 1.32 shows that the adjoint equation is equivalent to

$$\int_0^T \big(-\langle v(t), \bar{p}_t(t)\rangle_{V,V^*} + a(v(t), \bar{p}(t))\big)\,dt + (\bar{y}(T) - y_d + \bar{p}(T), v(T))_{L^2(\Omega)}$$
$$+ (v(0), \bar{q} - \bar{p}(0))_{L^2(\Omega)} = 0 \quad \forall v \in Y.$$

Using the fact that $C_c^\infty((0,T);V) \subset Y$ is dense in $L^2(0,T;V)$, we conclude that for $\bar{p} \in Y$ the adjoint equation is equivalent to

$$\int_0^T \big(-\langle v(t), \bar{p}_t(t)\rangle_{V,V^*} + a(v(t), \bar{p}(t))\big)\,dt = 0 \quad \forall v \in L^2(0,T;V),$$
$$\bar{p}(T) = -(\bar{y}(T) - y_d), \quad \bar{q} = \bar{p}(0).$$

But this variational equation is the weak formulation of

$$-\bar{p}_t - \Delta\bar{p} = 0, \qquad \bar{p}(T) = -(\bar{y}(T) - y_d), \qquad \frac{\partial \bar{p}}{\partial \nu}|_{(0,T)\times\partial\Omega} = 0$$

and has by Theorem 1.35 and (1.37) in fact a unique solution $\bar{p} \in Y$, which is together with $\bar{q} = \bar{p}(0)$ the unique adjoint state. By applying Lemma 1.12 we can summarize

Theorem 1.51 *If* $(\bar{y}, \bar{u})$ *is an optimal solution of* (1.125) *then there exist* $\bar{p} \in Y$, $\bar{\lambda}_a, \bar{\lambda}_b \in L^2((0,T)\times\partial\Omega)$ *such that the following optimality system holds in the weak sense.*

$$\bar{y}_t - \Delta\bar{y} = 0, \qquad \frac{\partial \bar{y}}{\partial \nu}|_{(0,T)\times\partial\Omega} = \bar{u}, \qquad \bar{y}(0) = y_0,$$
$$-\bar{p}_t - \Delta\bar{p} = 0, \qquad \frac{\partial \bar{p}}{\partial \nu}|_{(0,T)\times\partial\Omega} = 0, \qquad \bar{p}(T) = -(\bar{y}(T) - y_d),$$
$$\alpha\bar{u} - \bar{p} + \bar{\lambda}_b - \bar{\lambda}_a = 0,$$
$$\bar{u} \ge a, \quad \bar{\lambda}_a \ge 0, \quad \bar{\lambda}_a(\bar{u} - a) = 0,$$
$$\bar{u} \le b, \quad \bar{\lambda}_b \ge 0, \quad \bar{\lambda}_b(b - \bar{u}) = 0.$$

1.7.3 Optimality Conditions for Problems with General Constraints

We sketch now the theory of optimality conditions for general problems of the form

$$\min_{w\in W} J(w) \quad \text{subject to} \quad G(w) \in \mathcal{K}_G, \quad w \in \mathcal{C}. \tag{1.126}$$

Here, $J : W \to \mathbb{R}$, $G : W \to V$ are continuously Fréchet differentiable with Banach spaces W, V, $\mathcal{C} \subset V$ is non-empty, closed and convex, and $\mathcal{K}_G \subset V$ is a closed convex cone. Here, $\mathcal{K}_G$ is a cone if

$$\forall \lambda > 0: \quad v \in \mathcal{K}_G \quad \Longrightarrow \quad \lambda v \in \mathcal{K}_G.$$

We denote the feasible set by

$$F_{\text{ad}} := \{w \in W : G(w) \in \mathcal{K}_G, \ w \in \mathcal{C}\}.$$

Remark 1.20 It is no restriction not to include equality constraints. In fact

$$e(w) = 0, \qquad c(w) \in \mathcal{K}$$

is equivalent to

$$G(w) := \begin{pmatrix} e(w) \\ c(w) \end{pmatrix} \in \{0\} \times \mathcal{K} =: \mathcal{K}_G.$$

1.7.3.1 A Basic First Order Optimality Condition

Let $\bar{w}$ be a local solution of (1.126). To develop an extension of Theorem 1.48, we define the cone of feasible directions as follows.

Definition 1.31 Let $F_{\text{ad}} \subset W$ be nonempty. The *tangent cone* of F_{ad} at $w \in F_{\text{ad}}$ is defined by

$$T(F_{\text{ad}}; w) = \left\{ s \in W : \exists \eta_k > 0, \ w_k \in F_{\text{ad}} : \lim_{k\to\infty} w_k = w, \ \lim_{k\to\infty} \eta_k (w_k - w) = s \right\}.$$

Then we have the following optimality condition.

Theorem 1.52 *Let $J : W \to \mathbb{R}$ be continuously Fréchet differentiable. Then for any local solution $\bar{w}$ of* (1.126) *the following optimality condition holds.*

$$\bar{w} \in F_{\text{ad}} \quad \textit{and} \quad \langle J'(\bar{w}), s\rangle_{W^*,W} \ge 0 \quad \forall s \in T(F_{\text{ad}}; \bar{w}). \tag{1.127}$$

Proof $\bar{w} \in F_{\text{ad}}$ is obvious. Let $s \in T(F_{\text{ad}}; \bar{w})$ be arbitrary. Then there exist $(w_k) \subset F_{\text{ad}}$ and $\eta_k > 0$ with $w_k \to \bar{w}$ and $\eta_k(w_k - \bar{w}) \to s$. This yields for all sufficiently large k

$$\begin{aligned} 0 &\le \eta_k (J(w_k) - J(\bar{w})) = \langle J'(\bar{w}), \eta_k (w_k - \bar{w})\rangle_{W^*,W} + \eta_k o(\|w_k - \bar{w}\|_W) \\ &\to \langle J'(\bar{w}), s\rangle_{W^*,W} \end{aligned}$$

since $\eta_k o(\|w_k - \bar{w}\|_W) \to 0$, which follows from $\eta_k(w_k - \bar{w}) \to s$ and $w_k \to \bar{w}$.

1.7.3.2 Constraint Qualification and Robinson's Regularity Condition

We want to replace the tangent cone by a cone with a less complicated representation. Linearization of the constraints (assuming G is continuously differentiable) leads us to the *linearization cone* at a point $\bar{w} \in F_{\text{ad}}$ defined by

$$L(F_{\text{ad}}, G, \mathcal{K}_G, \mathcal{C}; \bar{w}) = \left\{\eta d : \eta > 0,\ d \in W,\ G(\bar{w}) + G'(\bar{w})d \in \mathcal{K}_G,\ \bar{w} + d \in \mathcal{C}\right\}.$$

Assume now that a local solution $\bar{w}$ of (1.126) satisfies the

Constraint Qualification:

$$L(F_{\text{ad}}, G, \mathcal{K}_G, \mathcal{C}; \bar{w}) \subset T(F_{\text{ad}}; \bar{w}) \tag{1.128}$$

Then the following result is obvious.

Theorem 1.53 *Let $J : W \to \mathbb{R}$, $G : W \to V$ be continuously Fréchet differentiable with Banach-spaces W, V. Further let $\mathcal{C} \subset W$ be non-empty, closed and convex, and let $\mathcal{K}_G \subset V$ be a closed convex cone. Then at every local solution $\bar{w}$ of* (1.126) *satisfying* (1.128) *the following optimality condition holds.*

$$\bar{w} \in F_{\text{ad}} \quad \textit{and} \quad \langle J'(\bar{w}), s\rangle_{W^*,W} \geq 0 \quad \forall s \in L(F_{\text{ad}}, G, \mathcal{K}_G, \mathcal{C}; \bar{w}). \tag{1.129}$$

Remark 1.21 If G is affine linear, then (1.128) is satisfied. In fact, let $s \in L(F_{\text{ad}}, G, \mathcal{K}_G, \mathcal{C}; \bar{w})$. Then $s = \eta d$ with $\eta > 0$ and $d \in W$,

$$G(\bar{w} + d) = G(\bar{w}) + G'(\bar{w})d \in \mathcal{K}_G, \quad \bar{w} + d \in \mathcal{C}.$$

Since $G(\bar{w}) \in \mathcal{K}_G$ and $\bar{w} \in \mathcal{C}$, the convexity of $\mathcal{K}_G$ and $\mathcal{C}$ yields $w_k := \bar{w} + \frac{1}{k}d \in F_{\text{ad}}$. Choosing $\eta_k = k\eta$ shows that $s \in T(F_{\text{ad}}; \bar{w})$.

In general, (1.128) can be ensured if $\bar{w}$ satisfies the

Regularity Condition of Robinson:

$$0 \in \operatorname{int}\left(G(\bar{w}) + G'(\bar{w})\,(\mathcal{C} - \bar{w}) - \mathcal{K}_G\right). \tag{1.130}$$

We have the following important and deep result by Robinson [116].

Theorem 1.54 *Robinson's regularity condition* (1.130) *implies the constraint qualification* (1.128).

Proof See [116, Thm. 1, Cor. 2].

In the convex case, i.e.,

$$G((1-t)w_1 + tw_2) - (1-t)G(w_1) - tG(w_2) \in \mathcal{K}_G \quad \forall t \in [0,1],\ w_1, w_2 \in W \tag{1.131}$$

Robinson's regularity condition is implied by the following Slater's condition.

Slater's condition:

There exists $\tilde{w} \in W$ such that

$$G(\tilde{w}) \in \operatorname{int} \mathcal{K}_G, \quad \tilde{w} \in \mathcal{C}. \tag{1.132}$$

Theorem 1.55 *Let as above $\mathcal{C} \subset W$ be closed and convex and $\mathcal{K}_G \subset V$ be a closed convex cone. If $G : W \to V$ is convex, i.e., if* (1.131) *holds, then Slater's condition* (1.132) *implies Robinson's regularity condition* (1.130) *for all $W \ni \bar{w} \in \mathcal{C}$ with $G(\bar{w}) \in \mathcal{K}_G$.*

Proof By (1.132) we have for sufficiently small $\varepsilon > 0$ that

$$G(\tilde{w}) + B_V(\varepsilon) \subset \mathcal{K}_G. \tag{1.133}$$

We show that Robinson's regularity condition holds for all $\bar{w} \in W$ with $G(\bar{w}) \in \mathcal{K}_G$, $\bar{w} \in \mathcal{C}$.

Let $w(t) = \bar{w} + t(\tilde{w} - \bar{w})$, $t \in [0, 1]$. Then by (1.131) we have for all $t \in (0, 1]$

$$\begin{aligned} \frac{1}{t}(G(w(t)) - (1-t)G(\bar{w}) - tG(\tilde{w})) &= \frac{G(w(t)) - G(\bar{w})}{t} + G(\bar{w}) - G(\tilde{w}) \\ &\in \frac{1}{t}\mathcal{K}_G = \mathcal{K}_G. \end{aligned}$$

Since $\mathcal{K}_G$ is closed, $t \searrow 0$ yields

$$G'(\bar{w})(\tilde{w} - \bar{w}) + G(\bar{w}) - G(\tilde{w}) \in \mathcal{K}_G.$$

Hence, there exists $d \in \mathcal{K}_G$ with

$$G(\bar{w}) + G'(\bar{w})(\tilde{w} - \bar{w}) - \mathcal{K}_G = G(\tilde{w}) + d - \mathcal{K}_G \supset G(\tilde{w}) - \mathcal{K}_G,$$

where the last inclusion follows form $d + \mathcal{K}_G \subset \mathcal{K}_G$ (note that $d + w = 2((d + w)/2) \in \mathcal{K}_G$ for all $d, w \in \mathcal{K}_G$, since $\mathcal{K}_G$ is a closed convex cone). We conclude with (1.133) that

$$\begin{aligned} G(\bar{w}) + G'(\bar{w})(\mathcal{C} - \bar{w}) - \mathcal{K}_G &\supset G(\bar{w}) + G'(\bar{w})(\tilde{w} - \bar{w}) \\ &- \mathcal{K}_G \supset G(\tilde{w}) - \mathcal{K}_G \supset B_V(\varepsilon). \end{aligned}$$

This shows that (1.130) holds.

1.7.3.3 Karush-Kuhn-Tucker Conditions

Using Robinson's regularity condition, we can write the optimality condition (1.129) in a more explicit form.

Theorem 1.56 (Zowe and Kurcyusz [150]) *Let $J : W \to \mathbb{R}$, $G : W \to V$ be continuously Fréchet differentiable with Banach-spaces W, V. Further let $\mathcal{C} \subset W$ be non-empty, closed and convex, and let $\mathcal{K}_G \subset V$ be a closed convex cone. Then for any local solution $\bar{w}$ of* (1.126) *at which Robinson's regularity condition* (1.130) *is satisfied, the following optimality condition holds*:

There exists a Lagrange multiplier $\bar{q} \in V^*$ with

$$G(\bar{w}) \in \mathcal{K}_G, \tag{1.134}$$

$$\bar{q} \in \mathcal{K}_G^\circ := \left\{q \in V^* : \langle q, v\rangle_{V^*,V} \le 0 \; \forall v \in \mathcal{K}_G\right\}, \tag{1.135}$$

$$\langle \bar{q}, G(\bar{w})\rangle_{V^*,V} = 0, \tag{1.136}$$

$$\bar{w} \in \mathcal{C}, \quad \langle J'(\bar{w}) + G'(\bar{w})^* \bar{q}, w - \bar{w}\rangle_{W^*,W} \ge 0 \quad \forall w \in \mathcal{C}. \tag{1.137}$$

Using the Lagrangian function

$$L(w, q) := J(w) + \langle q, G(w)\rangle_{V^*,V}$$

we can write (1.137) *in the compact form*

$$\bar{w} \in \mathcal{C}, \quad \langle L_w(\bar{w}, \bar{q}), w - \bar{w}\rangle_{W^*,W} \ge 0 \quad \forall w \in \mathcal{C}. \tag{1.137}$$

Proof Under Robinson's regularity condition (1.130), a separation argument can be used to derive (1.135)–(1.137), see [150].

A similar result can be shown if $\mathcal{K}_G$ is a closed convex set instead of a closed convex cone, see [15], but then (1.135), (1.136) have a more complicated structure.

1.7.3.4 Application to PDE-Constrained Optimization

In PDE-constrained optimization, we have usually a state equation and constraints on control and/or state. Therefore, we consider as a special case the problem

$$\min_{(y,u)\in Y\times U} J(y, u) \quad \text{subject to} \quad e(y, u) = 0, \qquad c(y) \in \mathcal{K}, \quad u \in U_{\text{ad}}, \tag{1.138}$$

where $e : Y \times U \to Z$ and $c : Y \to R$ are continuously Fréchet differentiable, $\mathcal{K} \subset R$ is a closed convex cone and $U_{\text{ad}} \subset U$ is a closed convex set. We set

$$G : \begin{pmatrix} y \\ u \end{pmatrix} \in W := Y \times U \mapsto \begin{pmatrix} e(y, u) \\ c(y) \end{pmatrix} \in Z \times R,$$

$$\mathcal{K}_G = \{0\} \times \mathcal{K}, \qquad \mathcal{C} = Y \times U_{\text{ad}}.$$

Then (1.138) has the form (1.126) and Robinson's regularity condition at a feasible point $\bar{w} = (\bar{y}, \bar{u})$ reads

$$0 \in \operatorname{int}\left(\begin{pmatrix} 0 \\ c(\bar{y}) \end{pmatrix} + \begin{pmatrix} e_y(\bar{w}) & e_u(\bar{w}) \\ c'(\bar{y}) & 0 \end{pmatrix} \begin{pmatrix} Y \\ U_{\text{ad}} - \bar{u} \end{pmatrix} - \begin{pmatrix} 0 \\ \mathcal{K} \end{pmatrix}\right). \tag{1.139}$$

We rewrite now (1.134)–(1.137) for our problem. The multiplier has the form $q = (p, \lambda) \in Z^* \times R^*$ and the Lagrangian function is given by

$$\begin{aligned}\mathcal{L}(y, u, p, \lambda) &= J(y, u) + \langle p, e(y, u)\rangle_{Z^*,Z} + \langle \lambda, c(y)\rangle_{R^*,R} \\ &= L(y, u, p) + \langle \lambda, c(y)\rangle_{R^*,R}\end{aligned}$$

with the Lagrangian

$$L(y, u, p) = J(y, u) + \langle p, e(y, u)\rangle_{Z^*,Z}$$

for the equality constraints.

Since $\mathcal{K}_G = \{0\} \times \mathcal{K}$, we have

$$\mathcal{K}_G^\circ = Z^* \times \mathcal{K}^\circ$$

and thus (1.134)–(1.137) read

$$\begin{aligned}&e(\bar{y}, \bar{u}) = 0, \qquad c(\bar{y}) \in \mathcal{K}, \\ &\bar{\lambda} \in \mathcal{K}^\circ, \quad \langle \bar{\lambda}, c(\bar{y})\rangle_{R^*,R} = 0, \\ &\langle L_y(\bar{y}, \bar{u}, \bar{p}) + c'(\bar{y})^*\bar{\lambda}, y - \bar{y}\rangle_{Y^*,Y} \geq 0 \quad \forall y \in Y, \\ &\bar{u} \in U_{\text{ad}}, \quad \langle L_u(\bar{y}, \bar{u}, \bar{p}), u - \bar{u}\rangle_{U^*,U} \geq 0 \quad \forall u \in U_{\text{ad}}.\end{aligned}$$

This yields finally

$$e(\bar{y}, \bar{u}) = 0, \qquad c(\bar{y}) \in \mathcal{K}, \tag{1.140}$$

$$\bar{\lambda} \in \mathcal{K}^\circ, \quad \langle \bar{\lambda}, c(\bar{y})\rangle_{R^*,R} = 0, \tag{1.141}$$

$$L_y(\bar{y}, \bar{u}, \bar{p}) + c'(\bar{y})^*\bar{\lambda} = 0, \tag{1.142}$$

$$\bar{u} \in U_{\text{ad}}, \quad \langle L_u(\bar{y}, \bar{u}, \bar{p}), u - \bar{u}\rangle_{U^*,U} \geq 0 \quad \forall u \in U_{\text{ad}}. \tag{1.143}$$

Remark 1.22 Without the state constraint $c(y) \in \mathcal{K}$ (which can formally be removed by omitting everything involving c or by making the constraint trivial, e.g., $c(y) = y$, $R = Y$, $\mathcal{K} = Y$), we recover exactly the optimality conditions (1.105)–(1.107) of Corollary 1.3.

We show next that the following Slater-type condition implies Robinson's regularity condition (1.139).

Lemma 1.14 *Let $\bar{w} \in F_{\text{ad}}$. If $e_y(\bar{w}) \in \mathcal{L}(Y, Z)$ is surjective and if there exist $\tilde{u} \in U_{\text{ad}}$ and $\tilde{y} \in Y$ with*

$$\begin{aligned}&e_y(\bar{w})(\tilde{y} - \bar{y}) + e_u(\bar{w})(\tilde{u} - \bar{u}) = 0, \\ &c(\bar{y}) + c'(\bar{y})(\tilde{y} - \bar{y}) \in \operatorname{int}(\mathcal{K})\end{aligned}$$

then Robinson's regularity condition (1.139) *is satisfied.*

Proof Let

$$\tilde{v} := c(\bar{y}) + c'(\bar{y})(\tilde{y} - \bar{y}).$$

Then there exists $\varepsilon > 0$ with

$$\tilde{v} + B_R(2\varepsilon) \subset \mathcal{K}.$$

Here $B_R(\varepsilon)$ is the open ε-ball in R. Furthermore, there exists $\delta > 0$ with

$$c'(\bar{y})B_Y(\delta) \subset B_R(\varepsilon).$$

Using that $\tilde{u} \in U_{\mathrm{ad}}$ and $\tilde{y} - \bar{y} + B_Y(\delta) \subset Y$ we have

$$\begin{aligned}
&\begin{pmatrix} 0 \\ c(\bar{y}) \end{pmatrix} + \begin{pmatrix} e_y(\bar{w}) & e_u(\bar{w}) \\ c'(\bar{y}) & 0 \end{pmatrix} \begin{pmatrix} Y \\ U_{\mathrm{ad}} - \bar{u} \end{pmatrix} - \begin{pmatrix} 0 \\ \mathcal{K} \end{pmatrix} \\
&\supset \begin{pmatrix} 0 \\ c(\bar{y}) \end{pmatrix} + \begin{pmatrix} e_y(\bar{w}) & e_u(\bar{w}) \\ c'(\bar{y}) & 0 \end{pmatrix} \begin{pmatrix} \tilde{y} - \bar{y} + B_Y(\delta) \\ \tilde{u} - \bar{u} \end{pmatrix} - \begin{pmatrix} 0 \\ \tilde{v} + B_R(2\varepsilon) \end{pmatrix} \\
&= \begin{pmatrix} e_y(\bar{w}) \\ c'(\bar{y}) \end{pmatrix} B_Y(\delta) + \begin{pmatrix} 0 \\ B_R(2\varepsilon) \end{pmatrix} \supset \begin{pmatrix} e_y(\bar{w})B_Y(\delta) \\ B_R(\varepsilon) \end{pmatrix}.
\end{aligned}$$

In the last step we have used $c'(\bar{y})B_Y(\delta) \subset B_R(\varepsilon)$ and that, for all $v \in B_R(\varepsilon)$, there holds $v + B_R(2\varepsilon) \supset B_R(\varepsilon)$. By the open mapping theorem $e_y(\bar{w})B_Y(\delta)$ is open in Z and contains 0. Therefore, the set on the right hand side is an open neighborhood of 0 in $Z \times R$.

1.7.3.5 Applications

Elliptic Problem with State Constraints

We consider the problem

$$\begin{aligned}
\min\ J(y,u) &:= \frac{1}{2}\|y - y_d\|^2_{L^2(\Omega)} + \frac{\alpha}{2}\|u\|^2_{L^2(\Omega)} \\
\text{subject to}\quad -\Delta y + y &= \gamma u \quad \text{on } \Omega, \\
\frac{\partial y}{\partial \nu} &= 0 \quad \text{on } \partial\Omega, \\
y &\geq 0 \quad \text{on } \Omega.
\end{aligned} \tag{1.144}$$

Let $n \leq 3$ and $\Omega \subset \mathbb{R}^n$ be open and bounded with Lipschitz boundary. We know from Theorem 1.22 that for $u \in U := L^2(\Omega)$ there exists a unique weak solution $y \in H^1(\Omega) \cap C(\bar{\Omega})$ of the state equation. We can write the problem in the form

$$\min J(y,u) \quad \text{subject to} \quad Ay + Bu = 0, \quad y \geq 0.$$

where $Bu = -\gamma u$, and A is induced by the bilinear form $a(y,v) = \int_\Omega \nabla y \cdot \nabla v\, dx + (y,v)_{L^2(\Omega)}$.

With appropriate spaces $Y \subset H^1(\Omega)$, $Z \subset H^1(\Omega)^*$ and $R \supset Y$ we set

$$e : \begin{pmatrix} y \\ u \end{pmatrix} \in Y \times U \mapsto Ay + Bu \in Z,$$
$$c(y) = y, \qquad \mathcal{K} = \{v \in R : v \geq 0\}, \quad U_{\mathrm{ad}} = U$$

and arrive at a problem of the form (1.138). For the naive choice $R = Y = H^1(\Omega)$, $Z = Y^*$, the cone $\mathcal{K}$ has no interior point. But since $Bu = -\gamma u \in L^2(\Omega)$, we know that all solutions y of the state equation live in the space

$$Y = \left\{ y \in H^1(\Omega) \cap C(\bar{\Omega}) : Ay \in L^2(\Omega) \right\}$$

and Y is a Banach space with the graph norm $\|y\|_{H^1(\Omega)} + \|y\|_{C(\bar{\Omega})} + \|Ay\|_{L^2(\Omega)}$. In fact, for a Cauchy sequence $(y_k) \subset Y$ there exists $y \in H^1(\Omega) \cap C(\bar{\Omega})$ and $z \in L^2(\Omega)$ with $y_k \to y$ in $H^1(\Omega) \cap C(\bar{\Omega})$ and $Ay_k \to z$ in $L^2(\Omega)$. Moreover, $Ay_k \to Ay$ in $H^1(\Omega)^*$ and therefore $Ay = z$.

By definition of Y the operator $A : Y \mapsto L^2(\Omega) =: Z$ is bounded and by Theorem 1.22 also surjective. Finally, we choose $R = C(\bar{\Omega})$, then $R \supset Y$ and $\mathcal{K} \subset R$ has an interior point. Summarizing, we work with the spaces

$$U = L^2(\Omega), \qquad Y = \left\{ y \in H^1(\Omega) \cap C(\bar{\Omega}) : Ay \in L^2(\Omega) \right\},$$
$$Z = L^2(\Omega), \qquad R = C(\bar{\Omega}).$$

Now assume that there exists $\tilde{y} \in Y$, $\tilde{y} > 0$ and $\tilde{u} \in U$ with (note that $e_y = A$, $e_u = B$)

$$A(\tilde{y} - \bar{y}) + B(\tilde{u} - \bar{u}) = 0.$$

For example in the case $\gamma \equiv 1$ the choice $\tilde{y} = \bar{y} + 1$, $\tilde{u} = \bar{u} + 1$ works. Then by Lemma 1.14 Robinson's regularity assumption is satisfied. Therefore, at a solution $(\bar{y}, \bar{u})$ the necessary conditions (1.140)–(1.143) are satisfied: Using that

$$L(y,u,p) = \frac{1}{2}\|y - y_d\|^2_{L^2(\Omega)} + \frac{\alpha}{2}\|u\|^2_{L^2(\Omega)} + (p, Ay + Bu)_{L^2(\Omega)}$$

we obtain

$$A\bar{y} + B\bar{u} = 0, \quad \bar{y} \geq 0,$$
$$\bar{\lambda} \in \mathcal{K}^\circ, \quad \langle \bar{\lambda}, \bar{y} \rangle_{C(\bar{\Omega})^*, C(\bar{\Omega})} = 0,$$
$$(\bar{y} - y_d, v)_{L^2(\Omega)} + (\bar{p}, Av)_{L^2(\Omega)} + \langle \bar{\lambda}, v \rangle_{C(\bar{\Omega})^*, C(\bar{\Omega})} = 0 \quad \forall v \in Y,$$
$$(\alpha \bar{u} - \gamma \bar{p}, u - \bar{u})_{L^2(\Omega)} \geq 0 \quad \forall u \in U.$$

One can show, see for example [35, 110], that the set $\mathcal{K}^\circ \subset C(\bar{\Omega})^*$ of nonpositive functionals on $C(\bar{\Omega})$ can be identified with nonpositive regular Borel measures, i.e.

$$\lambda \in \mathcal{K}^\circ \iff$$

$$\langle \lambda, v \rangle_{C(\bar{\Omega})^*, C(\bar{\Omega})} = -\int_\Omega v(x)\, d\mu_\Omega(x) - \int_{\partial\Omega} v(x)\, d\mu_{\partial\Omega}(x)$$

with nonneg. measures $\mu_\Omega, \mu_{\partial\Omega}$.

Therefore, the optimality system is formally a weak formulation of the following system.

$$-\Delta \bar{y} + \bar{y} = \gamma \bar{u} \quad \text{on } \Omega, \qquad \frac{\partial y}{\partial \nu} = 0 \quad \text{on } \partial\Omega,$$

$$\bar{y} \geq 0, \qquad \bar{\mu}_\Omega, \qquad \bar{\mu}_{\partial\Omega} \quad \text{nonnegative regular Borel measures,}$$

$$\int_\Omega \bar{y}(x)\, d\mu_\Omega(x) + \int_{\partial\Omega} \bar{y}(x)\, d\mu_{\partial\Omega}(x) = 0,$$

$$-\Delta \bar{p} + \bar{p} = -(\bar{y} - y_d) + \bar{\mu}_\Omega \quad \text{on } \Omega, \qquad \frac{\partial p}{\partial \nu} = \bar{\mu}_{\partial\Omega} \quad \text{on } \partial\Omega,$$

$$\alpha \bar{u} - \gamma \bar{p} = 0.$$

1.8 Optimal Control of Instationary Incompressible Navier-Stokes Flow

We conclude the chapter by providing the basic analytical foundations for optimization problems governed by the instationary, incompressible Navier-Stokes equations. These equations describe the flow of incompressible viscous fluid flow and are thus of central importance in science and engineering. The flow in the bounded domain with Lipschitz boundary $\Omega \subset \mathbb{R}^d$, $d = 2$ or 3, is characterized by the velocity field $y : [0, T] \times \Omega \to \mathbb{R}^d$ and by the pressure $p : [0, T] \times \Omega \to \mathbb{R}$. The viscosity of the fluid is characterized by a parameter $\nu > 0$, the kinetic viscosity. We denote by $x \in \Omega$ the spatial location, by $I = [0, T]$ the time horizon, and by $t \in I$ the time.

Let $f : [0, T] \times \Omega \to \mathbb{R}^d$ be the force per unit mass acting on the fluid and denote by $y_0 : \Omega \to \mathbb{R}^d$ the initial velocity of the fluid at $t = 0$. Then the Navier Stokes equations can be written as follows:

$$y_t - \nu \Delta y + (y \cdot \nabla) y + \nabla p = f \quad \text{on } \Omega_T := (0, T) \times \Omega, \tag{1.145}$$

$$\nabla \cdot y = 0 \quad \text{on } \Omega_T, \tag{1.146}$$

$$y(0, \cdot) = y_0 \quad \text{on } \Omega, \tag{1.147}$$

$$\text{suitable boundary conditions in } I \times \partial\Omega. \tag{1.148}$$

We have used the following standard notations for differential operators:

$$\Delta = \sum_{j=1}^{d} \frac{\partial^2}{\partial x_j^2}, \qquad (y \cdot \nabla) = \sum_{j=1}^{d} y_j \frac{\partial}{\partial x_j},$$

which are applied componentwise to the vector field y. Furthermore,

$$\nabla \cdot y = \operatorname{div} y = \sum_{j=1}^{d} \frac{\partial y_j}{\partial x_j}$$

is the divergence of y and y_t is a short notation for $\frac{\partial y}{\partial t}$.

The boundary conditions have to be chosen appropriately. Suitable is the prescription of the velocity field on the boundary, i.e.,

$$y = y_b \quad \text{on } I \times \partial\Omega,$$

where $y_b : I \times \partial\Omega \to \mathbb{R}^d$ is given.

1.8.1 Functional Analytic Setting

Since only for space dimension $d = 2$ a complete existence and uniqueness theory for the Navier-Stokes equations is available, we will consider the case $d = 2$ throughout, i.e., $\Omega \subset \mathbb{R}^2$. We assume throughout that Ω is bounded with Lipschitz boundary.

For simplicity, we consider in the following homogeneous Dirichlet conditions for the velocities, i.e.,

$$y = 0 \quad \text{on } I \times \partial\Omega.$$

It is advantageous to work in the space of divergence-free vector fields. To this end we introduce the following Hilbert spaces

$$V := \mathrm{cl}_{H_0^1(\Omega)^2}\{y \in C_c^\infty(\Omega)^2 : \nabla \cdot y = 0\},$$

$$H := \mathrm{cl}_{L^2(\Omega)^2}\{y \in C_c^\infty(\Omega)^2 : \nabla \cdot y = 0\},$$

$$(\cdot,\cdot)_V := (\cdot,\cdot)_{H_0^1(\Omega)^2}, \qquad (\cdot,\cdot)_H := (\cdot,\cdot)_{L^2(\Omega)^2}.$$

Here, cl_X denotes the closure in the space X. We choose the dual pairings such that we obtain the Gelfand triple

$$V \hookrightarrow H = H^* \hookrightarrow V^*$$

with continuous and dense imbeddings.

As introduced previously in the context of linear parabolic PDEs, we define the space

$$W(I) := W(I; H, V) = \{y \in L^2(I; V) : y_t \in L^2(I; V^*)\}.$$

Note that $L^2(I; V^*)$ is the dual space of $L^2(I; V)$, see Theorem 1.31.

Let the right hand side f in (1.145) satisfy $f \in L^2(I; V^*)$. Then testing the equation (1.145) with $v \in V$ as in the parabolic case yields

$$\langle y_t, v\rangle_{V^*,V} + \nu(\nabla y, \nabla v)_{L^2(\Omega)^{2\times 2}} + \langle (y \cdot \nabla)y, v\rangle_{V^*,V} = \langle f, v\rangle_{V^*,V} \quad \forall v \in V. \tag{1.149}$$

Here, the pressure term has disappeared since the weak form of $\langle \nabla p, v\rangle_{V^*,V}$ is $-(p, \nabla \cdot v)_{L^2(\Omega)}$, where integration by parts and $v|_{\partial\Omega} = 0$ has been used, and from $v \in V$ we see that

$$(p, \nabla \cdot v)_{L^2(\Omega)} = (p, 0)_{L^2(\Omega)} = 0.$$

We call $y \in W(I)$ a weak solution of the Navier-Stokes equations corresponding to the initial condition $y_0 \in H$ if

$$y(0, \cdot) = y_0 \quad \text{in } H \quad \text{and} \quad (1.149) \text{ holds a.e. on } (0, T).$$

Since the continuous embedding $W(I) \hookrightarrow C(I; H)$ can be shown to hold, see Theorem 1.32, the initial condition $y(0, \cdot) = y_0$ in H makes sense. In fact, $y \in W(I) \hookrightarrow C(I; H)$ can be interpreted as $C(I; H)$-function and evaluated at $t = 0$.

With data $\nu > 0$, $y_0 \in H$, and $f \in L^2(I; V^*)$, we can – analogous to the parabolic case – view the Navier-Stokes equations as a nonlinear operator equation: The velocity field $y \in W(I)$ satisfies

$$\begin{aligned} y_t + (y \cdot \nabla)y - \nu\Delta y - f &= 0 \quad \text{in } L^2(I; V^*), \\ y - y_0 &= 0 \quad \text{in } H. \end{aligned}$$

To justify the appropriateness of the image space $L^2(I; V^*)$, we consider all terms in the differential equation. Due to $y \in W(I)$ we have $y_t \in L^2(I; V^*)$. Furthermore,

$$\langle -\Delta w, v\rangle = (\nabla w, \nabla v)_{L^2(\Omega)^{2\times 2}} \le \|w\|_V \|v\|_V \quad \forall v, w \in V.$$

Thus

$$\begin{aligned} \langle -\Delta y, v\rangle_{L^2(I;V^*),L^2(I;V)} &= \int_0^T (\nabla y, \nabla v)_{L^2(\Omega)^2}\, dt \\ &\le \int_0^T \|y\|_V \|v\|_V\, dt \le \text{const}\big\| \|y\|_V \big\|_{L^2(I)} \big\| \|v\|_V \big\|_{L^2(I)} \\ &= \text{const}\|y\|_{L^2(I;V)} \|v\|_{L^2(I;V)} \le \text{const}\|y\|_{W(I)} \|v\|_{L^2(I;V)}. \end{aligned}$$

Hence, $-\Delta y \in L^2(I; V^*)$.

The most delicate term is $(y \cdot \nabla)y$, but the following can be shown (here $d = 2$ is needed)

Lemma 1.15

$$\|(y \cdot \nabla)y\|_{L^2(I;V^*)} \le 2^{1/2}\|y\|_{L^\infty(I;H)}\|y\|_{L^2(I;V)} \quad \forall y \in L^2(I;V) \cap L^\infty(I;H).$$

In particular, $y \in W(I) \mapsto (y \cdot \nabla)y \in L^2(I;V^*)$ *is a bounded bilinear operator.*

For a proof see [130, Lemma III.3.4]. The boundedness of the operator follows from the estimate and the fact that $W(I)$ is continuously embedded in both, $L^2(I;V)$ and $L^\infty(I;H)$.

Summarizing, we can argue exactly as in the parabolic case, see (1.58), (1.61) and Remark 1.16, that it is appropriate to consider the differential equation as an operator equation with range space $L^2(I;V^*)$ defined by (1.149), i.e.,

$$y_t + (y \cdot \nabla)y - \nu\Delta y = f \text{ in } L^2(I;V^*) \quad :\Longleftrightarrow \quad \text{(1.149) holds a.e. on } (0,T). \tag{1.150}$$

We now introduce a control on the right hand side, i.e., we replace the volume force term f by Bu, where $B \in \mathcal{L}(U, L^2(I;V^*))$ and U is a Hilbert space of controls. As an objective function we choose for example the tracking type functional

$$J(y,u) = \frac{1}{2}\|y - y_d\|^2_{L^2(I\times\Omega)^2} + \frac{\gamma}{2}\|u - u_d\|^2_U$$

with target state $y_d \in L^2(I \times \Omega)^2$ and reference control $u_d \in U$.

Let $U_{\text{ad}} \subset U$ be nonempty, convex and closed. We obtain the following optimal control problem

$$\min_{u\in U, y\in W(I)} J(y,u) \quad \text{s.t.} \quad e(y,u) = 0, \quad u \in U_{\text{ad}}, \tag{1.151}$$

where the state equation $e(y,u) = 0$ is the weak Navier-Stokes equation, i.e.,

$$\begin{aligned} &e : W(I) \times U \to L^2(I;V^*) \times H, \\ &e(y,u) = \begin{pmatrix} y_t + (y \cdot \nabla)y - \nu\Delta y - Bu \\ y(0,\cdot) - y_0 \end{pmatrix} \quad \text{is defined by (1.150).} \end{aligned} \tag{1.152}$$

1.8.2 Analysis of the Flow Control Problem

The following existence and uniqueness result concerning the Navier-Stokes equations can be shown:

Theorem 1.57 *For all* $y_0 \in H$ *and* $u \in U$ *the Navier-Stokes equation* $e(y,u) = 0$ *with* $e : W(I) \times U \to L^2(I;V^*) \times H$ *defined in* (1.150), (1.152) *possesses a unique solution* $y(u) \in W(I)$ *and it satisfies*

$$\|y(u)\|_{C(I;H)} + \|y(u)\|_{W(I)} \le \text{const}(\|y_0\|_H + \|u\|_U + \|y_0\|^2_H + \|u\|^2_U).$$

Proof See for example [130, Ch. III].

Next, we consider derivatives of the state equation. We start by considering a general bounded bilinear operator.

Lemma 1.16 *Let*

$$A : X \times X \to Y$$

be a bilinear operator between Banach spaces that is bounded, i.e.,

$$\|A(x_1, x_2)\|_Y \le \text{const}\|x_1\|_X\|x_2\|_X.$$

Then $H : X \to Y$, $H(x) = A(x, x)$ *is Fréchet differentiable with derivative*

$$H'(x) : d \in X \mapsto A(d, x) + A(x, d).$$

The operator $H' : X \mapsto \mathcal{L}(X, Y)$ *is bounded and linear. Hence,* H *is infinitely differentiable with constant second derivative* $H''(x)(d_1, d_2) = A(d_1, d_2) + A(d_2, d_1)$ *and* $H^{(k)}(x) = 0$, $k \ge 3$.

Proof With $H'(x)$ as stated, we obtain, using bilinearity

$$\begin{aligned} H(x+d) - H(x) - H'(x)d &= A(x+d, x+d) - A(x, x) - A(d, x) - A(x, d) \\ &= A(d, d). \end{aligned}$$

The remainder term $A(d, d)$ satisfies

$$\|A(d, d)\|_Y \le \text{const}\|d\|_X^2.$$

By definition of F-differentiability, H is thus F-differentiable with derivative as stated. The continuous linearity of H' is obvious and differentiation gives the stated formula for H'' as well as $H^{(k)} = 0$, $k \ge 3$.

We now can derive the following result.

Lemma 1.17 *The operator* $e : W(I) \times U \to L^2(I; V^*) \times H$ *is infinitely Fréchet differentiable with its derivatives given by*

$$\begin{aligned} e'(y, u)(v, w) &= \begin{pmatrix} v_t + (y \cdot \nabla)v + (v \cdot \nabla)y - \nu\Delta v - Bw \\ v(0, \cdot) \end{pmatrix} \\ &\quad \forall y, v \in W(I),\ u, w \in U, \\ e''(y, u)((v_1, w_1), (v_2, w_2)) &= \begin{pmatrix} (v_2 \cdot \nabla)v_1 + (v_1 \cdot \nabla)v_2 \\ 0 \end{pmatrix} \\ &\quad \forall y, v_{1/2} \in W(I),\ u, w_{1/2} \in U, \\ e^{(k)}(y, u) &= 0 \quad \forall y \in W(I),\ u \in U,\ k \ge 3. \end{aligned}$$

Proof Since $e(y,u)$ is a sum of bounded linear and bilinear operators, the operator $e : W(I) \times U \to L^2(I;V^*) \times H$ is infinitely Fréchet differentiable by Lemma 1.16. The derivatives are obtained by the rules of differentiating linear and bilinear operators.

The linearized equation

$$e_y(y,u)v = \begin{pmatrix} g \\ v_0 \end{pmatrix} \tag{1.153}$$

can be shown to have a unique solution $v(g, v_0) \in W(I)$ for any $g \in L^2(I;V^*)$, $v_0 \in H$. Therefore, $e_y(y,u) \in \mathcal{L}(W(I), L^2(I;V^*) \times H)$ is boundedly invertible. See for example [70].

Written in expanded form, (1.153) reads

$$\begin{aligned} v_t + (v \cdot \nabla)y + (y \cdot \nabla)v + \nu \Delta v &= g, \\ v(0,\cdot) &= v_0. \end{aligned}$$

Now set $Y := W(I)$, $Z := L^2(I;V^*) \times H$. Then $e_y(y,u) \in \mathcal{L}(Y,Z)$ has a bounded inverse and thus we have verified Assumption 1.47 for the optimal control problem (1.151). Thus, at any local solution $(\bar{y}, \bar{u})$ the optimality condition of Corollary 1.3 holds, which yields in our case

$$\begin{aligned} &e(\bar{y},\bar{u}) = 0, \\ &J_y(\bar{y},\bar{u}) + e_y(\bar{y},\bar{u})^* \begin{pmatrix} \bar{p}_1 \\ \bar{p}_2 \end{pmatrix} = 0, \\ &\bar{u} \in U_{\mathrm{ad}}, \quad \left\langle J_u(\bar{y},\bar{u}) + e_u(\bar{y},\bar{u})^* \begin{pmatrix} \bar{p}_1 \\ \bar{p}_2 \end{pmatrix}, u - \bar{u} \right\rangle_U \ge 0 \quad \forall u \in U_{\mathrm{ad}}. \end{aligned} \tag{1.154}$$

Here $(\bar{p}_1, \bar{p}_2) \in L^2(I;V) \times H$ is the adjoint state.

We now take a closer look at the adjoint operator $e_y(y,u)^*$ and its inverse.

Since $e_y(y,u)$ is boundedly invertible, the same holds true for $e_y(y,u)^*$. Therefore, the adjoint equation is uniquely solvable with respect to the adjoint state (p_1, p_2).

Now let $g \in W(I)^*$ and consider the adjoint equation

$$e_y(y,u)^* \begin{pmatrix} p_1 \\ p_2 \end{pmatrix} = g. \tag{1.155}$$

Writing (1.155) in variational form, we obtain

$$\begin{aligned} &\langle (e_1)_y(y,u)v, p_1 \rangle_{L^2(I;V^*),L^2(I;V)} + ((e_2)_y(y,u)v, p_2)_H = \langle g, v \rangle_{W(I)^*,W(I)} \\ &\forall v \in W(I). \end{aligned}$$

Expanding the operator e, this becomes

$$\langle v_t, p_1 \rangle_{L^2(I;V^*),L^2(I;V)} + \langle (v \cdot \nabla)y, p_1 \rangle_{L^2(I;V^*),L^2(I;V)}$$

$$+\langle (y\cdot\nabla)v, p_1\rangle_{L^2(I;V^*),L^2(I;V)} + \nu(\nabla v, \nabla p_1)_{L^2(I;L^2(\Omega)^{2\times 2})} + (v(0,\cdot), p_2)_H$$
$$= \langle g, v\rangle_{W(I)^*,W(I)} \quad \forall v\in W(I). \tag{1.156}$$

The following can be shown.

Lemma 1.18 *The adjoint equation possesses for each* $g\in W(I)^*$ *a unique solution* $(p_1, p_2)\in L^2(I;V)\times H$. *The following estimate holds*:

$$\|p_1\|_{L^2(I;V)} + \|p_2\|_H \le c(\|y\|_{W(I)})\|g\|_{W^*(I)}.$$

Here $c(\cdot)$ *is locally Lipschitz continuous. If* g *has additional regularity in the sense that* $g\in L^s(I;V^*)\cap W(I)^*$ *with* $s\in[1,4/3]$ *then* $p_1\in L^2(I;V)$ *satisfies* $p_1\in C(I;V^*)$, $(p_1)_t\in L^s(I;V^*)\cap W^*(I)$. *Furthermore, the following estimates hold*

$$\|(p_1)_t\|_{W^*(I)} \le c(\|y\|_{W(I)})\|g\|_{W^*(I)},$$
$$\|(p_1)_t\|_{L^s(I;V^*)} \le c(\|y\|_{W(I)})\|g\|_{W^*(I)} + \|g\|_{L^s(I;V^*)}.$$

Here $c(\cdot)$ *is locally Lipschitz continuous.*

The adjoint equation (1.155) *can be interpreted as the weak formulation of the PDE*

$$-(p_1)_t - (y\cdot\nabla)p_1 + (\nabla y)^T p_1 - \nu\Delta p_1 = g \quad \text{in } \Omega\times I,\ p_1|_{t=T}=0 \quad \text{in } \Omega \tag{1.157}$$

and $p_2 = p_1(0,\cdot)$. *Here,* ∇y *denotes the Jacobian matrix of* y *w.r.t.* x.

For a proof see [70, 72, 134, 137]. The PDE (1.157) is obtained from (1.156) by integrating the terms $\langle v_t, p\rangle_{L^2(I;V^*),L^2(I;V)}$ and $\langle (y\cdot\nabla)v, p\rangle_{L^2(I;V^*),L^2(I;V)}$ by parts and using $\nabla\cdot y = 0$. However, the justification of the validity of integration by parts is a bit tricky since the involved function spaces are quite weak.

1.8.3 Reduced Optimal Control Problem

Due to the unique solvability of the Navier-Stokes equations, there exists a uniquely defined control-to-state operator $u\in U\mapsto y(u)\in W(I)$. Since the state equation operator e is infinitely differentiable and $e_y(y,u)$ is boundedly invertible, the implicit function theorem (Theorem 1.41) yields that the control-to-state operator is infinitely F-differentiable. Since a continuously differentiable operator is Lipschitz continuous on bounded sets, we obtain

Lemma 1.19 *The Navier-Stokes equations* $e(y,u)=0$ *with* $e: W(I)\times U\to L^2(I;V^*)\times H$ *defined in* (1.150), (1.152) *defines a unique control-to-state operator* $u\in U\mapsto y(u)\in W(I)$. *The operator* $y(u)$ *is infinitely F-differentiable and, as a consequence, F-derivatives of all orders are Lipschitz continuous on bounded sets.*

This amounts to considering the reduced objective function $\hat{J}(u) = J(y(u), u)$, which is as smooth as J is. Also, by unique solvability of the adjoint equation, there exists a unique control-to-adjoint state operator $u \mapsto (p_1, p_2)(u)$, where $(p_1, p_2)(u)$ is the adjoint state corresponding to u and $y(u)$. Due to infinite differentiability of e the smoothness of the control-to-adjoint state operator only depends on the differentiability properties of J. Using Lemma 1.18 and the adjoint presentation of $\hat{J}(u)'$ according to (1.89), we have the following result

Theorem 1.58 *Let the objective function J be $k \geq 1$ times continuously F-differentiable. Then the reduced objective function $\hat{J}(u) = J(y(u), u)$ is k times continuously F-differentiable. The adjoint gradient representation is given by*

$$\hat{J}(u)' = J_u(y(u), u) - B^* p_1(u).$$

If $k \geq 2$ and if $J_y(y(u), u) \in W^*(I)$ has the property that $u \in U \mapsto J_y(y(u), u) \in L^s(I; V^*)$, $s \in [1, 4/3]$ is Lipschitz continuous on bounded sets and if $B^* \in \mathcal{L}(L^2(I; V), U)$ induces an operator $B^* \in \mathcal{L}(W^s(I), \tilde{U})$, $\tilde{U} \hookrightarrow U$ with

$$W^s(I) := \left\{ y \in L^2(I; V) : y_t \in L^s(I; V^*) \right\},$$

then the mapping

$$u \in U \mapsto B^* p_1(u) \in \tilde{U}$$

is Lipschitz continuous on bounded sets.

Remark 1.23 The second part of the theorem is important for semismooth Newton methods and other second order methods that require a smoothing operator in the gradient representation.

For example, consider the case $U = L^2(I \times \Omega)^2$, $B = I_{U, L^2(I; V^*)}$, and

$$J(y, u) = \frac{1}{2} \|y - y_d\|^2_{L^2(I \times \Omega)^2} + \frac{\alpha}{2} \|u - u_d\|^2_U$$

with $u_d, y_d \in L^2(I \times \Omega)^2$. Then $u \in U \mapsto J_y(y(u), u) = y(u) - y_d \in L^2(I \times \Omega)^2 \hookrightarrow L^{4/3}(I; V^*)$ is Lipschitz continuous on bounded sets. Moreover, the imbedding $W^s(I) \hookrightarrow L^q(I \times \Omega)^2 =: \tilde{U}, 2 \leq q < 7/2$ is continuous [137]. Therefore, the mapping

$$u \in U \mapsto B^* p_1(u) = p_1(u) \in L^q(I \times \Omega)^2$$

is for all $2 \leq q < 7/2$ Lipschitz continuous on bounded sets and thus the reduced gradient

$$\nabla \hat{J}(u) = \alpha(u - u_d) - p_1(u)$$

contains a more regular part $u \in U \mapsto -p_1(u) \in L^q(I \times \Omega)^2$. This *smoothing property* will be important for the fast convergence of semismooth Newton methods.

Chapter 2
Optimization Methods in Banach Spaces

Michael Ulbrich

Abstract In this chapter we present a selection of important algorithms for optimization problems with partial differential equations. The development and analysis of these methods is carried out in a Banach space setting. We begin by introducing a general framework for achieving global convergence. Then, several variants of generalized Newton methods are derived and analyzed. In particular, necessary and sufficient conditions for fast local convergence are derived. Based on this, the concept of semismooth Newton methods for operator equations is introduced. It is shown how complementarity conditions, variational inequalities, and optimality systems can be reformulated as semismooth operator equations. Applications to constrained optimal control problems are discussed, in particular for elliptic partial differential equations and for flow control problems governed by the incompressible instationary Navier-Stokes equations. As a further important concept, the formulation of optimality systems as generalized equations is addressed. We introduce and analyze the Josephy-Newton method for generalized equations. This provides an elegant basis for the motivation and analysis of sequential quadratic programming (SQP) algorithms. The chapter concludes with a short outline of recent algorithmic advances for state constrained problems and a brief discussion of several further aspects.

2.1 Synopsis

The aim of this chapter is to give an introduction to selected optimization algorithms that are well-suited for PDE-constrained optimization. For the development and analysis of such algorithms, a functional analytic setting is the framework of choice. Therefore, we will develop optimization methods in this abstract setting and then return to concrete problems later.

Optimization methods are iterative algorithms for finding (global or local) solutions of minimization problems. Usually, we are already satisfied if the method can be proved to converge to *stationary* points. These are points that satisfy the first-order necessary optimality conditions. Besides global convergence, which will not be the main focus of this chapter, fast local convergence is desired. All fast converging optimization methods use the idea of Newton's method in some sense. Therefore, our main focus will be on Newton-type methods for optimization problems in Banach spaces.

M. Ulbrich (✉)
Lehrstuhl für Mathematische Optimierung, TU München, Garching, Germany
e-mail: mulbrich@ma.tum.de

M. Hinze et al., *Optimization with PDE Constraints*,
Mathematical Modelling: Theory and Applications 23,
DOI 10.1007/978-1-4020-8839-1_2, © Springer Science + Business Media B.V. 2009

Optimization methods for minimizing an objective function $f : W \to \mathbb{R}$ on a feasible set $W_{\mathrm{ad}} \subset W$, where W is a Banach space, generate a sequence $(w^k) \subset W$ of iterates. Essentially, as already indicated, there are two desirable properties an optimization algorithm should have:

1. Global convergence:
   There are different flavors to formulate global convergence. Some of them use the notion of a stationarity measure. This is a function $\Sigma : W \to \mathbb{R}_+$ with $\Sigma(w) = 0$ if w is stationary and $\Sigma(w) > 0$, otherwise. In the unconstrained case, i.e., $W_{\mathrm{ad}} = W$, a common choice is $\Sigma(w) := \|f'(w)\|_{W^*}$. The following is a selection of global convergence assertions:
   (a) Every accumulation point of (w^k) is a stationary point.
   (b) For some continuous stationarity measure $\Sigma(w)$ there holds
   $$\lim_{k\to\infty} \Sigma(w^k) = 0.$$
   (c) There exists an accumulation point of (w^k) that is stationary.
   (d) For the continuous stationarity measure $\Sigma(w)$ there holds
   $$\liminf_{k\to\infty} \Sigma(w^k) = 0.$$
   Note that (b) implies (a) and (c) implies (d).
2. Fast local convergence:
   These are local results in a neighborhood of a stationary point $\bar{w}$:
   There exists $\delta > 0$ such that, for all $w^0 \in W$ with $\|w^0 - \bar{w}\|_W < \delta$, we have $w^k \to \bar{w}$ and
   $$\|w^{k+1} - \bar{w}\|_W = o(\|w^k - \bar{w}\|_W) \quad \text{(q-superlinear convergence)},$$
   or even, for $\alpha > 0$,
   $$\|w^{k+1} - \bar{w}\|_W = O(\|w^k - \bar{w}\|_W^{1+\alpha})$$
   $$\text{(q-superlinear convergence with order } 1+\alpha).$$
   The case $1 + \alpha = 2$ is called q-quadratic convergence.

We begin with a discussion of globalization concepts. Then, in the rest of this chapter, we present locally fast convergent methods that all can be viewed as Newton-type methods.

Notation If W is a Banach space, we denote by W^* its dual space. The Fréchet-derivative (F-derivative) of an operator $G : X \to Y$ between Banach spaces is denoted by $G' : X \to \mathcal{L}(X, Y)$, where $\mathcal{L}(X, Y)$ are the bounded linear operators $A : X \to Y$. In particular, the derivative of a real-valued function $f : W \to \mathbb{R}$ is denoted by $f' : W \to W^*$. In case of a Hilbert space W, the gradient $\nabla f : W \to W$ is the Riesz representation of f', i.e.,

$$(\nabla f(w), v)_W = \langle f'(w), v\rangle_{W^*,W} \quad \forall v \in W.$$

Here $\langle f'(w), v\rangle_{W^*,W}$ denotes the dual pairing between the dual space $W^* = \mathcal{L}(W,\mathbb{R})$ and W and $(\cdot,\cdot)_W$ is the inner product. Note that in Hilbert space we can do the identification $W^* = W$ via $\langle\cdot,\cdot\rangle_{W^*,W} = (\cdot,\cdot)_W$, but this is not always done.

2.2 Globally Convergent Methods in Banach Spaces

2.2.1 Unconstrained Optimization

For understanding how global convergence can be achieved, it is important to look at unconstrained optimization first:

$$\min_{w\in W} f(w)$$

with W a real Banach space and $f : W \to \mathbb{R}$ continuously F-differentiable.

The first-order optimality conditions for a local minimum $\bar{w} \in W$ are well-known:

$\bar{w} \in W$ satisfies

$$f'(\bar{w}) = 0.$$

We develop a general class of methods that is globally convergent: *Descent methods*.

The idea of descent methods is to find, at the current (kth) iterate $w^k \in W$, a direction $s^k \in W$ such that $\phi_k(t) \overset{\text{def}}{=} f(w^k + ts^k)$ is decreasing at $t = 0$:

$$\phi_k'(0) = \langle f'(w^k), s^k\rangle_{W^*,W} < 0.$$

Of course, this descent can be very small. However, from the (sharp) estimate

$$\phi_k'(0) = \langle f'(w^k), s^k\rangle_{W^*,W} \ge -\|f'(w^k)\|_{W^*}\|s^k\|_W$$

it is natural to derive the following quality requirement ("angle" condition)

$$\langle f'(w^k), s^k\rangle_{W^*,W} \le -\eta\|f'(w^k)\|_{W^*}\|s^k\|_W \tag{2.1}$$

for the descent direction. Here $\eta \in (0, 1)$ is fixed.

A second ingredient of a descent method is a step size rule to obtain a step size $\sigma_k > 0$ such that

$$\phi_k(\sigma_k) < \phi_k(0).$$

Then, the new iterate is computed as $w^{k+1} := w^k + \sigma_k s^k$. Overall, we obtain:

Algorithm 2.1 (General descent method)

0. Choose an initial point $w^0 \in W$.

For $k = 0, 1, 2, \ldots$:

1. If $f'(w^k) = 0$, STOP.
2. Choose a descent direction $s^k \in W$: $\langle f'(w^k), s^k \rangle_{W^*,W} < 0$.
3. Choose a step size $\sigma_k > 0$ such that $f(w^k + \sigma_k s^k) < f(w^k)$.
4. Set $w^{k+1} := w^k + \sigma_k s^k$.

In this generality, it is not possible to prove global convergence. We need additional requirements on the quality of the descent direction and the step sizes:

1. Admissibility of the search directions:

$$\frac{\langle f'(w^k), s^k \rangle_{W^*,W}}{\|s^k\|_W} \stackrel{k\to\infty}{\longrightarrow} 0 \quad \Longrightarrow \quad \|f'(w^k)\|_{W^*} \stackrel{k\to\infty}{\longrightarrow} 0.$$

2. Admissibility of the step sizes:

$$f(w^k + \sigma_k s^k) < f(w^k) \quad \forall k \quad \text{and}$$

$$f(w^k + \sigma_k s^k) - f(w^k) \stackrel{k\to\infty}{\longrightarrow} 0 \quad \Longrightarrow \quad \frac{\langle f'(w^k), s^k \rangle_{W^*,W}}{\|s^k\|_W} \stackrel{k\to\infty}{\longrightarrow} 0.$$

These conditions become more intuitive by realizing that the expression $\frac{\langle f'(w^k), s^k \rangle_{W^*,W}}{\|s^k\|_W}$ is the slope of f at w^k in the direction s^k:

$$\frac{d}{dt} f\left(w^k + t \frac{s^k}{\|s^k\|_W} \right)\bigg|_{t=0} = \frac{\langle f'(w^k), s^k \rangle_{W^*,W}}{\|s^k\|_W}.$$

Therefore, admissible step sizes mean that if the f-decreases become smaller and smaller then the slopes along the s^k have to become smaller and smaller. And admissible search directions mean that if the slopes along the s^k become smaller and smaller then the steepest possible slopes have to become smaller and smaller.

With these two conditions at hand, we can prove global convergence.

Theorem 2.2 *Let f be continuously F-differentiable and (w^k), (s^k), (σ_k) be generated by Algorithm 2.1. Assume that (σ_k) and (s^k) are admissible and that $(f(w^k))$ is bounded below. Then*

$$\lim_{k\to\infty} f'(w^k) = 0. \tag{2.2}$$

In particular, every accumulation point of (w^k) is a stationary point.

Proof Let $f^* = \inf_{k\ge 0} f(w^k) > -\infty$. Then, using $f(w^k + \sigma_k s^k) - f(w^k) < 0$, we see that $f(w^k) \to f^*$ and

$$f(w^0) - f^* = \sum_{k=0}^{\infty} (f(w^k) - f(w^{k+1})) = \sum_{k=0}^{\infty} |f(w^k + \sigma_k s^k) - f(w^k)|.$$

This shows $f(w^k + \sigma_k s^k) - f(w^k) \to 0$. By the admissibility of (σ_k), this implies

$$\frac{\langle f'(w^k), s^k \rangle_{W^*,W}}{\|s^k\|_W} \stackrel{k\to\infty}{\longrightarrow} 0.$$

Now the admissibility of (s^k) yields

$$\|f'(w^k)\|_{W^*} \stackrel{k\to\infty}{\longrightarrow} 0.$$

Next, consider the situation where $\bar{w}$ is an accumulation point of (w^k). Then there exists a subsequence $(w^k)_K \to \bar{w}$ and due to monotonicity of $f(w^k)$ we conclude $f(w^k) \geq f(\bar{w})$ for all k. Hence, we can apply the first part of the theorem and obtain (2.2). Now, by continuity,

$$f'(\bar{w}) = \lim_{k\to\infty} f'(w^k) = 0.$$

There are two questions open:

(a) How can we check in practice if a search direction is admissible or not?
(b) How can we compute admissible step sizes?

An answer to question (a) is provided by the following Lemma:

Lemma 2.1 *If the search directions* (s^k) *satisfy the angle condition* (2.1) *then they are admissible.*

Proof The angle condition yields

$$\|f'(w^k)\|_{W^*} \leq -\frac{1}{\eta} \frac{\langle f'(w^k), s^k \rangle_{W^*,W}}{\|s^k\|_W}.$$

A very important step size rule is the

2.2.1.1 Armijo Rule

Given a descent direction s^k of f at w^k, choose the maximum $\sigma_k \in \{1, 1/2, 1/4, \dots\}$ for which

$$f(w^k + \sigma_k s^k) - f(w^k) \leq \gamma \sigma_k \langle f'(w^k), s^k \rangle_{W^*,W}.$$

Here $\gamma \in (0, 1)$ is a constant. The next result shows that Armijo step sizes exist.

Lemma 2.2 *Let* f' *be uniformly continuous on* $N_0^\rho = \{w + s : f(w) \leq f(w^0),\ \|s\|_W \leq \rho\}$ *for some* $\rho > 0$. *Then, for every* $\varepsilon > 0$, *there exists* $\delta > 0$ *such that for all* $w^k \in W$ *with* $f(w^k) \leq f(w^0)$ *and all* $s^k \in W$ *that satisfy*

$$\frac{\langle f'(w^k), s^k \rangle_{W^*,W}}{\|s^k\|_W} \leq -\varepsilon,$$

there holds

$$f(w^k + \sigma s^k) - f(w^k) \le \gamma\sigma \langle f'(w^k), s^k\rangle_{W^*,W} \quad \forall \sigma \in [0, \delta/\|s^k\|_W].$$

Proof We have, with appropriate $\tau_\sigma \in [0, \sigma]$,

$$\begin{aligned} f(w^k + \sigma s^k) - f(w^k) &= \sigma \langle f'(w^k + \tau_\sigma s^k), s^k\rangle_{W^*,W} \\ &\le \sigma \langle f'(w^k), s^k\rangle_{W^*,W} + \sigma \| f'(w^k + \tau_\sigma s^k) \\ &\quad - f'(w^k)\|_{W^*} \|s^k\|_W \\ &= \gamma\sigma \langle f'(w^k), s^k\rangle_{W^*,W} + \rho_k(\sigma), \end{aligned}$$

where

$$\rho_k(\sigma) := (1-\gamma)\sigma \langle f'(w^k), s^k\rangle_{W^*,W} + \sigma \| f'(w^k + \tau_\sigma s^k) - f'(w^k)\|_{W^*} \|s^k\|_W.$$

Now we use the uniform continuity of f' to choose $\delta \in (0, \rho)$ so small that

$$\| f'(w^k + \tau_\sigma s^k) - f'(w^k)\|_{W^*} < (1-\gamma)\varepsilon \quad \forall \sigma \in [0, \delta/\|s^k\|_W].$$

This is possible since

$$\|\tau_\sigma s^k\|_W \le \sigma \|s^k\|_W \le \delta.$$

Then

$$\begin{aligned} \rho_k(\sigma) &= (1-\gamma)\sigma \langle f'(w^k), s^k\rangle_{W^*,W} + \sigma \| f'(w^k + \tau_\sigma s^k) - f'(w^k)\|_{W^*} \|s^k\|_W \\ &\le -(1-\gamma)\varepsilon\sigma \|s^k\|_W + (1-\gamma)\varepsilon\sigma \|s^k\|_W = 0. \end{aligned}$$

Next, we prove the admissibility of Armijo step sizes under mild conditions.

Lemma 2.3 *Let f' be uniformly continuous on $N_0^\rho = \{w + s : f(w) \le f(w^0), \|s\|_W \le \rho\}$ for some $\rho > 0$. We consider Algorithm 2.1, where (σ_k) is generated by the Armijo rule and the descent directions s^k are chosen such that they are not too short in the following sense:*

$$\|s^k\|_W \ge \phi\left(-\frac{\langle f'(w^k), s^k\rangle_{W^*,W}}{\|s^k\|_W} \right),$$

where $\phi : [0, \infty) \to [0, \infty)$ is monotonically increasing and satisfies $\phi(t) > 0$ for all $t > 0$. Then the step sizes (σ_k) are admissible.

Proof Assume that there exist an infinite set K and $\varepsilon > 0$ such that

$$\frac{\langle f'(w^k), s^k\rangle_{W^*,W}}{\|s^k\|_W} \le -\varepsilon \quad \forall k \in K.$$

Then

$$\|s^k\|_W \geq \phi\left(-\frac{\langle f'(w^k), s^k\rangle_{W^*,W}}{\|s^k\|_W}\right) \geq \phi(\varepsilon) =: \eta > 0 \quad \forall k \in K.$$

By Lemma 2.2, for $k \in K$ we have either $\sigma_k = 1$ or $\sigma_k \geq \delta/(2\|s^k\|)$. Hence,

$$\sigma_k \|s^k\|_W \geq \min\{\delta/2, \eta\} \quad \forall k \in K.$$

This shows

$$\begin{aligned} f(w^k + \sigma_k s^k) - f(w^k) &\leq \gamma \sigma_k \langle f'(w^k), s^k\rangle_{W^*,W} = \gamma\sigma_k \|s^k\|_W \frac{\langle f'(w^k), s^k\rangle_{W^*,W}}{\|s^k\|_W} \\ &\leq -\gamma \min\{\delta/2, \eta\}\varepsilon \quad \forall k \in K. \end{aligned}$$

Therefore

$$f(w^k + \sigma_k s^k) - f(w^k) \not\to 0.$$

In the Banach space setting, the computation of descent directions is not straightforward. Note that the negative derivative of f is *not* suitable, since $W^* \ni f'(w^k) \notin W$.

In the Hilbert space setting, however, we *can* choose $W^* = W$ and $\langle\cdot,\cdot\rangle_{W^*,W} = (\cdot,\cdot)_W$ by the Riesz representation theorem. Then we have $f'(w^k) = \nabla f(w^k) \in W$ and $-\nabla f(w^k)$ is the direction of steepest descent, as we will show below.

Certainly the most well-known descent method is the steepest descent method. In Banach space, the steepest descent directions of f at w are defined by $s = td_{sd}$, $t > 0$, where d_{sd} solves

$$\min_{\|d\|_W=1} \langle f'(w), d\rangle_{W^*,W}.$$

Now consider the case where $W = W^*$ is a Hilbert space. Then

$$d_{sd} = -\frac{\nabla f(w)}{\|\nabla f(w)\|_W}.$$

In fact, by the Cauchy-Schwarz inequality,

$$\begin{aligned} \min_{\|d\|_W=1} \langle f'(w), d\rangle_{W^*,W} &= \min_{\|d\|_W=1} (\nabla f(w), d)_W \geq -\|\nabla f(w)\|_W \\ &= \left(\nabla f(w), -\frac{\nabla f(w)}{\|\nabla f(w)\|_W}\right)_W. \end{aligned}$$

Therefore, $-\nabla f(w)$ is a steepest descent direction. This is the reason why the steepest descent method is also called gradient method.

It should be mentioned that the steepest descent method is usually very inefficient. Therefore, the design of efficient globally convergent methods works as follows: A locally fast convergent method (e.g., Newton's method) is used to generate trial steps. If the generated step satisfies a (generalized) angle test ensuring admissibility of the step, the step is selected. Otherwise, another search direction is chosen, e.g., the steepest descent direction.

2.2.2 Optimization on Closed Convex Sets

We now develop descent methods for simply constrained problems of the form

$$\min f(w) \quad \text{s.t.} \quad w \in S \tag{2.3}$$

with W a Hilbert space, $f : W \to \mathbb{R}$ continuously F-differentiable, and $S \subset W$ closed and convex.

Example 2.1 A scenario frequently found in practice is

$$W = L^2(\Omega), \quad S = \left\{u \in L^2(\Omega) : a(x) \le u(x) \le b(x) \text{ a.e. on } \Omega\right\}$$

with L^∞-functions a, b. It is then very easy to compute the projection P_S onto S, which will be needed in the following:

$$P_S(w)(x) = P_{[a(x),b(x)]}(w(x)) = \max(a(x), \min(w(x), b(x))).$$

The presence of the constraint set S requires to take care that we stay feasible with respect to S, or—if we think of an infeasible method—that we converge to feasibility. In the following, we consider a feasible algorithm, i.e., $w^k \in S$ for all k.

If w^k is feasible and we try to apply the unconstrained descent method, we have the difficulty that already very small step sizes $\sigma > 0$ can result in points $w^k + \sigma s^k$ that are infeasible. The backtracking idea of considering only those $\sigma \ge 0$ for which $w^k + \sigma s^k$ is feasible is not viable, since very small step sizes or even $\sigma_k = 0$ might be the result.

Therefore, instead of performing a line search along the ray $\{w^k + \sigma s^k : \sigma \ge 0\}$, we perform a line search along the projected path

$$\left\{P_S(w^k + \sigma s^k) : \sigma \ge 0\right\},$$

where P_S is the projection onto S. Of course, we have to ensure that along this path we achieve sufficient descent as long as w^k is not a stationary point. Unfortunately, not any descent direction is suitable here.

Example 2.2 Consider

$$S = \left\{w \in \mathbb{R}^2 : w_1 \ge 0, \ w_1 + w_2 \ge 3\right\}, \qquad f(w) = 5w_1^2 + w_2^2.$$

Then, at $w^k = (1,2)^T$, we have $\nabla f(w^k) = (10,4)^T$. Since f is convex quadratic with minimum $\bar{w} = 0$, the Newton step is

$$d^k = -w^k = -(1,2)^T.$$

This is a descent direction, since

$$\nabla f(w^k)^T d^k = -18.$$

But, for $\sigma \geq 0$, there holds

$$P_S(w^k - \sigma d^k) = P_S((1-\sigma)(1,2)^T) = (1-\sigma)\begin{pmatrix}1\\2\end{pmatrix} + \sigma\begin{pmatrix}3/2\\3/2\end{pmatrix} = \begin{pmatrix}1\\2\end{pmatrix} + \frac{\sigma}{2}\begin{pmatrix}1\\-1\end{pmatrix}.$$

From

$$\nabla f(w^k)^T \begin{pmatrix}1\\-1\end{pmatrix} = 6$$

we see that we are getting ascent, not descent, along the projected path, although d^k is a descent direction.

The example shows that care must be taken in choosing appropriate search directions for projected methods. Since the projected descent properties of a search direction are more complicated to judge than in the unconstrained case, it is out of the scope of this chapter to give a general presentation of this topic. In the finite dimensional setting, we refer to [84] for a detailed discussion. Here, we only consider the projected gradient method.

Algorithm 2.3 (Projected gradient method)

0. Choose $w^0 \in S$.

For $k = 0, 1, 2, 3, \ldots$:

1. Set $s^k = -\nabla f(w^k)$.
2. Choose σ_k by a projected step size rule such that $f(P_S(w^k + \sigma_k s^k)) < f(w^k)$.
3. Set $w^{k+1} := P_S(w^k + \sigma_k s^k)$.

For abbreviation, let

$$w_\sigma^k = w^k - \sigma \nabla f(w^k).$$

We will prove global convergence of this method. To do this, we need the facts about the projection operator P_S collected in Lemma 1.10.

The following result shows that along the projected steepest descent path we achieve a certain amount of descent:

Lemma 2.4 *Let W be a Hilbert space and let $f : W \to \mathbb{R}$ be continuously F-differentiable on a neighborhood of the closed convex set S. Let $w^k \in S$ and assume* *that* ∇f *is* α*-order Hölder-continuous with modulus* $L > 0$ *on*

$$\left\{(1-t)w^k + t P_S(w_\sigma^k) : 0 \le t \le 1\right\},$$

for some $\alpha \in (0, 1]$. *Then there holds*

$$f(P_S(w_\sigma^k)) - f(w^k) \le -\frac{1}{\sigma}\|P_S(w_\sigma^k) - w^k\|_W^2 + L\|P_S(w_\sigma^k) - w^k\|_W^{1+\alpha}.$$

Proof

$$\begin{aligned} f(P_S(w_\sigma^k)) - f(w^k) &= (\nabla f(v_\sigma^k), P_S(w_\sigma^k) - w^k)_W \\ &= (\nabla f(w^k), P_S(w_\sigma^k) - w^k)_W \\ &\quad + (\nabla f(v_\sigma^k) - \nabla f(w^k), P_S(w_\sigma^k) - w^k)_W \end{aligned}$$

with appropriate $v_\sigma^k \in \{(1-t)w^k + t P_S(w_\sigma^k) : 0 \le t \le 1\}$.

Now, since $w_\sigma^k - w^k = \sigma s^k = -\sigma \nabla f(w^k)$ and $w^k = P_S(w^k)$, we obtain

$$\begin{aligned} -\sigma(\nabla f(w^k), P_S(w_\sigma^k) - w^k)_W &= (w_\sigma^k - w^k, P_S(w_\sigma^k) - w^k)_W \\ &= (w_\sigma^k - P_S(w^k), P_S(w_\sigma^k) - P_S(w^k))_W \\ &= (P_S(w_\sigma^k) - P_S(w^k), P_S(w_\sigma^k) - P_S(w^k))_W \\ &\quad + \underbrace{(w_\sigma^k - P_S(w_\sigma^k), P_S(w_\sigma^k) - P_S(w^k))_W}_{\ge 0} \\ &\ge (P_S(w_\sigma^k) - P_S(w^k), P_S(w_\sigma^k) - P_S(w^k))_W \\ &= \|P_S(w_\sigma^k) - w^k\|_W^2. \end{aligned}$$

Next, we use

$$\|v_\sigma^k - w^k\|_W \le \|P_S(w_\sigma^k) - w^k\|_W.$$

Hence,

$$\begin{aligned} (\nabla f(v_\sigma^k) - \nabla f(w^k), P_S(w_\sigma^k) - w^k)_W &\le \|\nabla f(v_\sigma^k) - \nabla f(w^k)\|_W \|P_S(w_\sigma^k) - w^k\|_W \\ &\le L\|v_\sigma^k - w^k\|_W^\alpha \|P_S(w_\sigma^k) - w^k\|_W \\ &\le L\|P_S(w_\sigma^k) - w^k\|_W^{1+\alpha}. \end{aligned}$$

We now consider the following

2.2.2.1 Projected Armijo Rule

Choose the maximum $\sigma_k \in \{1, 1/2, 1/4, \dots\}$ for which

$$f(P_S(w^k + \sigma_k s^k)) - f(w^k) \le -\frac{\gamma}{\sigma_k}\|P_S(w^k + \sigma_k s^k) - w^k\|_W^2.$$

Here $\gamma \in (0, 1)$ is a constant.

In the unconstrained case, we recover the classical Armijo rule:

$$f(P_S(w^k + \sigma_k s^k)) - f(w^k) = f(w^k + \sigma_k s^k) - f(w^k),$$

$$\begin{aligned}-\frac{\gamma}{\sigma_k}\|P_S(w^k + \sigma_k s^k) - w^k\|_W^2 &= -\frac{\gamma}{\sigma_k}\|\sigma_k s^k\|_W^2 = -\gamma\sigma_k\|s^k\|_W^2 \\ &= \gamma\sigma_k(\nabla f(w^k), s^k)_W.\end{aligned}$$

As a stationarity measure $\Sigma(w) = \|p(w)\|_W$ we use the norm of the *projected gradient*

$$p(w) \stackrel{\text{def}}{=} w - P_S(w - \nabla f(w)).$$

In fact, the first-order optimality conditions for (2.3) are

$$w \in S, \quad (\nabla f(w), v - w)_W \ge 0 \quad \forall v \in S.$$

By Lemma 1.10, this is equivalent to

$$w - P_S(w - \nabla f(w)) = 0.$$

As a next result we show that projected Armijo step sizes exist.

Lemma 2.5 *Let W be a Hilbert space and let $f : W \to \mathbb{R}$ be continuously F-differentiable on a neighborhood of the closed convex set S. Then, for all $w^k \in S$ with $p(w^k) \ne 0$, the projected Armijo rule terminates successfully.*

Proof We proceed as in the proof of Lemma 2.4 and obtain (we have not assumed Hölder continuity of ∇f here)

$$f(P_S(w_\sigma^k)) - f(w^k) \le \frac{-1}{\sigma}\|P_S(w_\sigma^k) - w^k\|_W^2 + o(\|P_S(w_\sigma^k) - w^k\|_W).$$

It remains to show that, for all small $\sigma > 0$,

$$\frac{\gamma - 1}{\sigma}\|P_S(w_\sigma^k) - w^k\|_W^2 + o(\|P_S(w_\sigma^k) - w^k\|_W) \le 0.$$

But this follows easily from (Lemma 1.10(e)):

$$\frac{\gamma - 1}{\sigma}\|P_S(w_\sigma^k) - w^k\|_W^2 \le \underbrace{(\gamma - 1)\|p(w^k)\|_W}_{<0}\|P_S(w_\sigma^k) - w^k\|_W.$$

Theorem 2.4 *Let W be a Hilbert space, $f : W \to \mathbb{R}$ be continuously F-differentiable, and $S \subset W$ be nonempty, closed, and convex. Consider Algorithm 2.1 and assume that $f(w^k)$ is bounded below. Furthermore, let ∇f be α-order Hölder continuous on*

$$N_0^\rho = \left\{ w + s : f(w) \le f(w^0),\ \|s\|_W \le \rho \right\}$$

for some $\alpha > 0$ and some $\rho > 0$. Then

$$\lim_{k\to\infty} \|p(w^k)\|_W = 0.$$

Proof Set $p^k = p(w^k)$ and assume $p^k \not\to 0$. Then there exist $\varepsilon > 0$ and an infinite set K with $\|p^k\|_W \ge \varepsilon$ for all $k \in K$.

By construction we have that $f(w^k)$ is monotonically decreasing and by assumption the sequence is bounded below. For all $k \in K$, we obtain

$$f(w^k) - f(w^{k+1}) \ge \frac{\gamma}{\sigma_k} \|P_S(w^k + \sigma_k s^k) - w^k\|_W^2 \ge \gamma\sigma_k \|p^k\|_W^2 \ge \gamma\sigma_k\varepsilon^2,$$

where we have used the Armijo condition and Lemma 1.10(e). This shows $(\sigma_k)_K \to 0$ and $(\|P_S(w^k + \sigma_k s^k) - w^k\|_W)_K \to 0$.

For large $k \in K$ we have $\sigma_k \le 1/2$ and therefore, the Armijo condition did not hold for the step size $\sigma = 2\sigma_k$. Hence,

$$\begin{aligned}
&-\frac{\gamma}{2\sigma_k} \|P_S(w^k + 2\sigma_k s^k) - w^k\|_W^2 \\
&\le f(P_S(w^k + 2\sigma_k s^k)) - f(w^k) \\
&\le -\frac{1}{2\sigma_k} \|P_S(w^k + 2\sigma_k s^k) - w^k\|_W^2 + L\|P_S(w^k + 2\sigma_k s^k) - w^k\|_W^{1+\alpha}.
\end{aligned}$$

Here, we have applied Lemma 2.4 and the fact that by Lemma 1.10(e)

$$\|P_S(w^k + 2\sigma_k s^k) - w^k\|_W \le 2\|P_S(w^k + \sigma_k s^k) - w^k\|_W \overset{K\ni k\to\infty}{\longrightarrow} 0.$$

Hence,

$$\frac{1-\gamma}{2\sigma_k} \|P_S(w^k + 2\sigma_k s^k) - w^k\|_W^2 \le L\|P_S(w^k + 2\sigma_k s^k) - w^k\|_W^{1+\alpha}.$$

From this we derive

$$(1-\gamma)\|p^k\|_W \|P_S(w^k + 2\sigma_k s^k) - w^k\|_W \le L\|P_S(w^k + 2\sigma_k s^k) - w^k\|_W^{1+\alpha}.$$

Hence,

$$(1-\gamma)\varepsilon \le L\|P_S(w^k + 2\sigma_k s^k) - w^k\|_W^{\alpha} \le L2^\alpha \|P_S(w^k + \sigma_k s^k) - w^k\|_W^{\alpha} \overset{K\ni k\to\infty}{\longrightarrow} 0.$$

This is a contradiction.

A careful choice of search directions will allow to extend the convergence theory to more general classes of projected descent algorithms. For instance, in finite dimensions, q-superlinearly convergent projected Newton methods and their globalization are investigated in [14, 84]. In an L^2 setting, the superlinear convergence of projected Newton methods was investigated by Kelley and Sachs in [85].

2.2.3 General Optimization Problems

For more general optimization problems than we discussed so far, one usually globalizes by choosing step sizes based on an Armijo-type rule that is applied to a suitable merit function. For instance, if we consider problems of the form

$$\min_{w} f(w) \quad \text{s.t.} \quad e(w) = 0, \qquad c(w) \in \mathcal{K},$$

with functions $f : W \to \mathbb{R}$, $e : W \to Z$, and $c : W \to R$, where W, Z, and R are Banach spaces and $\mathcal{K} \subset R$ is a closed convex cone, a possible choice for a merit function is

$$m_\rho(w) = f(w) + \rho \|e(w)\|_Z + \rho \operatorname{dist}(c(w), \mathcal{K})$$

with penalty parameter $\rho > 0$. In the case of equality constraints, a global convergence result for reduced SQP methods based on this merit function is presented in [82]. Other merit functions can be constructed by taking the norm of the residual of the KKT system, the latter being reformulated as a nonsmooth operator equation, see Sect. 2.5. This residual-based type of globalization, however, does not take into account second-order information.

2.3 Newton-Based Methods—A Preview

To give an impression of modern Newton-based approaches for optimization problems, we first consider all these methods in the finite dimensional setting: $W = \mathbb{R}^n$.

2.3.1 Unconstrained Problems—Newton's Method

Consider

$$\min_{w \in \mathbb{R}^n} f(w) \tag{2.4}$$

with $f : \mathbb{R}^n \to \mathbb{R}$ twice continuously differentiable.

From analysis we know that the first-order optimality conditions are:

$$\nabla f(w) = 0. \tag{2.5}$$

Newton's method for (2.4) is obtained by applying Newton's method to (2.5).

This, again, is done by linearizing $G = \nabla f$ about the current iterate w^k and equating this linearization to zero:

$$G(w^k) + G'(w^k)s^k = 0, \qquad w^{k+1} = w^k + s^k.$$

It is well-known—and will be proved later in a much more general context—that Newton's method converges q-superlinearly if G is C^1 and $G'(\bar{w})$ is invertible.

2.3.2 *Simple Constraints*

Let $f : \mathbb{R}^n \to \mathbb{R}$ be C^2 and let $S \subset \mathbb{R}^n$ be a nonempty closed convex set.

We consider the problem

$$\min_{w \in \mathbb{R}^n} f(w) \quad \text{s.t.} \quad w \in S.$$

The optimality conditions, written in a form that directly generalizes to a Banach space setting, are: $w = \bar{w}$ solves

$$w \in S, \quad \nabla f(w)^T (v - w) \geq 0 \quad \forall v \in S. \tag{2.6}$$

This is a *Variational Inequality*, which we abbreviate VI(∇f, S).

Note that the necessity of VI(∇f, S) can be derived very easily: For all $v \in S$, the line segment $\{\bar{w} + t(v - \bar{w}) : 0 \leq t \leq 1\}$ connecting $\bar{w}$ and v is contained in S (convexity) and therefore, the function

$$\phi(t) := f(\bar{w} + t(v - \bar{w}))$$

is nondecreasing at $t = 0$:

$$0 \leq \phi'(0) = \nabla f(\bar{w})^T (v - \bar{w}).$$

Similarly, in the Banach space setting, we will have that $w = \bar{w}$ solves

$$w \in S, \quad \langle f'(w), v - w \rangle_{W^*,W} \geq 0 \quad \forall v \in S$$

with $S \subset W$ closed, convex and $f' : W \to W^*$.

Note that if $S = \mathbb{R}^n$, then (2.6) is equivalent to (2.5).

2.3.2.1 Nonsmooth Reformulation Approach and Generalized Newton Methods

In the development of projected descent methods we already used the important fact that the VI (2.6) is equivalent to

$$w - P_S(w - \theta \nabla f(w)) = 0, \tag{2.7}$$

where $\theta > 0$ is fixed.

Example 2.3 If S is a box, i.e.,

$$S = [a_1, b_1] \times \cdots \times [a_n, b_n],$$

then $P_S(w)$ can be computed very easily as follows:

$$P_S(w)_i = \max(a_i, \min(w_i, b_i)).$$

It is instructive (and not difficult) to check the equivalence of (2.6) and (2.7) by hand.

The function

$$\Phi(w) := w - P_S(w - \theta \nabla f(w))$$

is locally Lipschitz continuous (P_S is non-expansive and ∇f is C^1), but cannot be expected to be differentiable. Therefore, *at a first sight*, Newton's method is *not* applicable.

However, a second look shows that Φ has nice properties if S is sufficiently nice. To be more concrete, let

$$S = [a_1, b_1] \times \cdots \times [a_n, b_n]$$

be a box in the following. Then Φ is *piecewise* continuously differentiable, i.e., it consists of finitely many C^1-pieces $\Phi^j : \mathbb{R}^n \to \mathbb{R}^n$, $j = 1, \ldots, m$. More precisely, every component Φ_i of Φ consists of three pieces:

$$w_i - a_i, \qquad w_i - b_i, \qquad w_i - (w_i - \theta \nabla f(w)_i) = \theta \nabla f(w)_i,$$

hence Φ consists of (at most) 3^n pieces Φ^j.

Denote by

$$A(w) = \left\{ j : \Phi^j(w) = \Phi(w) \right\}$$

the active indices at w and by

$$I(w) = \left\{ j : \Phi^j(w) \neq \Phi(w) \right\}$$

the set of inactive indices at w.

By continuity, $I(w) \supset I(\bar{w})$ in a neighborhood U of $\bar{w}$. Now consider the following

Algorithm 2.5 (Generalized Newton's method for piecewise C^1 equations)

0. Chose w^0 (sufficiently close to $\bar{w}$).

For $k = 0, 1, 2, \ldots$:

1. Choose $M_k \in \{(\Phi^j)'(w^k) : j \in A(w^k)\}$ and solve

$$M_k s^k = -\Phi(w^k).$$

2. Set $w^{k+1} = w^k + s^k$.

For w^k close to $\bar{w}$, we have $A(w^k) \subset A(\bar{w})$ and thus s^k is the Newton step for the C^1 equation

$$\Phi^{j_k}(w) = 0,$$

where $j_k \in A(w^k) \subset A(\bar{w})$ is the active index with $M_k = (\Phi^{j_k})'(w^k)$.

Therefore, if all the finitely many Newton processes for

$$\Phi^j(w) = 0, \quad j \in A(\bar{w})$$

converge locally fast, our generalized Newton's method converges locally fast, too. In particular, this is the case if f is C^2 and all $(\Phi^j)'(\bar{w})$, $j \in A(\bar{w})$, are invertible.

2.3.2.2 SQP Methods

A further appealing idea is to obtain an iterative method by linearizing ∇f in VI(∇f, S) about the current iterate $w^k \in S$:

$$w \in S, \quad (\nabla f(w^k) + \nabla^2 f(w^k)(w - w^k))^T (v - w) \geq 0 \quad \forall v \in S.$$

The solution w^{k+1} of this VI is then the new iterate. The resulting method, of course, can just as well be formulated for general variational inequalities VI(F, S) with C^1-function $F : \mathbb{R}^n \to \mathbb{R}^n$. We obtain the following method:

Algorithm 2.6 (Josephy-Newton method for VI(F, S))

0. Choose $w^0 \in S$ (sufficiently close to the solution $\bar{w}$ of VI(F, S)).

For $k = 0, 1, 2 \ldots$:

1. STOP if w^k solves VI(F, S) (holds if $w^k = w^{k-1}$).
2. Compute the solution w^{k+1} of

$$\text{VI}(F(w^k) + F'(w^k)(\cdot - w^k), S):$$

$$w \in S, \quad (F(w^k) + F'(w^k)(w - w^k))^T (v - w) \geq 0 \quad \forall v \in S$$

that is closest to w^k.

In the case $F = \nabla f$, it is easily seen that VI($\nabla f(w^k) + \nabla^2 f(w^k)(\cdot - w^k)$, S) is the first-order necessary optimality condition of the problem

$$\min_{w \in \mathbb{R}^n} \nabla f(w^k)^T (w - w^k) + \frac{1}{2}(w - w^k)^T \nabla^2 f(w^k)(w - w^k) \quad \text{s.t.} \quad w \in S.$$

The objective function is quadratic, and in the case of box constraints, we have a box-constrained quadratic program.

This is why this approach is called sequential quadratic programming.

Algorithm 2.7 (Sequential Quadratic Programming for simple constraints)

0. Chose $w^0 \in \mathbb{R}^n$ (sufficiently close to $\bar{w}$).

For $k = 0, 1, 2, \dots$:

1. Compute the first-order optimal point s^k of the QP

$$\min_{s\in\mathbb{R}^n} \nabla f(w^k)^T s + \frac{1}{2} s^T \nabla^2 f(w^k) s \quad \text{s.t.} \quad w^k + s \in S$$

   that is closest to 0.
2. Set $w^{k+1} = w^k + s^k$.

The local convergence analysis of the Josephy-Newton method is intimately connected with the locally unique and Lipschitz-stable solvability of the parameterized VI

$$\begin{aligned}&\text{VI}(F(\bar{w}) + F'(\bar{w})(\cdot - \bar{w}) - p, S):\\ &w \in S, \quad (F(\bar{w}) + F'(\bar{w})(w - \bar{w}) - p)^T(v - w) \geq 0 \quad \forall v \in S.\end{aligned}$$

In fact, if there exist open neighborhoods $U_p \subset \mathbb{R}^n$ of 0, $U_w \subset \mathbb{R}^n$ of $\bar{w}$, and a Lipschitz continuous function $U_p \ni p \mapsto w(p) \in U_w$ such that $w(p)$ is the unique solution of $\text{VI}(F(\bar{w}) + F'(\bar{w})(\cdot - \bar{w}) - p, S)$ in U_w, then $\text{VI}(F, S)$ is called *strongly regular* at $\bar{w}$.

As we will see, strong regularity implies local q-superlinear convergence of the above SQP method if f is C^2.

In the case $S = \mathbb{R}^n$ we have

$$\text{VI}(F, \mathbb{R}^n): \quad F(w) = 0.$$

Hence, the Josephy-Newton method for $\text{VI}(F, \mathbb{R}^n)$ is Newton's method for $F(w) = 0$. Furthermore, from

$$\text{VI}(F(\bar{w}) + F'(\bar{w})(\cdot - \bar{w}) + p, \mathbb{R}^n): \quad F(\bar{w}) + F'(\bar{w})(w - \bar{w}) + p = 0$$

we see that in this case strong regularity is the same as the invertibility of $F'(\bar{w})$.

2.3.3 General Inequality Constraints

We now consider general nonlinear optimization in $\mathbb{R}^n$:

$$\min_{w\in\mathbb{R}^n} f(w) \quad \text{s.t.} \quad e(w) = 0, \qquad c(w) \leq 0, \tag{2.8}$$

where $f : \mathbb{R}^n \to \mathbb{R}$, $e : \mathbb{R}^n \to \mathbb{R}^p$, and $c : \mathbb{R}^n \to \mathbb{R}^m$ are C^2 and $\le$ is meant component-wise.

Denote by

$$L(w, \lambda, \mu) = f(w) + \lambda^T c(w) + \mu^T e(w)$$

the Lagrange function of problem (2.8).

Under a constraint qualification (CQ), the first-order optimality conditions (KKT conditions) hold at $(\bar{w}, \bar{\lambda}, \bar{\mu})$:

$$\begin{aligned} &\nabla_w L(\bar{w}, \bar{\lambda}, \bar{\mu}) = \nabla f(\bar{w}) + c'(\bar{w})^T \bar{\lambda} + e'(\bar{w})^T \bar{\mu} = 0, \\ &\bar{\lambda} \ge 0, \quad \nabla_\lambda L(\bar{w}, \bar{\lambda}, \bar{\mu})^T (z - \bar{\lambda}) = c(\bar{w})^T (z - \bar{\lambda}) \le 0 \quad \forall z \ge 0, \\ &\nabla_\mu L(\bar{w}, \bar{\lambda}, \bar{\mu}) = e(\bar{w}) = 0. \end{aligned} \tag{2.9}$$

Remark 2.1

(a) An easy way to remember these conditions is the following: $(\bar{w}, \bar{\lambda}, \bar{\mu})$ is a first-order saddle point of L on $\mathbb{R}^n \times (\mathbb{R}^m_+ \times \mathbb{R}^p)$.
(b) The second equation can be equivalently written in the following way:

$$\bar{\lambda} \ge 0, \quad c(\bar{w}) \le 0, \quad c(\bar{w})^T \bar{\lambda} = 0.$$

The KKT system consists of two equations and the variational inequality VI$(-c(\bar{w}), \mathbb{R}^m_+)$. This is a VI w.r.t. λ that is parameterized by $\bar{w}$. Also, since equations are special cases of variational inequalities, we have that (2.9) is in fact the same as VI$(-\nabla L, \mathbb{R}^n \times \mathbb{R}^m_+ \times \mathbb{R}^p)$.

We now can use the same techniques as for simple constraints.

2.3.3.1 Nonsmooth Reformulation Approach and Generalized Newton Methods

Using the projection, we rewrite the VI in (2.9) as a nonsmooth equation:

$$\Phi(w, \lambda) := \lambda - P_{\mathbb{R}^m_+}(\lambda + \theta c(w)) = 0,$$

where $\theta > 0$ is fixed. The reformulated KKT system

$$G(w, \lambda, \mu) := \begin{pmatrix} \nabla f(w) + c'(w)^T \lambda + e'(w)^T \mu \\ \Phi(w, \lambda) \\ e(w) \end{pmatrix} = 0$$

is a system of $n + m + p$ equations in $n + m + p$ unknowns.

The function on the left is C^1, except for the second row which is piecewise C^1. Therefore, the generalized Newton's method for piecewise smooth equations (Algorithm 2.5) can be applied. It is q-superlinearly convergent if $(G^j)'(\bar{w}, \bar{\lambda}, \bar{\mu})$ is invertible for all active indices $j \in A(\bar{w}, \bar{\lambda}, \bar{\mu})$.

2.3.3.2 SQP Methods

As already observed, the KKT system is identical to VI($-\nabla L, \mathbb{R}^n \times \mathbb{R}^m_+ \times \mathbb{R}^p$).

The SQP method for (2.8) can now be derived as in the simply constrained case by linearizing $-\nabla L$ about the current iterate $x^k := (w^k, \lambda^k, \mu^k)$: The resulting subproblem is VI($-\nabla L(x^k) - \nabla L(x^k)(\cdot - x^k), \mathbb{R}^n \times \mathbb{R}^m_+ \times \mathbb{R}^p$), or, in detail:

$$\begin{aligned}
&\nabla_w L(x^k) + \nabla_{wx} L(x^k)(x - x^k) = 0 \\
&\lambda \geq 0, \quad (c(w^k) + c'(w^k)(w - w^k))^T (z - \lambda) \leq 0 \quad \forall z \geq 0, \\
&e(w^k) + e'(w^k)(w - w^k) = 0.
\end{aligned} \tag{2.10}$$

As in the simply constrained case, it is straightforward to verify that (2.10) is equivalent to the KKT conditions of the following quadratic program:

$$\begin{aligned}
&\min_{w} \ \nabla f(w^k)^T (w - w^k) + \frac{1}{2}(w - w^k)^T \nabla_{ww} L(x^k)(w - w^k) \\
&\text{s.t.} \quad e(w^k) + e'(w^k)(w - w^k) = 0, \qquad c(w^k) + c'(w^k)(w - w^k) \leq 0.
\end{aligned}$$

2.4 Generalized Newton Methods

We have seen in the previous section that we can reformulate KKT systems of finite dimensional optimization problems as nonsmooth equations. This also holds true for PDE-constrained optimization with inequality constraints, as we will sketch below. In finite dimensions, we observed that a projection-based reformulation results in a piecewise C^1-function to which a Newton-type method can be applied. In order to develop similar approaches in a function space framework, it is important to find minimum requirements on the operator $G : X \to Y$ that allow us to develop and analyze a Newton-type method for the (possibly nonsmooth) operator equation

$$G(x) = 0. \tag{2.11}$$

2.4.1 Motivation: Application to Optimal Control

We will show now that the optimality conditions of constrained optimal control problems can be converted to nonsmooth operator equations.

Consider the following elliptic optimal control problem:

$$\begin{aligned}
&\min_{y \in H_0^1(\Omega), u \in L^2(\Omega)} \quad J(y, u) \stackrel{\text{def}}{=} \frac{1}{2}\|y - y_d\|^2_{L^2(\Omega)} + \frac{\alpha}{2}\|u\|^2_{L^2(\Omega)} \\
&\text{s.t.} \qquad Ay = u, \ \beta_l \leq u \leq \beta_r.
\end{aligned}$$

Here, $y \in H_0^1(\Omega)$ is the state, which is defined on the open bounded domain $\Omega \subset \mathbb{R}^n$, and $u \in L^2(\Omega)$ is the control. Furthermore, $A : H_0^1(\Omega) \to H^{-1}(\Omega) = H_0^1(\Omega)^*$ is a (for simplicity) linear elliptic partial differential operator, e.g., $A = -\Delta$.

The control is subject to pointwise bounds $\beta_l < \beta_r$. The objective is to drive the state as close to $y_d \in L^2(\Omega)$ as possible. The second part of the objective function penalizes excessive control costs; the parameter $\alpha > 0$ is typically small.

We eliminate the state y via the state equation, i.e., $y = y(u) = A^{-1}u$, and obtain the reduced problem

$$\min_{u \in L^2(\Omega)} \ \hat{J}(u) \stackrel{\text{def}}{=} J(y(u), u) \stackrel{\text{def}}{=} \frac{1}{2}\|A^{-1}u - y_d\|^2_{L^2(\Omega)} + \frac{\alpha}{2}\|u\|^2_{L^2(\Omega)}$$

$$\text{s.t.} \quad \beta_l \le u \le \beta_r.$$

The feasible set is

$$S = \left\{u \in L^2(\Omega) : \beta_l \le u \le \beta_r\right\}$$

and the optimality conditions are given by

$$\mathrm{VI}(\nabla \hat{J}, S): \quad u \in S, \quad (\nabla \hat{J}(u), v - u)_{L^2(\Omega)} \ge 0 \quad \forall v \in S.$$

Using the projection $P_S(u) = P_{[\beta_l, \beta_r]}(u(\cdot))$ onto S, this can be rewritten as

$$\Phi(u) \stackrel{\text{def}}{=} u - P_{[\beta_l, \beta_r]}(u - \theta \nabla \hat{J}(u)) = 0,$$

where $\theta > 0$ is fixed. This is a nonsmooth operator equation in the space $L^2(\Omega)$. Hence, we were able to convert the optimality system into a nonsmooth operator equation.

2.4.2 A General Superlinear Convergence Result

Consider the operator equation (2.11) with $G : X \to Y$, X, Y Banach spaces.

A general Newton-type method for (2.11) has the form

Algorithm 2.8 (Generalized Newton's method)

0. Choose $x^0 \in X$ (sufficiently close to the solution $\bar{x}$).

For $k = 0, 1, 2, \ldots$:

1. Choose an invertible operator $M_k \in \mathcal{L}(X, Y)$.
2. Obtain s^k by solving

$$M_k s = -G(x^k), \tag{2.12}$$

   and set $x^{k+1} = x^k + s^k$.

We now investigate the generated sequence (x^k) in a neighborhood of a solution $\bar{x} \in X$, i.e., $G(\bar{x}) = 0$.

For the distance $d^k := x^k - \bar{x}$ to the solution we have

$$\begin{aligned} M_k d^{k+1} &= M_k(x^{k+1} - \bar{x}) = M_k(x^k + s^k - \bar{x}) = M_k d^k - G(x^k) \\ &= G(\bar{x}) + M_k d^k - G(x^k). \end{aligned}$$

Hence, we obtain:

1. (x^k) converges q-linearly to $\bar{x}$ with rate $\gamma \in (0, 1)$ iff

$$\|M_k^{-1}(G(\bar{x} + d^k) - G(\bar{x}) - M_k d^k)\|_X \leq \gamma \|d^k\|_X \quad \forall k \text{ with } \|d^k\|_X \text{ suff. small.} \tag{2.13}$$

2. (x^k) converges q-superlinearly to $\bar{x}$ iff

$$\|M_k^{-1}(G(\bar{x} + d^k) - G(\bar{x}) - M_k d^k)\|_X = o(\|d^k\|_X) \quad \text{for } \|d^k\|_X \to 0. \tag{2.14}$$

3. (x^k) converges with q-order $1 + \alpha > 1$ to $\bar{x}$ iff

$$\|M_k^{-1}(G(\bar{x} + d^k) - G(\bar{x}) - M_k d^k)\|_X = O(\|d^k\|_X^{1+\alpha}) \quad \text{for } \|d^k\|_X \to 0. \tag{2.15}$$

In 1., the estimate is meant uniformly in k, i.e., there exists $\delta_\gamma > 0$ such that

$$\|M_k^{-1}(G(\bar{x} + d^k) - G(\bar{x}) - M_k d^k)\|_X \leq \gamma \|d^k\|_X \quad \forall k \text{ with } \|d^k\|_X < \delta_\gamma.$$

In 2., $o(\|d^k\|_X)$ is meant uniformly in k, i.e., for all $\eta \in (0, 1)$, there exists $\delta_\eta > 0$ such that

$$\|M_k^{-1}(G(\bar{x} + d^k) - G(\bar{x}) - M_k d^k)\|_X \leq \eta \|d^k\|_X \quad \forall k \text{ with } \|d^k\|_X < \delta_\eta.$$

The condition in 3. and those stated below are meant similarly.

It is convenient, and often done, to split the smallness assumption on

$$\|M_k^{-1}(G(\bar{x} + d^k) - G(\bar{x}) - M_k d^k)\|_X$$

in two parts:

1. *Regularity condition*:

$$\|M_k^{-1}\|_{Y \to X} \leq C \quad \forall k \geq 0. \tag{2.16}$$

2. *Approximation condition*:

$$\|G(\bar{x} + d^k) - G(\bar{x}) - M_k d^k\|_Y = o(\|d^k\|_X) \quad \text{for } \|d^k\|_X \to 0 \tag{2.17}$$

or

$$\|G(\bar{x} + d^k) - G(\bar{x}) - M_k d^k\|_Y = O(\|d^k\|_X^{1+\alpha}) \quad \text{for } \|d^k\|_X \to 0. \tag{2.18}$$

We obtain

Theorem 2.9 *Consider the operator equation* (2.11) *with* $G : X \to Y$, *where* X *and* Y *are Banach spaces. Let* (x^k) *be generated by the generalized Newton method* (*Algorithm* 2.8). *Then*:

1. *If* x^0 *is sufficiently close to* $\bar{x}$ *and* (2.13) *holds then* $x^k \to \bar{x}$ *q-linearly with rate* γ.
2. *If* x^0 *is sufficiently close to* $\bar{x}$ *and* (2.14) (*or* (2.16) *and* (2.17)) *holds then* $x^k \to \bar{x}$ *q-superlinearly.*
3. *If* x^0 *is sufficiently close to* $\bar{x}$ *and* (2.15) *holds* (*or* (2.16) *and* (2.18)) *then* $x^k \to \bar{x}$ *q-superlinearly with order* $1+\alpha$.

Proof 1. Let $\delta > 0$ be so small that (2.13) holds for all x^k with $\|d^k\|_X < \delta$. Then, for x^0 satisfying $\|x^0 - \bar{x}\|_X < \delta$, we have

$$\begin{aligned}\|x^1 - \bar{x}\|_X &= \|d^1\|_X = \|M_0^{-1}(G(\bar{x}+d^0) - G(\bar{x}) - M_0 d^0)\|_X \le \gamma\|d^0\|_X \\ &= \gamma\|x^0 - \bar{x}\|_X < \delta.\end{aligned}$$

Inductively, let $\|x^k - \bar{x}\|_X < \delta$. Then

$$\begin{aligned}\|x^{k+1} - \bar{x}\|_X &= \|d^{k+1}\|_X = \|M_k^{-1}(G(\bar{x}+d^k) - G(\bar{x}) - M_k d^k)\|_X \\ &\le \gamma\|d^k\|_X = \gamma\|x^k - \bar{x}\|_X < \delta.\end{aligned}$$

Hence, we have

$$\|x^{k+1} - \bar{x}\|_X \le \gamma\|x^k - \bar{x}\|_X \quad \forall k \ge 0.$$

2. Fix $\gamma \in (0,1)$ and let $\delta > 0$ be so small that (2.13) holds for all x^k with $\|d^k\|_X < \delta$. Then, for x^0 satisfying $\|x^0 - \bar{x}\|_X < \delta$, we can apply 1. to conclude $x^k \to \bar{x}$ with rate γ.

Now, (2.14) immediately yields

$$\begin{aligned}\|x^{k+1} - \bar{x}\|_X &= \|d^{k+1}\|_X = \|M_k^{-1}(G(\bar{x}+d^k) - G(\bar{x}) - M_k d^k)\|_X = o(\|d^k\|_X) \\ &= o(\|x^k - \bar{x}\|_X) \quad (k \to \infty).\end{aligned}$$

3. As in 2, but now

$$\begin{aligned}\|x^{k+1} - \bar{x}\|_X &= \|d^{k+1}\|_X = \|M_k^{-1}(G(\bar{x}+d^k) - G(\bar{x}) - M_k d^k)\|_X = O(\|d^k\|_X^{1+\alpha}) \\ &= O(\|x^k - \bar{x}\|_X^{1+\alpha}) \quad (k \to \infty).\end{aligned}$$

We emphasize that an inexact solution of the Newton system (2.12) can be interpreted as a solution of the same system, but with M_k replaced by a perturbed operator $\tilde{M}_k$. Since the condition (2.14) (or the conditions (2.16) and (2.17)) remain valid if M_k is replaced by a perturbed operator $\tilde{M}_k$ and the perturbation is sufficiently small, we see that the fast convergence of the generalized Newton's method is not affected if the system is solved inexactly and the accuracy of the solution is controlled suitably. The Dennis-Moré condition [36] characterizes perturbations that are possible without destroying q-superlinear convergence.

We will now specialize on particular instances of generalized Newton methods. The first one, of course, is Newton's method itself.

2.4.3 The Classical Newton's Method

In the classical Newton's method, we assume that G is continuously F-differentiable and choose $M_k = G'(x^k)$.

The regularity condition then reads

$$\|G'(x^k)^{-1}\|_{Y\to X} \le C \quad \forall k \ge 0.$$

By Banach's Lemma (asserting continuity of $M \mapsto M^{-1}$), this holds true if G' is continuous at $\bar{x}$ and

$$G'(\bar{x}) \in \mathcal{L}(X, Y) \quad \text{is continuously invertible.}$$

This condition is the textbook regularity requirement in the analysis of Newton's method.

Fréchet differentiability at $\bar{x}$ means

$$\|G(\bar{x} + d^k) - G(\bar{x}) - G'(\bar{x})d^k\|_Y = o(\|d^k\|_X).$$

Now, due to the continuity of G',

$$\begin{aligned}
&\|G(\bar{x} + d^k) - G(\bar{x}) - M_k d^k\|_Y \\
&\quad = \|G(\bar{x} + d^k) - G(\bar{x}) - G'(\bar{x} + d^k)d^k\|_Y \\
&\quad \le \|G(\bar{x} + d^k) - G(\bar{x}) - G'(\bar{x})d^k\|_Y + \|(G'(\bar{x}) - G'(\bar{x} + d^k))d^k\|_Y \\
&\quad \le o(\|d^k\|_X) + \|G'(\bar{x}) - G'(\bar{x} + d^k)\|_{X\to Y}\|d^k\|_X \\
&\quad = o(\|d^k\|_X) \quad \text{for } \|d^k\|_X \to 0.
\end{aligned}$$

Therefore, we have proved the superlinear approximation condition.

If G' is α-order Hölder continuous near $\bar{x}$, we even obtain the approximation condition of order $1 + \alpha$. In fact, let $L > 0$ be the modulus of Hölder continuity. Then

$$\begin{aligned}
&\|G(\bar{x} + d^k) - G(\bar{x}) - M_k d^k\|_Y \\
&\quad = \|G(\bar{x} + d^k) - G(\bar{x}) - G'(\bar{x} + d^k)d^k\|_Y \\
&\quad = \left\| \int_0^1 (G'(\bar{x} + td^k) - G'(\bar{x} + d^k))d^k \, dt \right\|_Y
\end{aligned}$$

$$\leq \int_0^1 \|G'(\bar{x}+td^k) - G'(\bar{x}+d^k)\|_{X\to Y}\, dt\, \|d^k\|_X$$

$$\leq L\int_0^1 (1-t)^\alpha \|d^k\|_X^\alpha\, dt\, \|d^k\|_X = \frac{L}{1+\alpha}\|d^k\|_X^{1+\alpha} = O(\|d^k\|_X^{1+\alpha}).$$

Summarizing, we have proved the following

Corollary 2.1 *Let $G : X \to Y$ be a continuously F-differentiable operator between Banach spaces and assume that $G'(\bar{x})$ is continuously invertible at the solution $\bar{x}$. Then Newton's method (i.e., Algorithm 2.8 with $M_k = G'(x^k)$ for all k) converges locally q-superlinearly. If, in addition, G' is α-order Hölder continuous near $\bar{x}$, the order of convergence is $1+\alpha$.*

Remark 2.2 The choice of M_k in the classical Newton's method, $M_k = G'(x^k)$, is *point-based*, since it depends on the point x^k.

2.4.4 Generalized Differential and Semismoothness

If G is nonsmooth, the question arises if a suitable substitute for G' can be found. We follow [134, 136] here; a related approach can be found in [87] and [69]. Thinking at subgradients of convex functions, which are set-valued, we consider set-valued generalized differentials $\partial G : X \rightrightarrows \mathcal{L}(X,Y)$. Then we will choose M_k point-based, i.e.,

$$M_k \in \partial G(x^k).$$

If we want every such choice M_k to satisfy the superlinear approximation condition, then we have to require

$$\sup_{M\in\partial G(\bar{x}+d)} \|G(\bar{x}+d) - G(\bar{x}) - Md\|_Y = o(\|d\|_X) \quad \text{for } \|d\|_X \to 0.$$

This approximation property is called semismoothness [134, 136]:

Definition 2.1 (Semismoothness) Let $G : X \to Y$ be a continuous operator between Banach spaces. Furthermore, let be given the set-valued mapping $\partial G : X \rightrightarrows Y$ with nonempty images (which we will call generalized differential in the sequel). Then

(a) G is called ∂G-semismooth at $x \in X$ if

$$\sup_{M\in\partial G(x+d)} \|G(x+d) - G(x) - Md\|_Y = o(\|d\|_X) \quad \text{for } \|d\|_X \to 0.$$

(b) G is called ∂G-semismooth of order $\alpha > 0$ at $x \in X$ if

$$\sup_{M\in\partial G(x+d)} \|G(x+d) - G(x) - Md\|_Y = O(\|d\|_X^{1+\alpha}) \quad \text{for } \|d\|_X \to 0.$$

Lemma 2.6 *If $G : X \to Y$ is continuously F-differentiable near x, then G is $\{G'\}$-semismooth at x. Furthermore, if G' is α-order Hölder continuous near x, then G is $\{G'\}$-semismooth at x of order α. Here, $\{G'\}$ denotes the setvalued operator $\{G'\} : X \rightrightarrows \mathcal{L}(X, Y)$, $\{G'\}(x) = \{G'(x)\}$.*

Proof

$$\begin{aligned}
&\|G(x+d) - G(x) - G'(x+d)d\|_Y \\
&\quad \le \|G(x+d) - G(x) - G'(x)d\|_Y + \|G'(x)d - G'(x+d)d\|_Y \\
&\quad \le o(\|d\|_X) + \|G'(x) - G'(x+d)\|_{X\to Y}\|d\|_X = o(\|d\|_X).
\end{aligned}$$

Here, we have used the definition of F-differentiability and the continuity of G'.

In the case of α-order Hölder continuity we have to work a little bit more:

$$\begin{aligned}
&\|G(x+d) - G(x) - G'(x+d)d\|_Y \\
&\quad = \left\| \int_0^1 (G'(x+td) - G'(x+d))d\,dt \right\|_Y \\
&\quad \le \int_0^1 \|G'(x+td) - G'(x+d)\|_{X\to Y}\,dt\,\|d\|_X \le \int_0^1 L(1-t)^\alpha \|d\|_X^\alpha\,dt\,\|d\|_X \\
&\quad = \frac{L}{1+\alpha}\|d\|_X^{1+\alpha} = O(\|d\|_X^{1+\alpha}).
\end{aligned}$$

Example 2.4 For locally Lipschitz-continuous functions $G : \mathbb{R}^n \to \mathbb{R}^m$, the standard choice for ∂G is Clarke's generalized Jacobian:

$$\partial^{cl} G(x) = \text{conv}\left\{M : x^k \to x,\ G'(x^k) \to M,\ G \text{ differentiable at } x^k\right\}. \tag{2.19}$$

This definition is justified since G' exists almost everywhere on $\mathbb{R}^n$ by Rademacher's theorem (which is a deep result).

Remark 2.3 The classical definition of semismoothness for functions $G : \mathbb{R}^n \to \mathbb{R}^m$ [105, 113] is equivalent to $\partial^{cl} G$-semismoothness, where $\partial^{cl} G$ is Clarke's generalized Jacobian defined in (2.19), in connection with directional differentiability of G.

Next, we give a concrete example of a semismooth function:

Example 2.5 Consider $\psi : \mathbb{R} \to \mathbb{R}$, $\psi(x) = P_{[a,b]}(x)$, $a < b$, then Clarke's generalized derivative is

$$\partial^{cl}\psi(x) = \begin{cases} \{0\} & x < a \text{ or } x > b, \\ \{1\} & a < x < b, \\ \text{conv}\{0, 1\} = [0, 1] & x = a \text{ or } x = b. \end{cases}$$

The $\partial^{cl}\psi$-semismoothness of ψ can be shown easily:

For all $x \notin \{a, b\}$ we have that ψ is continuously differentiable in a neighborhood of x with $\partial^{cl}\psi \equiv \{\psi'\}$. Hence, by Lemma 2.6, ψ is $\partial^{cl}\psi$-semismooth at x.

For $x = a$, we estimate explicitly: For small $d > 0$, we have $\partial^{cl}\psi(x) = \{\psi'(a+d)\} = \{1\}$ and thus

$$\sup_{M\in\partial^{cl}\psi(x+d)} |\psi(x+d) - \psi(x) - Md| = a + d - a - 1 \cdot d = 0.$$

For small $d < 0$, we have $\partial^{cl}\psi(x) = \{\psi'(a+d)\} = \{0\}$ and thus

$$\sup_{M\in\partial^{cl}\psi(x+d)} |\psi(x+d) - \psi(x) - Md| = a - a - 0 \cdot d = 0.$$

Hence, the semismoothness of ψ at $x = a$ is proved.

For $x = b$ we can do exactly the same.

The class of semismooth operators is closed with respect to a wide class of operations, see [134]:

Theorem 2.10 *Let X, Y, Z, X_i, Y_i be Banach spaces.*

(a) *If the operators $G_i : X \to Y_i$ are ∂G_i-semismooth at x then (G_1, G_2) is $(\partial G_1, \partial G_2)$-semismooth at x.*
(b) *If $G_i : X \to Y$, $i = 1, 2$, are ∂G_i-semismooth at x then $G_1 + G_2$ is $(\partial G_1 + \partial G_2)$-semismooth at x.*
(c) *Let $G_1 : Y \to Z$ and $G_2 : X \to Y$ be ∂G_i-semismooth at $G_2(x)$ and x, respectively. Assume that ∂G_1 is bounded near $y = G_2(x)$ and that G_2 is Lipschitz continuous near x. Then $G = G_1 \circ G_2$ is ∂G-semismooth with*

$$\partial G(x) = \{M_1 M_2 : M_1 \in \partial G_1(G_2(x)),\ M_2 \in \partial G_2(x)\}.$$

Proof Parts (a) and (b) are straightforward to prove.

Part (c):

Let $y = G_2(x)$ and consider $d \in X$. Let $h(d) = G_2(x+d) - y$. Then, for $\|d\|_X$ sufficiently small,

$$\|h(d)\|_Y = \|G_2(x+d) - G_2(x)\|_Y \le L_2 \|d\|_X.$$

Hence, for $M_1 \in \partial G_1(G_2(x+d))$ and $M_2 \in \partial G_2(x+d)$, we obtain

$$\begin{aligned}
&\|G_1(G_2(x+d)) - G_1(G_2(x)) - M_1 M_2 d\|_Z \\
&\quad = \|G_1(y + h(d)) - G_1(y) - M_1 h(d) + M_1(G_2(x+d) - G_2(x) - M_2 d)\|_Z \\
&\quad \le \|G_1(y + h(d)) - G_1(y) - M_1 h(d)\|_Z \\
&\qquad + \|M_1\|_{Y\to Z} \|G_2(x+d) - G_2(x) - M_2 d\|_Y.
\end{aligned}$$

By assumption, there exists C with $\|M_1\|_{Y\to Z} \le C$ if $\|d\|_X$ is sufficiently small. Taking the supremum with respect to M_1, M_2 and using the semismoothness of G_1 and G_2 gives

$$\begin{aligned}
&\sup_{M\in\partial G(x+d)} \|G(x+d)-G(x)-Md\|_Z \\
&\quad\le \sup_{M_1\in\partial G_1(y+h(d))} \|G_1(y+h(d))-G_1(y)-M_1h(d)\|_Z \\
&\qquad + C \sup_{M_2\in\partial G_2(x+d)} \|G_2(x+d)-G_2(x)-M_2d\|_Y \\
&\quad = o(\|h(d)\|_Y)+o(\|d\|_X)=o(\|d\|_X).
\end{aligned}$$

2.4.5 Semismooth Newton Methods

The semismoothness concept ensures the approximation property required for generalized Newton methods. In addition, we need a regularity condition, which can be formulated as follows:

There exist constants $C>0$ and $\delta>0$ such that

$$\|M^{-1}\|_{Y\to X} \le C \quad \forall M\in\partial G(x) \quad \forall x\in X, \quad \|x-\bar{x}\|_X<\delta. \tag{2.20}$$

Under these two assumptions, the following generalized Newton method for semismooth operator equations is q-superlinearly convergent:

Algorithm 2.11 (Semismooth Newton's method)

0. Choose $x^0\in X$ (sufficiently close to the solution $\bar{x}$).

For $k=0,1,2,\ldots$:

1. Choose $M_k\in\partial G(x^k)$.
2. Obtain s^k by solving

$$M_k s^k = -G(x^k),$$

and set $x^{k+1}=x^k+s^k$.

The local convergence result is a simple corollary of Theorem 2.9:

Theorem 2.12 *Let $G: X\to Y$ be continuous and ∂G-semismooth at a solution $\bar{x}$ of* (2.11). *Furthermore, assume that the regularity condition* (2.20) *holds. Then there exists $\delta>0$ such that for all $x^0\in X$, $\|x^0-\bar{x}\|_X<\delta$, the semismooth Newton method (Algorithm* 2.11) *converges q-superlinearly to $\bar{x}$.*

If G is ∂G-semismooth of order $\alpha>0$ at $\bar{x}$, then the convergence is of order $1+\alpha$.

Proof The regularity condition (2.20) implies (2.16) as long as x^k is close enough to $\bar{x}$. Furthermore, the semismoothness of G at $\bar{x}$ ensures the q-superlinear approximation condition (2.17).

In the case of α-order semismoothness, the approximation condition (2.18) with order $1+\alpha$ holds.

Therefore, Theorem 2.9 yields the assertions.

2.4.5.1 Semismooth Newton Method for Finite Dimensional KKT Systems

At the beginning of this chapter we have seen that we can rewrite the KKT conditions of the NLP

$$\min f(w) \quad \text{s.t.} \quad e(w) = 0, \qquad c(w) \le 0$$

in the following form:

$$G(x) \overset{\text{def}}{=} \begin{pmatrix} \nabla_w L(w, \lambda, \mu) \\ \lambda - P_{\mathbb{R}^p_+}(\lambda + c(w)) \\ e(w) \end{pmatrix} = 0,$$

where we have set $x = (w, \lambda, \mu)$. With the developed results, we now can show that the function G on the left is semismooth. In fact, $\nabla_w L$ is $\{\nabla_{wx} L\}$-semismooth and e is $\{e'\}$-semismooth.

Furthermore, as shown above, $\psi(t) = P_{\mathbb{R}_+}(t)$ is $\partial^{cl}\psi$-semismooth with

$$\partial^{cl}\psi(t) = \{0\} \quad (t < 0), \qquad \partial^{cl}\psi(t) = \{1\} \quad (t > 0), \qquad \partial^{cl}\psi(0) = [0, 1].$$

Hence, by the sum and chain rules from Theorem 2.10

$$\phi_i(w, \lambda_i) \overset{\text{def}}{=} \lambda_i - P_{\mathbb{R}_+}(\lambda_i + c_i(w)),$$

is semismooth with respect to

$$\partial\phi_i(w, \lambda_i) := \left\{(-g_i c_i'(w), 1 - g_i) : g_i \in \partial^{cl}\psi(\lambda_i + c_i(w))\right\}.$$

Therefore, the operator $\Phi(w, \lambda) = \lambda - P_{\mathbb{R}^p_+}(\lambda + c(w))$ is semismooth with respect to

$$\partial\Phi(w, \lambda) := \left\{(-D_g c_i'(w), I - D_g) : D_g = \text{diag}(g_i),\ g_i \in \partial^{cl}\psi(\lambda_i + c_i(w))\right\}.$$

This shows that G is semismooth with respect to

$$\partial G(x) \overset{\text{def}}{=} \left\{ \begin{pmatrix} \nabla_{ww} L(x) & c'(w)^T & e'(w)^T \\ -D_g c'(w) & I - D_g & 0 \\ e'(w) & 0 & 0 \end{pmatrix}; \right.$$
$$\left. D_g = \text{diag}(g_i),\ g_i \in \partial^{cl}\psi(\lambda_i + c_i(w)) \right\}.$$

Under the regularity condition

$$\|M^{-1}\| \le C \quad \forall M \in \partial G(x) \quad \forall x, \quad \|x - \bar{x}\| < \delta,$$

where $\bar{x} = (\bar{w}, \bar{\lambda}, \bar{\mu})$ is a KKT triple, Theorem 2.12 is applicable and yields the q-superlinear convergence of the semismooth Newton method.

Remark 2.4 The compact-valuedness and the upper semicontinuity of Clarke's generalized differential [34] even allows to reduce the regularity condition to

$$\|M^{-1}\| \le C \quad \forall M \in \partial G(\bar{x}).$$

Remark 2.5 We also can view G as a piecewise smooth equation and apply Algorithm 2.5. In fact, it can be shown that Clarke's generalized Jacobian is the convex hull of the Jacobians of all essentially active pieces [123, 134]. We are not going into details here.

2.4.5.2 Discussion

So far, we have looked at semismooth Newton methods from an abstract point of view. The main point, however, is to prove semismoothness for concrete instances of nonsmooth operators. In particular, we aim at reformulating KKT systems arising in PDE-constrained optimization in the same way as we did this in finite dimensions in the above section. We will investigate this in detail in Sect. 2.5.

It should be mentioned that the class of semismooth Newton method includes as a special case the *primal dual active set strategy*, see [13, 69].

2.5 Semismooth Newton Methods in Function Spaces

In the finite dimensional setting we have shown that variational inequalities and complementarity conditions can be reformulated as nonsmooth equations. We also described how generalized Newton methods can be developed that solve these nonsmooth equations.

In Sect. 2.4.5 we introduced the concept of semismoothness for nonsmooth operators and developed superlinearly convergent generalized Newton methods for semismooth operator equations. We now will show that, similar to the finite dimensional case, it is possible to reformulate variational inequalities and complementarity conditions in function space.

2.5.1 Pointwise Bound Constraints in L^2

Let $\Omega \subset \mathbb{R}^n$ be measurable with measure $0 < |\Omega| < \infty$. If boundary spaces are considered, Ω can also be a measurable surface, e.g., the boundary of an open Lipschitz domain, on which L^p-spaces can be defined.

We consider the problem

$$\min_{u\in L^2(\Omega)} \quad f(u) \quad a \le u \le b \quad \text{a.e. on } \Omega$$

with $f : L^2(\Omega) \to \mathbb{R}$ twice continuously F-differentiable. We can admit unilateral constraints ($a \le u$ or $u \le b$) just as well. To avoid distinguishing cases, we will focus on the bilateral case $a, b \in L^\infty(\Omega)$, $b - a \ge \nu > 0$ on Ω. We also could consider problems in $L^p(\Omega)$, $p \neq 2$. However, for the sake of compact presentation, we focus on the case $p = 2$, which is the most important situation.

It is convenient to transform the bounds to constant bounds, e.g., via

$$u \mapsto \frac{u-a}{b-a}.$$

Hence, we will consider the problem

$$\min_{u\in L^2(\Omega)} \quad f(u), \quad \beta_l \le u \le \beta_r \quad \text{a.e. on } \Omega \tag{2.21}$$

with constants $\beta_l < \beta_r$. Let $U = L^2(\Omega)$ and $S = \{u \in L^2(\Omega) : \beta_l \le u \le \beta_r\}$. We choose the standard dual pairing $\langle \cdot, \cdot \rangle_{U^*,U} = (\cdot, \cdot)_{L^2(\Omega)}$ and then have $U^* = U = L^2(\Omega)$. The optimality conditions are

$$u \in S, \quad (\nabla f(u), v - u)_{L^2(\Omega)} \ge 0 \quad \forall v \in S.$$

We now use the projection P_S onto S, which is given by

$$P_S(v)(x) = P_{[\beta_l,\beta_r]}(v(x)), \quad x \in \Omega.$$

Then the optimality conditions can be written as

$$\Phi(u) := u - P_S(u - \theta\nabla f(u)) = 0, \tag{2.22}$$

where $\theta > 0$ is arbitrary, but fixed. Note that, since P_S coincides with the pointwise projection onto $[\beta_l, \beta_r]$, we have

$$\Phi(u)(x) = u(x) - P_{[\beta_l,\beta_r]}(u(x) - \theta\nabla f(u)(x)).$$

Our aim now is to define a generalized differential $\partial\Phi$ for Φ in such a way that Φ is semismooth.

By the chain rule and sum rule that we developed, this reduces to the question how a suitable differential for the superposition $P_{[\beta_l,\beta_r]}(v(\cdot))$ can be defined.

2.5.2 Semismoothness of Superposition Operators

More generally than the superposition operator in the previous subsection, we look at the superposition operator

$$\Psi : L^p(\Omega)^m \to L^q(\Omega), \quad \Psi(w)(x) = \psi(w_1(x), \ldots, w_m(x))$$

with $1 \le q \le p \le \infty$.

Here, $\psi : \mathbb{R}^m \to \mathbb{R}$ is assumed to be Lipschitz continuous. Since we aim at semismoothness of Ψ, it is more than natural to assume semismoothness of ψ. As differential we choose Clarke's generalized differential $\partial^{cl}\psi$. Now it is reasonable to define $\partial\Psi$ in such a way that, for all $M \in \partial\Psi(w+d)$, the remainder

$$|(\Psi(u+d) - \Psi - Md)(x)| = |\psi(w(x)+d(x)) - \psi(w(x)) - (Md)(x)|$$

becomes pointwise small if $|d(x)|$ is small. By semismoothness of ψ, this, again, holds true if $(Md)(x) \in \partial^{cl}\psi(w(x)+d(x))d(x)$ is satisfied.

Hence, we define:

Definition 2.2 Let $\psi : \mathbb{R}^m \to \mathbb{R}$ be Lipschitz continuous and ($\partial^{cl}\psi$-) semismooth. For $1 \le q \le p \le \infty$, consider

$$\Psi : L^p(\Omega)^m \to L^q(\Omega), \quad \Psi(w)(x) = \psi(w_1(x), \ldots, w_m(x)).$$

We define the differential

$$\partial\Psi : L^p(\Omega)^m \rightrightarrows \mathcal{L}(L^p(\Omega)^m, L^q(\Omega)),$$
$$\partial\Psi(w) = \left\{M : Mv = g^T v,\ g \in L^\infty(\Omega)^m,\ g(x) \in \partial^{cl}\psi(w(x)) \text{ for a.a. } x \in \Omega\right\}.$$

The operator Φ in (2.22) is naturally defined as a mapping from $L^2(\Omega)$ to $L^2(\Omega)$. Therefore, since ∇f maps to $L^2(\Omega)$, we would like the superposition $v \mapsto P_{[\beta_l,\beta_r]}(v(\cdot))$ to be semismooth from $L^2(\Omega)$ to $L^2(\Omega)$. But this is not true, as the following Lemma shows in great generality.

Lemma 2.7 *Let $\psi : \mathbb{R} \to \mathbb{R}$ be any Lipschitz continuous function that is* not *affine linear. Furthermore, let $\Omega \subset \mathbb{R}^n$ be nonempty, open and bounded. Then, for all $q \in [1, \infty)$, the operator*

$$\Psi : L^q(\Omega) \ni u \mapsto \psi(u(\cdot)) \in L^q(\Omega)$$

is not *$\partial\Psi$-semismooth.*

Proof Fix $b \in \mathbb{R}$ and choose $g_b \in \partial\psi(b)$ Since ψ is not affine linear, there exists $a \in \mathbb{R}$ with

$$\psi(a) \neq \psi(b) + g_b(a-b).$$

Hence,

$$\rho := |\psi(b) - \psi(a) - g_b(b-a)| > 0.$$

Let $x_0 \in \Omega$ and $U_\varepsilon = (x_0 - h_\varepsilon, x_0 + h_\varepsilon)^n$, $h_\varepsilon = \varepsilon^{1/n}/2$. Define

$$u(x) = a, \quad x \in \Omega, \qquad d_\varepsilon(x) = \begin{cases} b-a & x \in U_\varepsilon, \\ 0 & x \notin U_\varepsilon. \end{cases}$$

Then

$$\|d_\varepsilon\|_{L^q} = \left(\int_\Omega |d_\varepsilon(x)|^q\, dx\right)^{1/q} = \left(\int_{U_\varepsilon} |b-a|^q dx\right)^{1/q} = \varepsilon^{1/q}|b-a|.$$

Choose some $g_a \in \partial\psi(a)$ and define

$$g_\varepsilon(x) = \begin{cases} g_b & x \in U_\varepsilon, \\ g_a & x \notin U_\varepsilon. \end{cases}$$

Then $M : L^q(\Omega) \ni v \mapsto g_\varepsilon \cdot v \in L^q(\Omega)$ is an element of $\partial\Psi(u+d_\varepsilon)$. Now, for all $x \in \Omega$,

$$\begin{aligned} &|\psi(u(x)+d_\varepsilon(x)) - \psi(u(x)) - g_\varepsilon(x)d_\varepsilon(x)| \\ &= \begin{cases} |\psi(b) - \psi(a) - g_b(b-a)| = \rho > 0, & x \in U_\varepsilon, \\ |\psi(a) - \psi(a) - g_a(a-a)| = 0, & x \notin U_\varepsilon. \end{cases} \end{aligned}$$

Therefore,

$$\begin{aligned} &\|\Psi(u+d_\varepsilon) - \Psi(u) - Md_\varepsilon\|_{L^q} \\ &= \left(\int_\Omega |\psi(u(x)+d_\varepsilon(x)) - \psi(u(x)) - g_\varepsilon(x)d_\varepsilon(x)|^q\, dx\right)^{1/q} \\ &= \left(\int_{U_\varepsilon} \rho^q\, dx\right)^{1/q} = \varepsilon^{1/q}\rho = \frac{\rho}{|b-a|}\|d_\varepsilon\|_{L^q}. \end{aligned}$$

Note that the trouble is not caused by the nonsmoothness of ψ, but by the nonlinearity of ψ.

Fortunately, Ulbrich [134, 136] proved a result that helps us. See also [69]. To formulate the result in its full generality, we extend our definition of generalized differentials to superposition operators of the form $\psi(G(\cdot))$, where G is a continuously F-differentiable operator.

Definition 2.3 Let $\psi : \mathbb{R}^m \to \mathbb{R}$ be Lipschitz continuous and ($\partial^{cl}\psi$-) semismooth. Furthermore, let $1 \le q \le p \le \infty$ be given, consider

$$\Psi_G : Y \to L^q(\Omega), \quad \Psi_G(y)(x) = \psi(G(y)(x)),$$

where $G : Y \to L^p(\Omega)^m$ is continuously F-differentiable and Y is a Banach space. We define the differential

$$\begin{aligned} &\partial\Psi_G : Y \rightrightarrows \mathcal{L}(Y, L^q(\Omega)), \\ &\partial\Psi_G(y) = \big\{M : Mv = g^T(G'(y)v),\ g \in L^\infty(\Omega)^m, \\ &\qquad\qquad g(x) \in \partial^{cl}\psi(G(y)(x)) \text{ for a.a. } x \in \Omega\big\}. \end{aligned} \tag{2.23}$$

Note that this is just the differential that we would obtain by the construction in part (c) of Theorem 2.10.

Now we can state the following semismoothness result.

Theorem 2.13 *Let $\Omega \subset \mathbb{R}^n$ be measurable with $0 < |\Omega| < \infty$. Furthermore, let $\psi : \mathbb{R}^m \to \mathbb{R}$ be Lipschitz continuous and semismooth. Let Y be a Banach space, $1 \le q < p \le \infty$, and assume that the operator $G : Y \to L^q(\Omega)^m$ is continuously F-differentiable and that G maps Y locally Lipschitz continuously to $L^p(\Omega)$. Then, the operator*

$$\Psi_G : Y \to L^q(\Omega), \quad \Psi_G(y)(x) = \psi(G(y)(x)),$$

is $\partial\Psi_G$-semismooth, where $\partial\Psi_G$ is defined in (2.23).

Addition: Under additional assumptions, the operator Ψ_G is $\partial\Psi_G$-semismooth of order $\alpha > 0$ with α appropriate.

A proof can be found in [134, 136].

2.5.3 Pointwise Bound Constraints in L^2 Revisited

We return to the operator Φ defined in (2.22). To be able to prove the semismoothness of $\Phi : L^2(\Omega) \to L^2(\Omega)$ defined in (2.22), we thus need some kind of smoothing property of the mapping

$$u \mapsto u - \theta\nabla f(u).$$

Therefore, we assume that ∇f has the following structure:

$$\begin{aligned}
&\text{There exist } \alpha > 0 \text{ and } p > 2 \text{ such that}\\
&\nabla f(u) = \alpha u + H(u),\\
&H : L^2(\Omega) \to L^2(\Omega) \text{ continuously F-differentiable,}\\
&H : L^2(\Omega) \to L^p(\Omega) \text{ locally Lipschitz continuous.}
\end{aligned} \tag{2.24}$$

This structure is met by many optimal control problems, as illustrated in Sect. 2.5.4.

If we now choose $\theta = 1/\alpha$, then we have

$$\Phi(u) = u - P_{[\beta_l,\beta_r]}(u - (1/\alpha)(\alpha u + B(u))) = u - P_{[\beta_l,\beta_r]}(-(1/\alpha)B(u)).$$

Therefore, we have achieved that the operator inside the projection satisfies the requirements of Theorem 2.13. We obtain:

Theorem 2.14 *Consider the problem* (2.21) *with $\beta_l < \beta_r$ and let the continuously F-differentiable function $f : L^2(\Omega) \to \mathbb{R}$ satisfy condition* (2.24). *Then, for*

$\theta = 1/\alpha$, *the operator* Φ *in the reformulated optimality conditions* (2.22) *is* $\partial\Phi$-*semismooth with*

$$\partial\Phi : L^2(\Omega) \rightrightarrows \mathcal{L}(L^2(\Omega), L^2(\Omega)),$$

$$\partial\Phi(u) = \Big\{M\ ;\ M = I + \frac{g}{\alpha}\cdot H'(u),\ g \in L^\infty(\Omega),$$
$$g(x) \in \partial^{cl} P_{[\beta_l,\beta_r]}(-(1/\alpha)H(u)(x))\ \textit{for a.a.}\ x \in \Omega\Big\}.$$

Here,

$$\partial^{cl} P_{[\beta_l,\beta_r]}(t) = \begin{cases} \{0\} & t < \beta_l \text{ or } t > \beta_r, \\ \{1\} & \beta_l < t < \beta_r, \\ [0,1] & t = \beta_l \text{ or } t = \beta_r. \end{cases}$$

Proof Setting $q = 2$, $\psi = P_{[\beta_l,\beta_r]}$ and $G = -(1/\alpha)H$, we can apply Theorem 2.13 and obtain that the operator $\Psi_G : L^2(\Omega) \to L^2(\Omega)$ is $\partial\Psi_G$-semismooth. Therefore, $\Phi = I - \Psi_G$ is $(I - \partial\Psi_G)$-semismooth by Theorem 2.10. Since $\partial\Phi = I - \partial\Psi_G$, the proof is complete.

For the applicability of the semismooth Newton method (Algorithm 2.11) we need, in addition, the following regularity condition:

$$\|M^{-1}\|_{L^2(\Omega)\to L^2(\Omega)} \le C \quad \forall M \in \partial\Phi(u) \quad \forall u \in L^2(\Omega), \quad \|u - \bar{u}\|_{L^2(\Omega)} < \delta.$$

Sufficient conditions for this regularity assumption in the flavor of second order sufficient optimality conditions can be found in [134, 135].

2.5.4 Application to Optimal Control

Consider the following elliptic optimal control problem:

$$\min_{y\in H_0^1(\Omega), u\in L^2(\Omega)} \quad J(y,u) \overset{\text{def}}{=} \frac{1}{2}\|y - y_d\|^2_{L^2(\Omega)} + \frac{\alpha}{2}\|u\|^2_{L^2(\Omega)}$$
$$\text{s.t.} \quad Ay = r + Bu, \quad \beta_l \le u \le \beta_r. \tag{2.25}$$

Here, $y \in H_0^1(\Omega)$ is the state, which is defined on the open bounded domain $\Omega \subset \mathbb{R}^n$, and $u \in L^2(\Omega_c)$ is the control, which is defined on the open bounded domain $\Omega_c \subset \mathbb{R}^m$. Furthermore, $A : H_0^1(\Omega) \to H^{-1}(\Omega) = H_0^1(\Omega)^*$ is a (for simplicity) linear elliptic partial differential operator, e.g., $A = -\Delta$, and $r \in H^{-1}(\Omega)$ is given.

The control operator $B : L^{p'}(\Omega_c) \to H^{-1}(\Omega)$ is continuous and linear, with $p' \in [1,2)$ (the reason why we do not choose $p' = 2$ here will become clear later; note however, that $L^2(\Omega_c)$ is continuously embedded in $L^{p'}(\Omega_c)$). For instance, distributed control on the whole domain Ω would correspond to the choice $\Omega_c = \Omega$

and $B : u \in L^{p'}(\Omega) \mapsto u \in H^{-1}(\Omega)$, where p' is chosen in such a way that $H_0^1(\Omega)$ is continuously embedded in the dual space $L^p(\Omega)$, $p = p'/(p'-1)$, of $L^{p'}(\Omega)$.

The control is subject to pointwise bounds $\beta_l < \beta_r$. The objective is to drive the state as close to $y_d \in L^2(\Omega)$ as possible. The second part penalizes excessive control costs; the parameter $\alpha > 0$ is typically small.

We eliminate the state y via the state equation, i.e., $y = y(u) = A^{-1}(r + Bu)$, and obtain the reduced problem

$$\min_{u \in L^2(\Omega)} \hat{J}(u) \stackrel{\text{def}}{=} J(y(u), u) \stackrel{\text{def}}{=} \frac{1}{2}\|y(u) - y_d\|^2_{L^2(\Omega)} + \frac{\alpha}{2}\|u\|^2_{L^2(\Omega)}$$
$$\text{s.t.} \quad \beta_l \le u \le \beta_r.$$

This problem is of the form (2.21).

For the gradient we obtain

$$\begin{aligned}(\nabla \hat{J}(u), d)_{L^2(\Omega)} &= (y(u) - y_d, y'(u)d)_{L^2(\Omega)} + \alpha(u, d)_{L^2(\Omega_c)} \\ &= (y'(u)^*(y(u) - y_d) + \alpha u, d)_{L^2(\Omega_c)}.\end{aligned}$$

Therefore,

$$\begin{aligned}\nabla \hat{J}(u) &= y'(u)^*(y(u) - y_d) + \alpha u = B^*(A^{-1})^*(A^{-1}(r + Bu) - y_d) + \alpha u \\ &= \alpha u + B^*(A^{-1})^*(A^{-1}(r + Bu) - y_d) \stackrel{\text{def}}{=} \alpha u + H(u).\end{aligned}$$

Since $B \in \mathcal{L}(L^{p'}(\Omega_c), H^{-1}(\Omega))$, we have $B^* \in \mathcal{L}(H_0^1(\Omega), L^p(\Omega_c))$ with $p = p'/(p'-1) > 2$. Hence, the affine linear operator

$$H(u) = B^*(A^{-1})^*(A^{-1}(r + Bu) - y_d)$$

is a continuous affine linear mapping $L^2(\Omega_c) \to L^p(\Omega)$.

Therefore, we can apply Theorem 2.13 to rewrite the optimality conditions as a semismooth operator equation

$$\Phi(u) \stackrel{\text{def}}{=} u - P_{[\beta_l, \beta_r]}(-(1/\alpha)H(u)) = 0.$$

The Newton system reads

$$\left(I + \frac{1}{\alpha} g^k \cdot H'(u^k)\right) s^k = -\Phi(u^k), \tag{2.26}$$

where $g \cdot H'(u)$ stands for $v \mapsto g \cdot (H'(u)v)$ and $g^k \in L^\infty(\Omega_c)$ is chosen such that

$$g^k(x) \begin{cases} = 0 & -(1/\alpha)H(u^k)(x) \notin [\beta_l, \beta_r], \\ = 1 & -(1/\alpha)H(u^k)(x) \in (\beta_l, \beta_r), \\ \in [0, 1] & -(1/\alpha)H(u^k)(x) \in \{\beta_l, \beta_r\}. \end{cases}$$

The linear operator on the left has the form

$$M_k \stackrel{\text{def}}{=} I + \frac{1}{\alpha} g^k \cdot H'(u^k) = I + \frac{1}{\alpha} g^k \cdot B^*(A^{-1})^* A^{-1} B.$$

For solving (2.26), it can be advantageous to note that s^k solves (2.26) if and only if $s^k = d_u^k$ and $(d_y^k, d_u^k, d_\mu^k)^T$ solves

$$\begin{pmatrix} I & 0 & A^* \\ 0 & I & -\frac{1}{\alpha} g^k \cdot B^* \\ A & -B & 0 \end{pmatrix} \begin{pmatrix} d_y^k \\ d_u^k \\ d_\mu^k \end{pmatrix} = \begin{pmatrix} 0 \\ -\Phi(u^k) \\ 0 \end{pmatrix}. \tag{2.27}$$

As we will see later in Sect. 2.8.2, this is system is amenable to multigrid methods.

2.5.5 General Optimization Problems with Inequality Constraints in L^2

We now consider problems of the form

$$\min_{w \in W} f(w) \quad \text{s.t.} \quad e(w) = 0, \qquad c_j(w) \le 0 \quad \text{a.e. on } \Omega_j, \quad j = 1, \ldots, m.$$

Here W and Z are Banach spaces, $f : W \to \mathbb{R}$, $e : W \to Z$, and $c_j : W \to L^2(\Omega_j)$ are twice continuously F-differentiable. The sets $\Omega_j \subset \mathbb{R}^{n_j}$ are assumed to be measurable and bounded.

This, in particular, includes control-constrained optimal control problems with L^2-control u and state $y \in Y$:

$$\min_{y \in Y, u \in L^2(\Omega)} J(y,u) \quad \text{s.t.} \quad e(y,u) = 0, \quad a_i \le u_i \le b_i, \quad i = 1, \ldots, l,$$

with $y \in Y$ denoting the state, $u \in L^2(\Omega_1) \times \cdots \times L^2(\Omega_l)$ denoting the controls, and $a_i, b_i \in L^\infty(\Omega_i)$.

In this case, we have

$$w = (y,u), \qquad m = 2l, \qquad c_{2i-1}(y,u) = a_i - u_i,$$
$$c_{2i}(y,u) = u_i - b_i, \quad i = 1, \ldots, l.$$

To simplify the presentation, consider the case $m = 1$, i.e.,

$$\min_{w \in W} f(w) \quad \text{s.t.} \quad e(w) = 0, \qquad c(w) \le 0 \quad \text{a.e. on } \Omega. \tag{2.28}$$

The Lagrange function is given by

$$L : W \times L^2(\Omega) \times Z^* \to \mathbb{R},$$
$$L(w, \lambda, \mu) = f(w) + (\lambda, c(w))_{L^2(\Omega)} + \langle \mu, e(w) \rangle_{Z^*,Z}.$$

Assuming that a CQ holds at the solution $\bar{w} \in W$, the KKT conditions hold:

There exist $\bar{\lambda} \in L^2(\Omega)$ and $\bar{\mu} \in Z^*$ such that $(\bar{w}, \bar{\lambda}, \bar{\mu})$ satisfies

$$L_w(\bar{w}, \bar{\lambda}, \bar{\mu}) = 0, \tag{2.29}$$
$$e(\bar{w}) = 0, \tag{2.30}$$
$$c(\bar{w}) \le 0, \ \bar{\lambda} \ge 0, \quad (\bar{\lambda}, c(\bar{w}))_{L^2(\Omega)} = 0. \tag{2.31}$$

The last line can equivalently be written as $\mathrm{VI}(-c(\bar{w}), \mathcal{K})$ with $\mathcal{K} = \{u \in L^2(\Omega) : u \ge 0\}$ and this VI can again be rewritten using the projection onto $\mathcal{K}$:

$$\bar{\lambda} - P_{\mathcal{K}}(\bar{\lambda} + \theta c(\bar{w})) = 0$$

with fixed $\theta > 0$. Since $P_{\mathcal{K}}(u) = P_{[0,\infty)}(u(\cdot))$, we again have to deal with a superposition operator.

To make the whole KKT system a semismooth equation, we need to get a smoothing operator inside of the projection.

We need additional structure to achieve this. Since it is not very enlightening to define this structure in full generality without giving a motivation, we look at an example first.

2.5.6 Application to Elliptic Optimal Control Problems

2.5.6.1 Distributed Control

Very similar as in Sect. 2.5.4, we consider the following control-constrained elliptic optimal control problem

$$\min_{y \in H_0^1(\Omega), u \in L^2(\Omega)} \quad J(y, u) \overset{\text{def}}{=} \frac{1}{2}\|y - y_d\|^2_{L_2(\Omega)} + \frac{\alpha}{2}\|u\|^2_{L^2(\Omega)}$$
$$\text{s.t.} \quad Ay = r + Bu, \quad u \le b. \tag{2.32}$$

Here $\Omega \subset \mathbb{R}^n$ is an open bounded domain and $A : H_0^1(\Omega) \to H^{-1}(\Omega)$ is a second order linear elliptic operator, e.g., $A = -\Delta$. Furthermore, $b \in L^\infty(\Omega)$ is an upper bound on the control, $r \in H^{-1}(\Omega)$ is a source term, and $B \in \mathcal{L}(L^{p'}(\Omega_c), H^{-1}(\Omega))$, $p' \in [1, 2)$ is the control operator. For a more detailed explanation of the problem setting, see Sect. 2.5.4.

We convert this control problem into the form (2.28) by setting

$$w=(y,u),\qquad W=Y\times U,\qquad Y=H_0^1(\Omega),\qquad U=L^2(\Omega),$$
$$Z=H^{-1}(\Omega),\qquad e(y,u)=Ay-Bu-r,\qquad c(y,u)=u-b.$$

Note that e and c are continuous affine linear operators. Hence,

$$e_y(y,u)=A,\qquad e_u(y,u)=-B,\qquad c_y(y,u)=0,\qquad c_u(y,u)=I.$$

The Lagrange function is

$$L(y,u,\lambda,\mu)=J(y,u)+(\lambda,c(y,u)_{L^2(\Omega)})+\langle\mu,e(y,u)\rangle_{H_0^1(\Omega),H^{-1}(\Omega)}.$$

We write down the optimality conditions:

$$L_y(y,u,\lambda,\mu)=J_y(y,u)+c_y(y,u)^*\lambda+e_y(y,u)^*\mu=y-y_d+A^*\mu=0,$$
$$L_u(y,u,\lambda,\mu)=J_u(y,u)+c_u(y,u)^*\lambda+e_u(y,u)^*\mu=\alpha u+\lambda-B^*\mu=0,$$
$$\lambda\geq 0,\qquad c(y,u)=u-b\leq 0,\qquad (\lambda,c(y,u))_{L^2(\Omega)}=(\lambda,u-b)_{L^2(\Omega)}=0,$$
$$e(y,u)=Ay-Bu-r=0.$$

The second equation yields $\lambda=B^*\mu-\alpha u$ and inserting this, we arrive at

$$A^*\mu=-(y-y_d),\qquad\text{(adjoint equation)}$$
$$B^*\mu-\alpha u\geq 0,\quad u\leq b,\quad (B^*\mu-\alpha u,u-b)_{L^2(\Omega)}=0,$$
$$Ay=r+Bu.\qquad\text{(state equation)}$$

We can reformulate the complementarity condition by using the projection $P_{[0,\infty)}$ as follows:

$$b-u-P_{[0,\infty)}(b-u-\theta(B^*\mu-\alpha u))=0.$$

If we choose $\theta=1/\alpha$, this simplifies to

$$\Phi(u,\mu):=u-b+P_{[0,\infty)}(b-(1/\alpha)B^*\mu)=0.$$

Since $B^*\in\mathcal{L}(H_0^1(\Omega),L^p(\Omega))$ with $p=p'/(p'-1)>2$, we see that

$$(u,\mu)\in L^2(\Omega)\times H_0^1(\Omega)\mapsto b-(1/\alpha)B^*\mu\in L^p(\Omega)$$

is continuous and affine linear, and thus Φ is $\partial\Phi$-semismooth w.r.t.

$$\partial\Phi: L^2(\Omega)\times H_0^1(\Omega)\rightrightarrows\mathcal{L}(L^2(\Omega)\times H_0^1(\Omega),L^2(\Omega)),$$
$$\partial\Phi(u,\mu)=\big\{M;\ M=(I,-(g/\alpha)\cdot B^*),\ g\in L^\infty(\Omega),$$
$$g(x)\in\partial^{cl}P_{[0,\infty)}(b(x)-(1/\alpha)(B^*\mu)(x))\text{ for a.a. }x\in\Omega\big\}.$$

Here,

$$\partial^{cl} P_{[0,\infty)}(t) = \begin{cases} \{0\} & t < 0, \\ \{1\} & t > 0, \\ [0,1] & t = 0. \end{cases} \tag{2.33}$$

The semismooth Newton system looks as follows

$$\begin{pmatrix} I & 0 & A^* \\ 0 & I & -(g^k/\alpha) \cdot B^* \\ A & -B & 0 \end{pmatrix} \begin{pmatrix} s_y \\ s_u \\ s_\mu \end{pmatrix} = - \begin{pmatrix} y^k - y_d + A^* \mu^k \\ u^k - b + P_{[0,\infty)}(b - (1/\alpha) B^* \mu^k) \\ Ay^k - Bu^k - r \end{pmatrix}. \tag{2.34}$$

It is important to note that this equation has exactly the same linear operator on the left as the extended system in (2.27). In particular, the regularity condition for the Newton system (2.34) is closely connected to the regularity condition for (2.26).

2.5.6.2 Neumann Boundary Control

We now consider a similar problem as before, but with Neumann boundary control:

$$\begin{aligned} \min_{y \in H^1(\Omega), u \in L^2(\partial\Omega)} \quad & J(y,u) \stackrel{\text{def}}{=} \frac{1}{2} \|y - y_d\|^2_{L_2(\Omega)} + \frac{\alpha}{2} \|u\|^2_{L^2(\partial\Omega)} \\ \text{s.t.} \quad & -\Delta y + cy = r \quad \text{in } \Omega, \\ & \frac{\partial y}{\partial \nu} = u \quad \text{in } \partial\Omega, \\ & u \le b \quad \text{in } \partial\Omega. \end{aligned} \tag{2.35}$$

Here $\Omega \subset \mathbb{R}^n$ is an open bounded Lipschitz domain and $c \in L^\infty(\Omega)$, $c > 0$. Furthermore, $b \in L^\infty(\partial\Omega)$ is an upper bound on the control and $r \in H^1(\Omega)^*$ is a source term.

The weak formulation of the state equation including boundary condition is

$$\int_\Omega (\nabla y \cdot \nabla v + cyv)\, dx = \int_\Omega rv\, dx + \int_{\partial\Omega} uv\, dS(x) \quad \forall v \in H^1(\Omega),$$

which in operator form can be written as

$$Ay = r + Bu,$$

where

$$B \in \mathcal{L}(L^2(\partial\Omega), H^1(\Omega)^*), \quad \langle Bu, v \rangle_{H^1(\Omega)^*, H^1(\Omega)} = \int_{\partial\Omega} uv\, dS(x),$$

$$A \in \mathcal{L}(H^1(\Omega), H^1(\Omega)^*),$$

$$\langle Ay, v\rangle_{H^1(\Omega)^*,H^1(\Omega)} = \int_\Omega (\nabla y \cdot \nabla v + cyv)\,dx \quad \forall v \in H^1(\Omega).$$

The adjoint $B^* \in \mathcal{L}(H^1(\Omega), L^2(\partial\Omega))$ of B is given by $B^*v = v|_{\partial\Omega}$. In fact,

$$(B^*v, w)_{L^2(\partial\Omega)} = \langle Bw, v\rangle_{H^1(\Omega)^*,H^1(\Omega)} = \int_{\partial\Omega} wv\,dS(x) = (v, w)_{L^2(\partial\Omega)}.$$

This control problem assumes the form (2.28) by setting

$$w = (y, u), \qquad W = Y \times U, \qquad Y = H^1(\Omega), \qquad U = L^2(\partial\Omega),$$
$$Z = H^1(\Omega)^*, \qquad e(y, u) = Ay - Bu - r, \qquad c(y, u) = u - b.$$

The operators e and c are continuous and affine linear with derivatives

$$e_y(y, u) = A, \qquad e_u(y, u) = -B, \qquad c_y(y, u) = 0, \qquad c_u(y, u) = I.$$

The Lagrange function reads

$$L(y, u, \lambda, \mu) = J(y, u) + (\lambda, c(y, u))_{L^2(\partial\Omega)} + \langle \mu, e(y, u)\rangle_{H^1(\Omega),H^1(\Omega)^*}.$$

We write down the optimality conditions:

$$L_y(y, u, \lambda, \mu) = J_y(y, u) + c_y(y, u)^*\lambda + e_y(y, u)^*\mu = y - y_d + A^*\mu = 0,$$

$$L_u(y, u, \lambda, \mu) = J_u(y, u) + c_u(y, u)^*\lambda + e_u(y, u)^*\mu = \alpha u + \lambda - B^*\mu = 0,$$

$$\lambda \geq 0, \qquad c(y, u) = u - b \leq 0, \qquad (\lambda, c(y, u))_{L^2(\partial\Omega)} = (\lambda, u - b)_{L^2(\partial\Omega)} = 0,$$

$$e(y, u) = Ay - Bu - r = 0.$$

The second equation yields $\lambda = B^*\mu - \alpha u$ and using this to eliminate λ, we arrive at

$$A^*\mu = -(y - y_d), \qquad \text{(adjoint equation)}$$
$$B^*\mu - \alpha u \geq 0, \quad u \leq b, \quad (B^*\mu - \alpha u, u - b)_{L^2(\partial\Omega)} = 0, \qquad (2.36)$$
$$Ay = r + Bu. \qquad \text{(state equation)}$$

Inserting $Av = A^*v = -\Delta v + cv$, $B^*v = v|_{\partial\Omega}$, and the definition of B, we can express this system as a coupled system of elliptic partial differential equations:

$$-\Delta\mu + c\mu = -(y - y_d) \quad \text{in } \Omega,$$

$$\frac{\partial\mu}{\partial\nu} = 0 \quad \text{in } \partial\Omega,$$

$$\mu|_{\partial\Omega} - \alpha u \geq 0, \quad u \leq b, \quad (\mu|_{\partial\Omega} - \alpha u)(u - b) = 0 \quad \text{in } \partial\Omega,$$

$$-\Delta y + cy = r \quad \text{in } \Omega,$$
$$\frac{\partial y}{\partial \nu} = u \quad \text{in } \partial\Omega.$$

Here, we have written the complementarity condition pointwise. Note that in the adjoint equation we have homogeneous Neumann boundary conditions since a Neumann boundary condition $\frac{\partial y}{\partial \nu} = h$ would result in the term Bh on the right hand side of the differential equation. Since no such term is present in the adjoint equation, we must have $h = 0$.

We return to the more compact notation of (2.36) and reformulate the complementarity condition by using the projection $P_{[0,\infty)}$ as follows:

$$b - u - P_{[0,\infty)}(b - u - \theta(B^*\mu - \alpha u)) = 0 \quad \text{in } L(\partial\Omega).$$

If we choose $\theta = 1/\alpha$, this simplifies to

$$\Phi(u, \mu) := u - b + P_{[0,\infty)}(b - (1/\alpha)B^*\mu) = 0 \quad \text{in } L(\partial\Omega).$$

From $B^*v = v|_{\partial\Omega}$ we see that B^* is a bounded linear operator from $H^1(\Omega)$ not only to $L^2(\partial\Omega)$, but even to $H^{1/2}(\partial\Omega)$. By the Sobolev embedding theorem, we can find $p > 2$ with $H^{1/2}(\partial\Omega) \hookrightarrow L^p(\partial\Omega)$. We then have $B^* \in \mathcal{L}(H^1(\Omega), L^p(\partial\Omega))$ with $p > 2$. Hence,

$$(u, \mu) \in L^2(\partial\Omega) \times H^1(\Omega) \mapsto b - (1/\alpha)B^*\mu \in L^p(\partial\Omega)$$

is continuous and affine linear, and thus Φ is $\partial\Phi$-semismooth w.r.t.

$$\partial\Phi : L^2(\partial\Omega) \times H^1(\Omega) \rightrightarrows \mathcal{L}(L^2(\partial\Omega) \times H^1(\Omega), L^2(\partial\Omega)),$$
$$\partial\Phi(u, \mu) = \big\{M;\ M = (I, -(g/\alpha)\cdot B^*),\ g \in L^\infty(\partial\Omega),$$
$$g(x) \in \partial^{cl} P_{[0,\infty)}(b(x) - (1/\alpha)(B^*\mu)(x)) \text{ for a.a. } x \in \partial\Omega\big\}.$$

Here, $\partial^{cl} P_{[0,\infty)}(t)$ is as in (2.33). The semismooth Newton system then is

$$\begin{pmatrix} I & 0 & A^* \\ 0 & I & -(g^k/\alpha)\cdot B^* \\ A & -B & 0 \end{pmatrix} \begin{pmatrix} s_y \\ s_u \\ s_\mu \end{pmatrix} = -\begin{pmatrix} y^k - y_d + A^*\mu^k \\ u^k - b + P_{[0,\infty)}(b - (1/\alpha)B^*\mu^k) \\ Ay^k - Bu^k - r \end{pmatrix}. \tag{2.37}$$

2.5.7 Optimal Control of the Incompressible Navier-Stokes Equations

We now discuss how an optimal control problem governed by the 2d incompressible instationary Navier-Stokes equations can be solved by a semismooth Newton method. We use exactly the notation of Sect. 1.8. In particular, $\Omega \subset \mathbb{R}^2$ is the open bounded flow domain and $I = [0, T]$ is the time horizon. By V we denote the closure of $\{y \in C_0^\infty(\Omega)^2 : \nabla \cdot y = 0\}$ in $H_0^1(\Omega)^2$ and by H its closure in $L^2(\Omega)^2$. Given the resulting Gelfand triple $V \hookrightarrow H \hookrightarrow V^*$ we can write the state equation of the flow control problem as follows: The velocity field $y \in W(I)$ satisfies

$$\begin{aligned} & y_t - \nu \Delta y + (y \cdot \nabla) y = Bu \quad \text{in } L^2(I; V^*), \\ & y|_{t=0} = y_0 \quad \text{in } H. \end{aligned} \tag{2.38}$$

Here, $B \in \mathcal{L}(U, L^2(I; V^*))$ is the control operator and U is a Hilbert space of controls. To be more concrete, we will consider time-dependent control on the right hand side on a subdomain Ω_c of the flow domain Ω. We achieve this by choosing $B \in \mathcal{L}(L^2(I \times \Omega_c)^2, L^2(I; V^*))$,

$$\langle Bu, w \rangle_{L^2(I;V^*), L^2(I;V)} = (u, w)_{L^2(I \times \Omega_c)^2}.$$

This is well defined, since $L^2(I; L^2(\Omega)) = L^2(I \times \Omega)$.

We consider an objective function of the form

$$J(y, u) = \frac{1}{2} \int_0^T \|Ny - q_d\|_{L^2(\Omega_d)^2}^2 \, dt + \frac{\alpha}{2} \|u\|_{L^2(I \times \Omega_c)^2}^2.$$

Here, $N \in \mathcal{L}(V, L^2(\Omega_d)^2)$ is an operator that maps the velocity field to the corresponding observation on the set $\Omega_d \subset \Omega$. For instance, $N = I$ or $N = \text{curl}$ are possible choices. On the control we will pose a pointwise constraint

$$u \in C \quad \text{on } I \times \Omega_c,$$

where $C \subset \mathbb{R}^2$ is a closed convex set such that the projection P_C onto C is semismooth.

We thus consider the problem

$$\min_{y,u} J(y, u) \quad \text{s.t.} \quad (y, u) \text{ satisfy (2.38)} \quad \text{and} \quad u \in C \quad \text{on } I \times \Omega_c.$$

The analysis of this problem was discussed in Sect. 1.8. In particular, for any $u \in U$ the state equation possesses a unique solution $y(u) \in W(I)$ and the operator $u \mapsto y(u)$ is infinitely F-differentiable. Since the objective function $J(y, u)$ is continuous and quadratic, it is infinitely F-differentiable. Therefore, the reduced objective function $\hat{J}(u) = J(y(u), u)$ is infinitely F-differentiable. The gradient of $\hat{J}(u)$ can be represented using the adjoint state in the form

$$\nabla \hat{J}(u) = \alpha u - B^* p_1,$$

where $p_1 = p_1(u) \in L^2(I; V)$ is the adjoint state corresponding to $(y, u) = (y(u), u)$ given by the weak solution of the adjoint equation

$$\begin{aligned} & -(p_1)_t - (y \cdot \nabla) p_1 + (\nabla y)^T p_1 - \nu \Delta p_1 = -N^*(Ny - q_d) \quad \text{in } I \times \Omega, \\ & p_1|_{t=T} = 0 \quad \text{in } \Omega. \end{aligned}$$

Due to the structure of B we see that

$$\langle Bu, w\rangle_{L^2(I;V^*),L^2(I;V)} = (u, w)_{L^2(I\times\Omega_c)^2} = (u, B^*w)_{L^2(I\times\Omega_c)^2}.$$

Therefore, $B^*w = w|_{I\times\Omega_c}$.

Since $N \in \mathcal{L}(V, L^2(\Omega_d)^2)$, we have $N^* \in \mathcal{L}(L^2(\Omega_d)^2, V^*)$ and thus the right hand side $-N^*(Ny(u) - q_d)$ maps $u \in U = L^2(I \times \Omega_c)^2 = L^2(I, L^2(\Omega_c)^2)$ infinitely F-differentiable to $L^2(I; V^*)$. From the imbedding $L^2(I; V^*) \hookrightarrow W(I)^* \cap L^{4/3}(I; V^*)$ and Theorem 1.58 we conclude that the operator

$$u \in U \mapsto p_1(u) \in W^{4/3}(I)$$

is well-defined and Lipschitz continuous on bounded sets.

Furthermore, it can be shown, see [134, 137], that

$$W^{4/3}(I) \hookrightarrow L^q(I \times \Omega)^2, \quad \forall 1 \le q < \frac{7}{2}.$$

Thus fixing $q \in (2, 7/2)$ we obtain that

$$u \in U \mapsto p_1(u) \in L^q(I \times \Omega)$$

is well-defined and Lipschitz continuous on bounded sets.

We collect what we have found so far

- $\hat{J} : U \to \mathbb{R}$ is infinitely F-differentiable.
- The reduced gradient has the following structure:

$$\nabla\hat{J}(u) = \alpha u + H(u)$$

with

$$H(u) = -B^*p_1(u) = -p_1(u)|_{I\times\Omega_c},$$

where $p_1(u) \in L^2(I; V)$ is the adjoint state.
- The operator $u \in U \mapsto p_1(u) \in L^2(I; V)$ is infinitely F-differentiable. Furthermore, the operator

$$u \in U \mapsto p_1(u) \in W^{4/3}(I) \hookrightarrow L^q(I \times \Omega)$$

is Lipschitz continuous on bounded sets for $q \in (2, 7/2)$. From this, it follows that $H : U \to U$ is infinitely F-differentiable and that the operator

$$u \in U \mapsto H(u) \in L^q(I \times \Omega_c)$$

is Lipschitz continuous on bounded sets.

We can write the first order optimality conditions in the form

$$u - P_C(u - \theta\nabla\hat{J}(u)) = 0$$

with $\theta > 0$ fixed. Choosing $\theta = 1/\alpha$ and inserting the adjoint representation of $\nabla \hat{J}(u)$, we obtain

$$u - P_C(-(1/\alpha)H(u)) = 0. \tag{2.39}$$

We made the assumption that P_C is semismooth. Due to the properties of the operator H it now follows from Theorem 2.13 that the operator in equation (2.39) is semismooth from U to U. Hence, a semismooth Newton's method can be applied to this optimal control problem. For further details, we refer to [134, 137].

2.6 Sequential Quadratic Programming

2.6.1 Lagrange-Newton Methods for Equality Constrained Problems

We consider

$$\min_{w \in W} f(w) \quad \text{s.t.} \quad e(w) = 0 \tag{2.40}$$

with $f : W \to \mathbb{R}$ and $e : W \to Z$ twice continuously F-differentiable.

If $\bar{w}$ is a local solution and a CQ holds (e.g., $e'(\bar{w})$ is surjective), then the KKT conditions hold:

There exists a Lagrange multiplier $\bar{\mu} \in Z^*$ such that $(\bar{w}, \bar{\mu})$ satisfies

$$\begin{aligned} L_w(\bar{w}, \bar{\mu}) &= f'(\bar{w}) + e'(\bar{w})^* \bar{\mu} = 0, \\ L_\mu(\bar{w}, \bar{\mu}) &= e(\bar{w}) = 0. \end{aligned}$$

Setting

$$x = (w, \mu), \qquad G(w, \mu) = \begin{pmatrix} L_w(w, \mu) \\ e(w) \end{pmatrix},$$

the KKT conditions form a nonlinear equation

$$G(x) = 0.$$

To this equation we can apply Newton's method:

$$G'(x^k)s^k = -G(x^k).$$

Written in detail,

$$\begin{pmatrix} L_{ww}(w^k, \mu^k) & e'(w^k)^* \\ e'(w^k) & 0 \end{pmatrix} \begin{pmatrix} s_w^k \\ s_\mu^k \end{pmatrix} = -\begin{pmatrix} L_w(w^k, \mu^k) \\ e(w^k) \end{pmatrix}. \tag{2.41}$$

The resulting method is called *Lagrange-Newton method*. We need a regularity condition:

$$\begin{pmatrix} L_{ww}(\bar{w}, \bar{\mu}) & e'(\bar{w})^* \\ e'(\bar{w}) & 0 \end{pmatrix} \quad \text{is boundedly invertible.} \tag{2.42}$$

Theorem 2.15 *Let f and e be twice continuously F-differentiable. Let $(\bar{w}, \bar{\mu})$ be a KKT pair of* (2.40) *at which the regularity condition* (2.42) *holds. Then there exists $\delta > 0$ such that, for all $(w^0, \mu^0) \in W \times Z^*$ with $\|(w^0, \mu^0) - (\bar{w}, \bar{\mu})\|_{W \times Z^*} < \delta$, the Lagrange-Newton iteration converges q-superlinearly to $(\bar{w}, \bar{\mu})$.*

If the second derivatives of f and e are locally Lipschitz continuous, then the rate of convergence is q-quadratic.

Proof We just have to apply the convergence theory of Newton's method.

If the second derivatives of f and e are locally Lipschitz continuous, then G' is locally Lipschitz continuous, and thus we have q-quadratic convergence.

So far, it is not clear what the connection is between the Lagrange-Newton method and sequential quadratic programming.

However, the connection is very close. Consider the following quadratic program:

SQP subproblem:

$$\begin{aligned} &\min_{d \in W} \langle f'(w^k), d \rangle_{W^*,W} + \frac{1}{2} \langle L_{ww}(w^k, \mu^k) d, d \rangle_{W^*,W} \\ &\text{s.t.} \quad e(w^k) + e'(w^k) d = 0. \end{aligned} \tag{2.43}$$

The constraint is linear with derivative $e'(w^k)$. As we will show below, $e'(w^k)$ is surjective for w^k close to $\bar{w}$ if $e'(\bar{w})$ is surjective.

Therefore, at a solution d^k of (2.43), the KKT conditions hold:

There exists $\mu_{qp}^k \in Z^*$ such that (d^k, μ_{qp}^k) solves

$$\begin{aligned} &f'(w^k) + L_{ww}(w^k, \mu^k) d^k + e'(w^k)^* \mu_{qp}^k = 0 \\ &e(w^k) + e'(w^k) d^k = 0. \end{aligned} \tag{2.44}$$

It is now easily seen that (d^k, μ_{qp}^k) solves (2.44) if and only if $(s_w^k, s_\mu^k) = (d^k, \mu_{qp}^k - \mu^k)$ solves (2.41).

Hence, locally, the Lagrange-Newton method is equivalent to the following method:

Algorithm 2.16 (SQP method for equality constrained problems)

0. Choose (w^0, μ^0) (sufficiently close to $(\bar{w}, \bar{\mu})$).

For $k = 0, 1, 2, \ldots$:

1. If (w^k, μ^k) is a KKT pair of (2.40), STOP.
2. Compute the KKT pair (d^k, μ^{k+1}) of

$$\begin{aligned} &\min_{d \in W} \langle f'(w^k), d \rangle_{W^*,W} + \frac{1}{2} \langle L_{ww}(w^k, \mu^k) d, d \rangle_{W^*,W} \\ &\text{s.t.} \quad e(w^k) + e'(w^k) d = 0, \end{aligned}$$

that is closest to $(0, \mu^k)$.

3. Set $w^{k+1} = w^k + d^k$.

For solving the SQP subproblems in step 2, it is important to know if for w^k close to $\bar{w}$, the operator $e'(w^k)$ is indeed surjective and if there exists a unique solution to the QP.

Lemma 2.8 *Let W be a Hilbert space and Z be a Banach space. Furthermore, let $e : W \to Z$ be continuously F-differentiable and let $e'(\bar{w})$ be surjective. Then $e'(w)$ is surjective for all w close to $\bar{w}$.*

Proof We set $B = e'(\bar{w})$, and $B(w) = e'(w)$, and do the splitting $W = W_0 \perp W_1$ with $W_0 = \text{Kern}(B)$. We then see that $B|_{W_1} \in \mathcal{L}(W_1, Z)$ is bijective and thus continuously invertible (open mapping theorem). Now, by continuity, for $w \to \bar{w}$ we have $B(w) \to B$ in $\mathcal{L}(W, Z)$ and thus also $B(w)|_{W_1} \to B|_{W_1}$ in $\mathcal{L}(W_1, Z)$. Therefore, by the Lemma of Banach, $B(w)|_{W_1}$ is continuously invertible for w close to $\bar{w}$ and thus $B(w)$ is onto.

Next, we show a second-order sufficient condition for the QP.

Lemma 2.9 *Let W be a Hilbert space and Z be a Banach space. Furthermore, let $f : W \to \mathbb{R}$ and $e : W \to Z$ be twice continuously F-differentiable. Let $e(\bar{w}) = 0$ and assume that $e'(\bar{w})$ is surjective. In addition, let the following second-order sufficient condition hold at $(\bar{w}, \bar{\mu})$:*

$$\langle d, L_{ww}(\bar{w}, \bar{\mu})d\rangle_{W,W^*} \geq \alpha \|d\|_W^2 \quad \forall d \in W \textit{ with } e'(\bar{w})d = 0,$$

where $\alpha > 0$ is a constant. Then, there exists $\delta > 0$ such that for all $(w, \mu) \in W \times Z^*$ with $\|(w, \mu) - (\bar{w}, \bar{\mu})\|_{W \times Z^*} < \delta$ the following holds:

$$\langle d, L_{ww}(w, \mu)d\rangle_{W,W^*} \geq \frac{\alpha}{2} \|d\|_W^2 \quad \forall d \in W \textit{ with } e'(w)d = 0.$$

Proof Set $B = e'(\bar{w})$, $B(w) = e'(w)$, $W_0 = \text{Kern}(B)$ and split $W = W_0 \perp W_1$. Remember that $B|_{W_1} \in \mathcal{L}(W_1, Z)$ is continuously invertible.

For any $d \in \text{Kern}(B(w))$ there exist unique $d_0 \in W_0$ and $d_1 \in W_1$ with $d = d_0 + d_1$. Our first aim is to show that d_1 is small. In fact,

$$\|Bd_1\|_Z = \|Bd\|_Z = \|(B - B(w))d\|_Z \leq \|B - B(w)\|_{W \to Z} \|d\|_W.$$

Hence,

$$\begin{aligned} \|d_1\|_W &= \|(B|_{W_1})^{-1} B d_1\|_W \leq \|(B|_{W_1})^{-1}\|_{Z \to W_1} \|B - B(w)\|_{W \to Z} \|d\|_W \\ &\overset{\text{def}}{=} \xi(w) \|d\|_W. \end{aligned}$$

Therefore, setting $x = (w, \mu)$,

$$\begin{aligned}
&\langle L_{ww}(x)d, d\rangle_{W^*,W} \\
&\quad = \langle L_{ww}(\bar{x})d, d\rangle_{W^*,W} + \langle (L_{ww}(x) - L_{ww}(\bar{x}))d, d\rangle_{W^*,W} \\
&\quad = \langle L_{ww}(\bar{x})d_0, d_0\rangle_{W^*,W} + \langle L_{ww}(\bar{x})(d + d_0), d_1\rangle_{W^*,W} \\
&\qquad + \langle (L_{ww}(x) - L_{ww}(\bar{x}))d, d\rangle_{W^*,W} \\
&\quad \ge \alpha\|d_0\|_W^2 - \|L_{ww}(\bar{x})\|_{W\to W^*}(\|d\|_W + \|d_0\|_W)\|d_1\|_W \\
&\qquad - \|L_{ww}(x) - L_{ww}(\bar{x})\|_{W\to W^*}\|d\|_W^2 \\
&\quad \ge (\alpha(1 - \xi^2(w)) - 2\|L_{ww}(\bar{x})\|_{W\to W^*}\xi(w) \\
&\qquad - \|L_{ww}(x) - L_{ww}(\bar{x})\|_{W\to W^*})\|d\|_W^2 \\
&\quad =: \alpha(x)\|d\|_W^2.
\end{aligned}$$

By continuity, $\alpha(x) \to \alpha$ for $x \to \bar{x}$.

A sufficient condition for the regularity condition (2.42) is the following:

Lemma 2.10 *Let W be a Hilbert space, let $e'(\bar{w})$ be surjective (this is a CQ), and assume that the following second order sufficient condition holds:*

$$\langle d, L_{ww}(\bar{w}, \bar{\mu})d\rangle_{W,W^*} \ge \alpha\|d\|_W^2 \quad \forall d \in W \text{ with } e'(\bar{w})d = 0,$$

where $\alpha > 0$ is a constant. Then the regularity condition (2.42) *holds.*

Proof For brevity, set $A = L_{ww}(\bar{w}, \bar{\mu})$ and $B = e'(\bar{w})$. We consider the unique solvability of

$$\begin{pmatrix} A & B^* \\ B & 0 \end{pmatrix}\begin{pmatrix} w \\ \mu \end{pmatrix} = \begin{pmatrix} r_1 \\ r_2 \end{pmatrix}.$$

Denote by W_0 the null space of B and by W_1 its orthogonal complement. Then $W = W_0 \perp W_1$ and W_0, W_1 are Hilbert spaces.

Since B is surjective, the equation $Bw = r_2$ is solvable and the set of all solutions is $w_1(r_2) + W_0$, where $w_1(r_2) \in W_1$ is uniquely determined.

We have

$$\langle d, Ad\rangle_{W,W^*} \ge \alpha\|d\|_W^2 \quad \forall d \in W_0.$$

Hence, by the Lax-Milgram Lemma 1.8, there exists a unique solution $w_0(r_1, r_2) \in W_0$ to the problem

$$w_0 \in W_0, \quad \langle Aw_0, d\rangle_{W^*,W} = \langle r_1 - Aw_1(r_2), d\rangle_{W^*,W} \quad \forall d \in W_0.$$

Since B is surjective, we have for all $z^* \in \mathrm{Kern}(B^*)$:

$$\langle z^*, Z\rangle_{Z^*,Z} = \langle z^*, BW\rangle_{Z^*,Z} = \langle B^*z^*, W\rangle_{W^*,W} = \langle \{0\}, W\rangle_{W^*,W} = \{0\}.$$

Hence, $\text{Kern}(B^*) = \{0\}$ and thus B^* is injective. Also, since $BW = Z$ is closed, the closed range theorem yields

$$B^* Z^* = \text{Kern}(B)^\perp = W_0^\perp.$$

Here, for $S \subset X$

$$S^\perp = \left\{x' \in X^* : \langle x', s\rangle_{X^*,X} = 0 \; \forall s \in S\right\}.$$

By construction, $r_1 - Aw_0(r_1, r_2) - Aw_1(r_2) \in W_0^\perp$. Hence, there exists a unique $\mu(r_1, r_2) \in Z^*$ such that

$$B^* \mu(r_1, r_2) = r_1 - Aw_0(r_1, r_2) - Aw_1(r_2).$$

Therefore, we have found the unique solution

$$\begin{pmatrix} w \\ \mu \end{pmatrix} = \begin{pmatrix} w_0(r_1, r_2) + w_1(r_2) \\ \mu(r_1, r_2) \end{pmatrix}.$$

2.6.2 The Josephy-Newton Method

In the previous section, we were able to derive the SQP method for equality-constrained problems by applying Newton's method to the KKT system.

For inequality constrained problems this is not directly possible since the KKT system consists of operator equations and a variational inequality. As we will see, such a combination can be most elegantly written as a

2.6.2.1 Generalized Equation

$$\text{GE}(G, N): \quad 0 \in G(x) + N(x).$$

Here, $G : X \to Y$ is assumed to be continuously F-differentiable and $N : X \rightrightarrows Y$ is a set-valued mapping with closed graph.

For instance, the variational inequality VI(F, S), with $F : W \to W^*$ and $S \subset W$ closed and convex, can be written as

$$0 \in F(w) + N_S(w),$$

where N_S is the normal cone mapping of S:

Definition 2.4 Let $S \subset W$ be a nonempty closed convex subset of the Banach space W. The *normal cone* $N_S(w)$ of S at $w \in W$ is defined by

$$N_S(w) = \begin{cases} \left\{y \in W^* : \langle y, z - w\rangle_{W^*,W} \le 0 \; \forall z \in S\right\}, & w \in S, \\ \emptyset, & w \notin S. \end{cases}$$

This defines a set-valued mapping $N_S : W \rightrightarrows W^*$.

The Josephy-Newton method for generalized equations looks as follows:

Algorithm 2.17 (Josephy-Newton method for GE(G, N))

0. Choose $x^0 \in X$ (sufficiently close to the solution $\bar{x}$ of GE(G, N)).

For $k = 0, 1, 2 \ldots$:

1. STOP if x^k solves GE(G, N) (holds if $x^k = x^{k-1}$).
2. Compute the solution x^{k+1} of

$$\mathrm{GE}(G(x^k) + G'(x^k)(\cdot - x^k), N):$$
$$0 \in G(x^k) + G'(x^k)(x - x^k) + N(x)$$

that is closest to x^k.

In the classical Newton's method, which corresponds to $N(x) = \{0\}$ for all x, an essential ingredient is the regularity condition that $G'(\bar{x})$ is continuously invertible. This means that the linearized equation

$$p = G(\bar{x}) + G'(\bar{x})(x - \bar{x})$$

possesses the unique solution $x(p) = \bar{x} + G'(\bar{x})^{-1} p$, which of course depends linearly and thus Lipschitz continuously on $p \in Y$.

The appropriate generalization of this regularity condition is the following:

Definition 2.5 (Strong regularity) The generalized equation GE(G, N) is called *strongly regular* at a solution $\bar{x}$ if there exist $\delta > 0$, $\varepsilon > 0$ and $L > 0$ such that, for all $p \in Y$, $\|p\|_Y < \delta$, there exists a unique $x = x(p) \in X$ with $\|x(p) - \bar{x}\|_X < \varepsilon$ such that

$$p \in G(\bar{x}) + G'(\bar{x})(x - \bar{x}) + N(x)$$

and $x(p)$ is Lipschitz continuous:

$$\|x(p_1) - x(p_2)\|_X \le L\|p_1 - p_2\|_Y \quad \forall p_1, p_2 \in Y, \ \|p_i\|_X < \delta, \ i = 1, 2.$$

It is a milestone result of Robinson [117] that then the following holds:

Theorem 2.18 *Let X, Y, and Z be Banach spaces. Furthermore, let $\bar{z} \in Z$ be fixed and assume that $\bar{x}$ is a solution of*

$$GE(G(\bar{z}, \cdot), N): \quad 0 \in G(\bar{z}, x) + N(x)$$

at which the GE is strongly regular with Lipschitz modulus L. Assume that G is F-differentiable with respect to x near $(\bar{z}, \bar{x})$ and that G and G_x are continuous at $(\bar{z}, \bar{x})$.

Then, for every $\varepsilon > 0$*, there exist neighborhoods* $Z_\varepsilon(\bar{z})$ *of* $\bar{z}$*,* $X_\varepsilon(\bar{x})$ *of* $\bar{x}$*, and a mapping* $x : Z_\varepsilon(\bar{z}) \to X_\varepsilon(\bar{x})$ *such that, for all* $z \in Z_\varepsilon(\bar{z})$*,* $x(z)$ *is the (locally) unique solution of the generalized equation*

$$0 \in G(z,x) + N(x), \quad x \in X_\varepsilon(\bar{x}).$$

In addition,

$$\|x(z_1) - x(z_2)\|_X \le (L+\varepsilon)\|G(z_1, x(z_2)) - G(z_2, x(z_2))\|_Y \quad \forall z_1, z_2 \in Z_\varepsilon(\bar{z}).$$

From this, it is not difficult to derive fast local convergence of the Josephy-Newton method:

Theorem 2.19 *Let* X*,* Y *be Banach spaces,* $G : X \to Y$ *continuously F-differentiable, and let* $N : X \rightrightarrows Y$ *be set-valued with closed graph. If* $\bar{x}$ *is a strongly regular solution of* GE(G, N)*, then the Josephy-Newton method* (*Algorithm* 2.17) *is locally q-superlinearly convergent in a neighborhood of* $\bar{x}$*. If, in addition,* G' *is* α*-Hölder continuous near* $\bar{x}$*, then the order of convergence is* $1+\alpha$*.*

Proof For compact notation, we set $B_\delta(x) = \{y \in X : \|y - x\|_X < \delta\}$.

Let L be the Lipschitz modulus of strong regularity. We set $Z = X$, $\bar{z} = \bar{x}$ and consider

$$\bar{G}(z,x) \stackrel{\text{def}}{=} G(z) + G'(z)(x-z).$$

Since $\bar{G}(\bar{z}, \cdot)$ is affine linear, we have

$$\bar{G}(\bar{z},\bar{x}) + \bar{G}_x(\bar{z},\bar{x})(x-\bar{x}) = \bar{G}(\bar{z},x) = G(\bar{z}) + G'(\bar{z})(x-\bar{z}) = G(\bar{x}) + G'(\bar{x})(x-\bar{x}).$$

Therefore, GE$(\bar{G}(\bar{z},\cdot), N)$ is strongly regular at $\bar{x}$ with Lipschitz constant L. Theorem 2.18 is applicable and thus, for $\varepsilon > 0$, there exist neighborhoods $Z_\varepsilon(\bar{x})$ of $\bar{z} = \bar{x}$ and $X_\varepsilon(\bar{x})$ of $\bar{x}$ such that, for all $z \in Z_\varepsilon(\bar{x})$,

$$0 \in \bar{G}(z,x) + N(x) = G(z) + G'(z)(x-z) + N(x), \quad x \in X_\varepsilon(\bar{x})$$

has a unique solution $x(z)$ that satisfies

$$\begin{aligned}
&\forall z_1, z_2 \in Z_\varepsilon(\bar{z}) = Z_\varepsilon(\bar{x}) : \\
&\quad \|x(z_1) - x(z_2)\|_X \le (L+\varepsilon)\|\bar{G}(z_1, x(z_2)) - \bar{G}(z_2, x(z_2))\|_Y \\
&\qquad = (L+\varepsilon)\|G(z_1) - G(z_2) + G'(z_1)(x(z_2) - z_1) - G'(z_2)(x(z_2) - z_2)\|_Y.
\end{aligned}$$

If we choose $z_1 = z \in Z_\varepsilon(\bar{x})$ and $z_2 = \bar{x}$, we obtain $x(z_2) = \bar{x}$ and thus for all $z \in Z_\varepsilon(\bar{x})$:

$$\begin{aligned}
\|x(z) - \bar{x}\|_X &\le (L+\varepsilon)\|G(z) - G(\bar{x}) + G'(z)(\bar{x} - z) - G'(\bar{x})(\bar{x} - \bar{x})\|_Y \\
&= (L+\varepsilon)\|G(z) - G(\bar{x}) - G'(z)(z - \bar{x})\|_Y
\end{aligned}$$

$$
\begin{aligned}
&\le (L+\varepsilon)\|G(z)-G(\bar{x})-G'(\bar{x})(z-\bar{x})\|_Y \\
&\quad + (L+\varepsilon)\|(G'(\bar{x})-G'(z))(z-\bar{x})\|_Y \\
&\le (L+\varepsilon)\|G(z)-G(\bar{x})-G'(\bar{x})(z-\bar{x})\|_Y \\
&\quad + (L+\varepsilon)\|G'(\bar{x})-G'(z)\|_{X\to Y}\|z-\bar{x}\|_X \\
&= o(\|z-\bar{x}\|_X) \quad (z\to\bar{x}). \qquad (2.45)
\end{aligned}
$$

In the last estimate, we have used the F-differentiability of G and the continuity of G'.

Now choose $\delta > 0$ such that $B_\delta(\bar{x}) \subset X_\varepsilon(\bar{x})$ and $B_{5\delta/2}(\bar{x}) \subset Z_\varepsilon(\bar{x})$. By possibly reducing δ, we achieve, using (2.45),

$$\|x(z)-\bar{x}\|_X \le \frac{1}{2}\|z-\bar{x}\|_X \quad \forall z \in B_\delta(\bar{x}).$$

In particular, this implies

$$x(z) \in B_{\delta/2}(\bar{x}) \subset B_\delta(\bar{x}) \quad \forall\, z \in B_\delta(\bar{x}).$$

Now observe that, for $x^k \in B_\delta(\bar{x})$, the unique solution of GE($G(x^k)+G'(x^k)(\cdot - x^k), N$) in $X_\varepsilon(\bar{x})$ is given by $x(x^k) \in B_{\delta/2}(\bar{x})$.

From

$$\|x(x^k)-x^k\| \le \|x(x^k)-\bar{x}\|_X + \|\bar{x}-x^k\|_X < \frac{\delta}{2}+\delta = \frac{3}{2}\delta$$

and $B_{5\delta/2}(\bar{x}) \subset X_\varepsilon(\bar{x})$ we conclude that $x(x^k)$ is the solution of GE($G(x^k)+G'(x^k)(\cdot - x^k), N$) that is closest to x^k. Hence, for $x^k \in B_\delta(\bar{x})$, we have

$$x^{k+1} = x(x^k) \in B_{\delta/2}(\bar{x}) \subset B_\delta(\bar{x}), \qquad \|x^{k+1}-\bar{x}\|_X \le \frac{1}{2}\|x^k-\bar{x}\|_X.$$

Thus, if we choose $x^0 \in B_\delta(\bar{x})$, we obtain by induction $x^k \to \bar{x}$.

Furthermore, from (2.45) it follows that

$$\|x^{k+1}-\bar{x}\|_X = \|x(x^k)-\bar{x}\|_X = o(\|x^k-\bar{x}\|_X) \quad (k\to\infty).$$

This proves the q-superlinear convergence.

If G' is α-order Hölder continuous near $\bar{x}$ with modulus $L_\alpha > 0$, then we can improve the estimate (2.45):

$$
\begin{aligned}
\|x(z)-\bar{x}\|_X &\le (L+\varepsilon)\|G(z)-G(\bar{x})-G'(z)(z-\bar{x})\|_Y \\
&= (L+\varepsilon)\left\|\int_0^1 (G'(\bar{x}+t(z-\bar{x}))-G'(z))(z-\bar{x})\,dt\right\|_Y \\
&\le (L+\varepsilon)\int_0^1 \|G'(\bar{x}+t(z-\bar{x}))-G'(z)\|_{X\to Y}\,dt\,\|z-\bar{x}\|_X
\end{aligned}
$$

$$
\begin{aligned}
&\leq (L+\varepsilon)\int_0^1 L_\alpha (1-t)^\alpha \|z-\bar{x}\|_X^\alpha \, dt \, \|z-\bar{x}\|_X \\
&= \frac{L+\varepsilon}{1+\alpha} L_\alpha \|z-\bar{x}\|_X^{1+\alpha} \\
&= O(\|z-\bar{x}\|_X^{1+\alpha}) \quad (z \to \bar{x}).
\end{aligned}
$$

Hence,

$$
\|x^{k+1}-\bar{x}\|_X = \|x(x^k)-\bar{x}\|_X = O(\|x^k-\bar{x}\|_X^{1+\alpha}) \quad (k \to \infty).
$$

2.6.3 SQP Methods for Inequality Constrained Problems

We consider the problem

$$
\min_{w \in W} f(w) \quad \text{s.t.} \quad e(w) = 0, \qquad c(w) \in \mathcal{K}, \tag{2.46}
$$

with $f : W \to \mathbb{R}$, $e : W \to Z$, and $c : W \to R$ twice continuously F-differentiable. Furthermore, W, Z, R are Banach spaces, R is reflexive (i.e., $R^{**} = R$), and $\mathcal{K} \subset R$ is a nonempty closed convex cone.

For this problem, we define the Lagrange function

$$
L(w, \lambda, \mu) = f(w) + \langle \lambda, c(w) \rangle_{R^*,R} + \langle \mu, e(w) \rangle_{Z^*,Z}.
$$

We will need the notion of the polar cone.

Definition 2.6 Let X be a Banach space and let $\mathcal{K} \subset X$ be a nonempty closed convex cone. Then the *polar cone* of $\mathcal{K}$ is defined by

$$
\mathcal{K}^\circ = \left\{ y \in X^* : \langle y, x \rangle_{X^*,X} \leq 0 \ \forall x \in \mathcal{K} \right\}.
$$

Obviously, $\mathcal{K}^\circ$ is a closed convex cone.

Recall also the definition of the normal cone mapping (Def. 2.4).

Under a constraint qualification, see Sect. 1.7.3.2, the following KKT conditions hold:

There exist Lagrange multipliers $\bar{\lambda} \in \mathcal{K}^\circ$ and $\bar{\mu} \in Z^*$ such that $(\bar{w}, \bar{\lambda}, \bar{\mu})$ satisfies

$$
\begin{aligned}
&L_w(\bar{w}, \bar{\lambda}, \bar{\mu}) = 0, \\
&c(\bar{w}) \in \mathcal{K}, \qquad \bar{\lambda} \in \mathcal{K}^\circ, \qquad \langle \bar{\lambda}, c(\bar{w}) \rangle_{R^*,R} = 0, \\
&e(\bar{w}) = 0.
\end{aligned}
$$

The second condition can be shown to be equivalent to VI$(-c(\bar{w}), \mathcal{K}^\circ)$. This is a VI w.r.t. λ with a constant operator parameterized by $\bar{w}$.

Now comes the trick, see, e.g., [5]:

By means of the normal cone $N_{\mathcal{K}^\circ}$, it is easily seen that VI$(-c(w), \mathcal{K}^\circ)$ is equivalent to the generalized equation

$$0 \in -c(w) + N_{\mathcal{K}^\circ}(\lambda).$$

Therefore, we can write the KKT system as a generalized equation:

$$0 \in \begin{pmatrix} L_w(w,\lambda,\mu) \\ -c(w) \\ e(w) \end{pmatrix} + \begin{pmatrix} \{0\} \\ N_{\mathcal{K}^\circ}(\lambda) \\ \{0\} \end{pmatrix}. \tag{2.47}$$

Setting

$$N(w,\lambda,\mu) = \begin{pmatrix} \{0\} \\ N_{\mathcal{K}^\circ}(\lambda) \\ \{0\} \end{pmatrix},$$

and noting $L_\lambda(w,\lambda,\mu) = c(w)$, $L_\mu(w,\lambda,\mu) = e(w)$, we can write (2.47) very compactly as GE$(-L', N)$.

The closed graph of the normal cone mapping is proved in the next lemma.

Lemma 2.11 *Let X be a Banach spaces and $S \subset X$ be nonempty, closed, and convex. Then the normal cone mapping N_S has closed graph.*

Proof Let $\mathrm{graph}(N_S) \ni (x^k, y^k) \to (\bar{x}, \bar{y})$. Then $y^k \in N_S(x^k)$ and thus $x^k \in S$, since otherwise $N_S(x^k)$ would be empty. Since S is closed, $\bar{x} \in S$ follows. Now, for all $z \in S$, by continuity

$$\langle \bar{y}, z - \bar{x}\rangle_{X^*,X} = \lim_{k\to\infty} \underbrace{\langle y^k, z - x^k\rangle_{X^*,X}}_{\le 0} \le 0,$$

hence $\bar{y} \in N_S(\bar{x})$. Therefore, $(\bar{x}, \bar{y}) \in \mathrm{graph}(N_S)$.

If we now apply the Josephy-Newton method to (2.47), we obtain the following subproblem (we set $x^k = (w^k, \lambda^k, \mu^k)$):

$$0 \in \begin{pmatrix} L_w(x^k) \\ -c(w^k) \\ e(w^k) \end{pmatrix} + \begin{pmatrix} L_{ww}(x^k) & c'(w^k)^* & e'(w^k)^* \\ -c'(w^k) & 0 & 0 \\ e'(w^k) & 0 & 0 \end{pmatrix} \begin{pmatrix} w - w^k \\ \lambda - \lambda^k \\ \mu - \mu^k \end{pmatrix} + \begin{pmatrix} \{0\} \\ N_{\mathcal{K}^\circ}(\lambda) \\ \{0\} \end{pmatrix}. \tag{2.48}$$

It is not difficult to see that (2.48) are exactly the KKT conditions of the following quadratic optimization problem:

2.6.3.1 SQP Subproblem

$$\begin{aligned}
&\min_{w \in W} \langle f'(w^k), w - w^k \rangle_{W^*,W} + \frac{1}{2} \langle L_{ww}(x^k)(w - w^k), w - w^k \rangle_{W^*,W} \\
&\text{s.t.} \quad e(w^k) + e'(w^k)(w - w^k) = 0, \quad c(w^k) + c'(w^k)(w - w^k) \in \mathcal{K}.
\end{aligned}$$

In fact, the Lagrange function of the QP is

$$\begin{aligned}
L^{qp}(x) = \langle f'(w^k), w - w^k \rangle_{W^*,W} &+ \frac{1}{2} \langle L_{ww}(x^k)(w - w^k), w - w^k \rangle_{W^*,W} \\
&+ \langle \lambda, c(w^k) + c'(w^k)(w - w^k) \rangle_{W^*,W} \\
&+ \langle \mu, e(w^k) + e'(w^k)(w - w^k) \rangle_{Z^*,Z}.
\end{aligned}$$

Since

$$\begin{aligned}
L_w^{qp}(x) &= f'(w^k) + L_{ww}(x^k)(w - w^k) + c'(w^k)^* \lambda + e'(w^k)^* \mu \\
&= L_w(x^k) + L_{ww}(x^k)(w - w^k) + c'(w^k)^*(\lambda - \lambda^k) + e'(w^k)^*(\mu - \mu^k),
\end{aligned}$$

we see that writing down the KKT conditions for the QP in the form (2.47) gives exactly the generalized equation (2.48).

We obtain:

Algorithm 2.20 (SQP method for inequality constrained problems)

0. Choose (w^0, λ^0, μ^0) (sufficiently close to $(\bar{w}, \bar{\lambda}, \bar{\mu})$).

For $k = 0, 1, 2, \ldots$:

1. If (w^k, λ^k, μ^k) is a KKT triple of (2.46), STOP.
2. Compute the KKT triple $(d^k, \lambda^{k+1}, \mu^{k+1})$ of

$$\begin{aligned}
&\min_{d \in W} \langle f'(w^k), d \rangle_{W^*,W} + \frac{1}{2} \langle L_{ww}(w^k, \lambda^k, \mu^k) d, d \rangle_{W^*,W} \\
&\text{s.t.} \quad e(w^k) + e'(w^k) d = 0, \qquad c(w^k) + c'(w^k) d \in \mathcal{K},
\end{aligned}$$

   that is closest to $(0, \lambda^k, \mu^k)$.
3. Set $w^{k+1} = w^k + d^k$.

Since this method is the Josephy-Newton algorithm applied to (2.47), we can derive local convergence results immediately if Robinson's strong regularity condition is satisfied. This condition has to be verified from case to case and is connected to second order sufficient optimality conditions. As an example where strong regularity is verified for an optimal control problem, we refer to [56].

2.6.3.2 Application to Optimal Control

For illustration, we consider the nonlinear elliptic optimal control problem

$$\min_{y \in H_0^1(\Omega), u \in L^2(\Omega)} \quad J(y,u) \stackrel{\text{def}}{=} \frac{1}{2}\|y - y_d\|_{L^2(\Omega)}^2 + \frac{\alpha}{2}\|u\|_{L^2(\Omega)}^2$$

$$\text{s.t.} \qquad Ay + y^3 + y = u, \quad u \le b. \tag{2.49}$$

Here, $y \in H_0^1(\Omega)$ is the state, which is defined on the open bounded domain $\Omega \subset \mathbb{R}^n$, $n \le 3$, and $u \in L^2(\Omega)$ is the control. Furthermore, $A : H_0^1(\Omega) \to H^{-1}(\Omega) = H_0^1(\Omega)^*$ is a linear elliptic partial differential operator, e.g., $A = -\Delta$. Finally $b \in L^\infty(\Omega)$ is an upper bound on the control. We convert this control problem into the form (2.46) by setting

$$Y = H_0^1(\Omega), \qquad U = L^2(\Omega), \qquad Z = H^{-1}(\Omega),$$

$$e(y,u) = Ay + y^3 + y - u, \qquad c(y,u) = u - b,$$

$$\mathcal{K} = \left\{ u \in L^2(\Omega) : u \le 0 \text{ a.e. on } \Omega \right\}.$$

One can show (note $n \le 3$) that the operator e is twice continuously F-differentiable with

$$e_y(y,u) = A + 3y^2 \cdot I + I, \quad e_{yy}(y,u)(h_1,h_2) = 6yh_1h_2$$

(the other derivatives are obvious due to linearity). Therefore, given $x^k = (y^k, u^k, \lambda^k, \mu^k)$, the SQP subproblem reads

$$\min_{d_y,d_u} (y^k - y_d, d_y)_{L^2(\Omega)} + \alpha(u^k, d_u)_{L^2(\Omega)} + \frac{1}{2}\|d_y\|_{L^2(\Omega)}^2$$

$$+ \frac{1}{2}\langle \mu_k, 6y^k d_y^2 \rangle_{H_0^1(\Omega), H^{-1}(\Omega)} + \frac{\alpha}{2}\|d_u\|_{L^2(\Omega)}^2$$

$$\text{s.t.} \quad Ay^k + (y^k)^3 + y^k - u^k + Ad_y + 3(y^k)^2 d_y + d_y - d_u = 0,$$

$$u_k + d_u \le b.$$

2.7 State-Constrained Problems

So far, we focused on optimization problems with control constraints. Only very recently, advances in the analysis of Newton based algorithms for state constrained problems have been made and much is to be done yet. We cannot go into a detailed discussion of this topic here. Rather, we just briefly sketch a couple of promising approaches that are suitable for state constrained optimization problems.

2.7.1 SQP Methods

In the case of SQP methods, state constraints do not pose direct conceptual difficulties, at least not at a first glance. In fact, sequential quadratic programming is applicable to very general problem settings. The constraints are linearized to generate the QP subproblems, i.e., state constraints arise as linearized state constraints in the subproblems and the difficulties of dealing with state constraints is thus transported to the subproblems. However, the efficient solution of the QP subproblems is not the only challenge. In fact, it is important to emphasize that second order optimality theory is challenging in the case of state constraints. Second order sufficient optimality conditions are closely linked to strong regularity of the generalized equation corresponding to the KKT conditions. Therefore, proving fast local convergence of SQP methods for problems with state constraints is challenging. Recent progress in second order optimality theory, e.g., [31, 57] may help paving the ground for future progress in this field.

2.7.2 Semismooth Newton Methods

The application of semismooth reformulation techniques for state constraints poses principal difficulties. In fact, consider for illustration the following model problem:

$$\begin{aligned}
&\min_{y,u}\ J(y,u) := \frac{1}{2}\|y-y_d\|^2_{L^2(\Omega)} + \frac{\alpha}{2}\|u\|^2_{L^2(\Omega)} \\
&\text{s.t.}\quad -\Delta y = u \quad \text{on } \Omega, \\
&\qquad\quad y = 0 \quad \text{on } \partial\Omega, \\
&\qquad\quad y \le b \quad \text{on } \Omega.
\end{aligned} \tag{2.50}$$

Here, $n \le 3$ and $\Omega \subset \mathbb{R}^n$ is open and bounded with C^2 boundary. Furthermore, $b \in H^2(\Omega)$, $b > 0$, $\alpha > 0$, and $y_d \in L^2(\Omega)$. From regularity results, see Theorem 1.28, we know that for $u \in U := L^2(\Omega)$ there exists a unique weak solution $y \in Y := H_0^1(\Omega) \cap H^2(\Omega) \hookrightarrow C(\bar{\Omega})$ of the state equation. The existence and uniqueness of a solution $(\bar{y},\bar{u}) \in Y \times U$ is easy to show by standard arguments.

Similar to the analysis of problem (1.144) it can be shown, see, e.g., [12], that the following optimality conditions hold at the solution: There exists a regular Borel measure $\bar{\mu} \in \mathcal{M}(\Omega)$ and an adjoint state $\bar{p} \in L^2(\Omega)$ such that

$$-\Delta\bar{y} = \bar{u} \quad \text{on } \Omega, \tag{2.51}$$

$$\bar{y} = 0 \quad \text{on } \partial\Omega, \tag{2.52}$$

$$(\bar{p}, -\Delta v)_{L^2(\Omega)} + \langle\bar{\mu}, v\rangle_{\mathcal{M}(\Omega),C(\bar{\Omega})} = (y_d - \bar{y}, v)_{L^2(\Omega)} \quad \forall v \in Y, \tag{2.53}$$

$$\bar{y} \le b, \qquad \langle\bar{\mu}, v - \bar{y}\rangle_{\mathcal{M}(\Omega),C(\bar{\Omega})} \le 0 \quad \forall v \in C(\bar{\Omega}),\ v \le b, \tag{2.54}$$

$$\alpha\bar{u} - \bar{p} = 0 \quad \text{in } \Omega. \tag{2.55}$$

The difficulty now is that the complementarity condition (2.54) between the function $\bar{y}$ and the measure $\bar{\mu}$ cannot be written in a pointwise fashion. Hence, nonsmooth pointwise reformulations as needed for semismooth Newton methods are not possible.

To avoid this difficulty, several approaches were presented recently.

2.7.2.1 Moreau-Yosida Regularization

One possibility is to treat the state constraint by a Moreau-Yosida regularization. The state constraint is converted to a penalty term, resulting in the following Moreau-Yosida regularized problem:

$$\begin{aligned} \min\ & \frac{1}{2}\|y - y_d\|^2_{L^2(\Omega)} + \frac{\alpha}{2}\|u\|^2_{L^2(\Omega)} + \frac{1}{2\gamma}\|\max(0, \hat{\mu} + \gamma(y - b))\|^2_{L^2(\Omega)} \\ \text{s.t.}\ & -\Delta y = u \quad \text{on } \Omega, \\ & y = 0 \quad \text{on } \partial\Omega. \end{aligned} \tag{2.56}$$

Here $\gamma > 0$ is a penalty parameter and $\hat{\mu} \geq 0$, $\hat{\mu} \in L^2(\Omega)$ is a shift parameter. For this problem without inequality constraints, the optimality conditions are

$$-\Delta\bar{y}_\gamma = \bar{u}_\gamma \quad \text{on } \Omega, \tag{2.57}$$

$$\bar{y}_\gamma = 0 \quad \text{on } \partial\Omega, \tag{2.58}$$

$$-\Delta\bar{p}_\gamma = y_d - \bar{y}_\gamma - \max(0, \hat{\mu} + \gamma(\bar{y}_\gamma - b)) \quad \text{on } \Omega \tag{2.59}$$

$$\bar{p}_\gamma = 0 \quad \text{on } \partial\Omega, \tag{2.60}$$

$$\alpha\bar{u}_\gamma - \bar{p}_\gamma = 0 \quad \text{on } \Omega. \tag{2.61}$$

To make this system more similar to the optimality conditions (2.51)–(2.55), we introduce

$$\bar{\mu}_\gamma = \max(0, \hat{\mu} + \gamma(\bar{y}_\gamma - b)).$$

We then can write the KKT conditions (2.57)–(2.61) in the form

$$-\Delta\bar{y}_\gamma = \bar{u}_\gamma \quad \text{on } \Omega, \tag{2.62}$$

$$\bar{y}_\gamma = 0 \quad \text{on } \partial\Omega, \tag{2.63}$$

$$-\Delta\bar{p}_\gamma + \bar{\mu}_\gamma = y_d - \bar{y}_\gamma \quad \text{on } \Omega, \tag{2.64}$$

$$\bar{p}_\gamma = 0 \quad \text{on } \partial\Omega, \tag{2.65}$$

$$\bar{\mu}_\gamma = \max(0, \hat{\mu} + \gamma(\bar{y}_\gamma - b)) \quad \text{on } \Omega, \tag{2.66}$$

$$\alpha\bar{u}_\gamma - \bar{p}_\gamma = 0 \quad \text{on } \Omega. \tag{2.67}$$

For further discussion, we rewrite (2.66) as follows

$$0 = \bar{\mu}_\gamma - \max(0, \hat{\mu} + \gamma(\bar{y}_\gamma - b))$$
$$= \bar{\mu}_\gamma - \max\left(0, \bar{\mu}_\gamma + \gamma\left(\bar{y}_\gamma - b + \frac{1}{\gamma}(\hat{\mu} - \bar{\mu}_\gamma)\right)\right). \tag{2.68}$$

If, just for an informal motivation, we suppose for a moment that $(\hat{\mu} - \bar{\mu}_\gamma)/\gamma$ becomes small for large γ, then we can interpret (2.68) as an approximation of

$$\bar{\mu}_\gamma = \max(0, \bar{\mu}_\gamma + \gamma(\bar{y}_\gamma - b)). \tag{2.69}$$

From earlier considerations we know that (2.69) is equivalent to

$$\bar{\mu}_\gamma \geq 0, \quad \bar{y}_\gamma - b \leq 0, \quad \bar{\mu}_\gamma(\bar{y}_\gamma - b) = 0,$$

which can be interpreted as a strong formulation of (2.54). This demonstrates the role of (2.66) as an approximation of (2.54).

We collect some results concerning the regularized solution tuple $(\bar{y}_\gamma, \bar{u}_\gamma, \bar{p}_\gamma, \bar{\mu}_\gamma)$, which we call primal dual path. The details can be found in [66, 67]:

For any $\gamma_0 > 0$, the primal dual path $\gamma \in [\gamma_0, \infty) \mapsto (\bar{y}_\gamma, \bar{u}_\gamma, \bar{p}_\gamma, \bar{\mu}_\gamma)$ can be shown to be bounded in $Y \times U \times L^2(\Omega) \times Y^*$ and Lipschitz continuous. In addition, $\gamma \in (0, \infty) \to \bar{\mu}_\gamma \in L^2(\Omega)$ is locally Lipschitz continuous. Moreover $(\bar{y}_\gamma, \bar{u}_\gamma, \bar{p}_\gamma, \bar{\mu}_\gamma)$ converges weakly to $(\bar{y}, \bar{u}, \bar{p}, \bar{\mu})$ as $\gamma \to \infty$ and the convergence $(\bar{y}_\gamma, \bar{u}_\gamma) \to (\bar{y}, \bar{u})$ is even strong in $Y \times U$.

The idea is now to apply a semismooth Newton method to (2.62)–(2.67) for solving (2.56) approximately and to drive γ to infinity in an outer iteration. The analysis of this approach was carried out in, e.g., [66, 67]. The adaption of the parameter γ can be controlled by models of the optimal value function along the path.

2.7.2.2 Lavrentiev Regularization

A second approach to state constrained problems is Lavrentiev regularization [103, 104]. We again consider the problem (2.50). The idea is to replace the constraint

$$y \leq b$$

by

$$y + \varepsilon u \leq b$$

with a parameter $\varepsilon > 0$. If we then introduce the new artificial control $w = y + \varepsilon u$, we have $u = (w - y)/\varepsilon$ and thus can express u in terms of w. The Lavrentiev regularized problem, transformed to w, then is given by

$$\begin{aligned}
\min\ & J(y,w) := \frac{1}{2}\|y-y_d\|^2_{L^2(\Omega)} + \frac{\alpha}{2\varepsilon^2}\|w-y\|^2_{L^2(\Omega)} \\
\text{s.t.}\ & -\varepsilon\Delta y + y = w \quad \text{on } \Omega, \\
& y = 0 \quad \text{on } \partial\Omega, \\
& w \le b \quad \text{on } \Omega.
\end{aligned} \tag{2.70}$$

Except for the modified L^2-regularization, this problem has the form of a control-constrained elliptic optimal control problem. It is not difficult to see that it is uniquely solvable and can be handled by semismooth Newton techniques.

Under suitable assumptions, it can be shown, see [104], that the regularized solution $(\bar{y}_\varepsilon, \bar{u}_\varepsilon)$ converges strongly to the solution $(\bar{y}, \bar{u})$ of (2.50) as $\varepsilon \to 0^+$.

2.8 Further Aspects

2.8.1 Mesh Independence

For numerical computations, we have to discretize the problem (Finite elements, finite differences, ...) and to apply the developed optimization methods to the discretized, finite dimensional problem. One such situation would be, for instance, to apply an SQP method to the discretization (P_h) of the infinite dimensional problem (P). If this is properly done, we can interpret the discrete SQP method as an inexact (i.e. perturbed) version of the SQP method applied to (P).

Abstractly speaking, we have an infinite dimensional problem (P) and an algorithm A for its solution. Furthermore, we have a family of finite dimensional approximations (P_h) of (P), and discrete versions A_h of algorithm A. Here $h > 0$ denotes the accuracy of discretization (with increasing accuracy as $h \to 0$). Starting from x^0 and the corresponding discrete point x_h^0, respectively, the algorithms A and A_h will generate sequences (x^k) and (x_h^k), respectively. Mesh independence means that the convergence behavior of (x^k) and (x_h^k) become more and more alike as the discretization becomes more and more accurate, i.e., as $h \to 0$. This means, for instance, that q-superlinear convergence of Alg. A on a δ-neighborhood of the solution implies the same rate of convergence for Alg. A_h on a δ-neighborhood of the corresponding discrete solution as soon as h is sufficiently small.

Mesh independence results for Newton's method were established in, e.g., [3, 44]. The mesh independence of SQP methods and Josephy-Newton methods was shown, e.g., in [6, 45]. Furthermore, the mesh independence of semismooth Newton methods was established in [68].

2.8.2 Application of Fast Solvers

An important ingredient in PDE constrained optimization is the combination of optimization methods with efficient solvers (sparse linear solvers, multigrid, preconditioned Krylov subspace methods, etc.). It is by far out of the scope of this chapter to give details. Instead, we focus on just two simple examples.

For both semismooth reformulations of the elliptic control problems (2.25) and (2.32), we showed that the semismooth Newton system is equivalent to

$$\begin{pmatrix} I & 0 & A^* \\ 0 & I & -\frac{1}{\alpha} g^k \cdot B^* \\ A & -B & 0 \end{pmatrix} \begin{pmatrix} s_y^k \\ s_u^k \\ s_\mu^k \end{pmatrix} = \begin{pmatrix} r_1^k \\ r_2^k \\ r_3^k \end{pmatrix} \tag{2.71}$$

with appropriate right hand side. Here $A \in \mathcal{L}(H_0^1(\Omega), H^{-1}(\Omega))$ is an elliptic operator, $B \in \mathcal{L}(L^{p'}(\Omega_c), H^{-1}(\Omega))$ with $p' \in [1, 2)$, and $g^k \in L^\infty(\Omega_c)$ with $g^k \in [0, 1]$ almost everywhere. We can do block elimination to obtain

$$\begin{pmatrix} I & A^* & 0 \\ A & -\frac{1}{\alpha} B(g^k \cdot B^*) & 0 \\ 0 & -\frac{g^k}{\alpha} \cdot B^* & I \end{pmatrix} \begin{pmatrix} s_y^k \\ s_\mu^k \\ s_u^k \end{pmatrix} = \begin{pmatrix} r_1^k \\ Br_2^k + r_3^k \\ r_2^k \end{pmatrix}.$$

The first two rows form a 2×2 elliptic system for which very efficient fast solvers (e.g., multigrid [62]) exist.

Similar techniques can successfully be used, e.g., for elastic contact problems [139].

2.8.3 Other Methods

Our treatment of Newton-type methods is not at all complete. There exist, for instance, interior point methods that are very well suited for optimization problems in function spaces, see, e.g., [121, 122, 138, 140, 145].

Chapter 3
Discrete Concepts in PDE Constrained Optimization

Michael Hinze

Abstract In the present chapter we give an introduction to discrete concepts for optimization problems with PDE constraints. As models for the state we consider elliptic and parabolic PDEs which are well understood from the analytical point of view. This allows to focus on structural aspects in discretization. We discuss and compare the approaches *First discretize, then optimize* and *First optimize, then discretize*, and introduce a variational discrete concept which avoids explicit discretization of the controls. We investigate problems with general constraints on the control, and also consider pointwise bounds on the state, and on the gradient of the state. We present error analysis for the variational discrete concept and accomplish our analytical findings with numerical examples which confirm our analytical results.

3.1 Introduction

This chapter presents an introduction to discrete concepts in PDE constrained optimization including control and state constraints. So far, concepts without constraints are fairly well understood, and theory and praxis for control constraints are strongly emerging. Currently there is a strong focus on the development of reliable numerical approaches for state constraints. This field in many respects requires further intensive research.

To approach an optimal control problem of the form (1.138) numerically one may either discretize this problem by substituting all appearing function spaces by finite dimensional spaces, and all appearing operators by suitable approximate counterparts which allow their numerical evaluation on a computer, say. Denoting by h the discretization parameter, one ends up with the problem

$$\min_{(y_h,u_h)\in Y_h\times U_h} J_h(y_h,u_h)$$

$$\text{subject to} \quad e_h(y_h,u_h)=0 \quad \text{and} \quad c_h(y_h)\in\mathcal{K}_h, \quad u_h\in U_{\text{ad}}^h, \tag{3.1}$$

The material presented in this chapter in parts is based on joint work with Klaus Deckelnick (Universität Magdeburg), Andreas Günther (Universität Hamburg), and Ulrich Matthes (Technische Universität Dresden).

M. Hinze (✉)
Department Mathematik, Universität Hamburg, Hamburg, Germany
e-mail: michael.hinze@uni-hamburg.de

M. Hinze et al., *Optimization with PDE Constraints*,
Mathematical Modelling: Theory and Applications 23,
DOI 10.1007/978-1-4020-8839-1_3,

where $J_h : Y_h \times U_h \to \mathbb{R}$, $e_h : Y_h \times U_h \to Z$, and $c_h : Y_h \to R$ with $\mathcal{K}_h \subset R_h$. For the finite dimensional subspaces one may require $Y_h \subset Y$, $U_h \subset U$, say, and $\mathcal{K}_h \subseteq R_h$ a closed and convex cone, $U^h_{\mathrm{ad}} \subseteq U_h$ closed and convex. This approach in general is referred to as first discretize, then optimize and is discussed in Sect. 3.2.2. On the other hand one may switch to the associated Karush-Kuhn-Tucker system (1.140)–(1.143) and substitute all appearing function spaces and operators accordingly. This leads to solving

$$e_h(y_h, u_h) = 0, \qquad c_h(y_h) \in \mathcal{K}_h, \tag{3.2}$$

$$\lambda_h \in \mathcal{K}_h{}^{\circ}, \qquad \langle \lambda_h, c_h(y_h)\rangle_{R^*,R} = 0, \tag{3.3}$$

$$L_{h_y}(y_h, u_h, p_h) + c_h'(y_h)^* \lambda_h = 0, \tag{3.4}$$

$$\bar{u}_h \in U^h_{\mathrm{ad}}, \qquad \langle L_{h_u}(y_h, u_h, p_h), u - u_h\rangle_{U^*,U} \geq 0 \quad \forall u \in U^h_{\mathrm{ad}} \tag{3.5}$$

for $\bar{y}_h, \bar{u}_h, \bar{p}_h, \bar{\lambda}_h$. This approach in general is referred to as first optimize, then discretize and is discussed in Sect. 3.2.3. As is shown in Sect. 3.2.5 the special structure of optimization problems of the form (1.138) allows for discrete concepts which avoid the (explicit) discretization of the control variables.

Instead of applying discrete concepts to problem (1.138) or (1.140)–(1.143) directly we may first apply an SQP approach on the continuous level and then apply first discretize, then optimize to the related linear quadratic constrained subproblems (compare step 2. of the Josephy-Newton method (2.17)), or first optimize, then discretize to the SQP systems (2.48) appearing in each iteration of the Josephy-Newton method on the infinite dimensional level. This is one of our motivations to illustrate all discrete concepts at hand of linear model PDEs which are well understood w.r.t. analysis and discretization concepts. This allows us to focus the presentation on structural aspects inherent in optimal control problems with PDE constraints.

3.2 Control Constraints

3.2.1 Stationary Model Problem

We consider the *Mother Problem* with control constraints;

$$(\mathbb{P}) \quad \begin{cases} \min_{(y,u)\in Y\times U} J(y,u) := \frac{1}{2}\|y - z\|^2_{L^2(\Omega)} + \frac{\alpha}{2}\|u\|^2_U \\ \text{s.t.} \\ -\Delta y = Bu \quad \text{in } \Omega, \\ y = 0 \qquad\quad \text{on } \partial\Omega, \\ \text{and} \\ u \in U_{\mathrm{ad}} \subseteq U. \end{cases} \tag{3.6}$$

Here, $\alpha > 0$ denotes a constant, $\Omega \subset \mathbb{R}^n$ denotes an open, bounded and sufficiently smooth (or convex polyhedral) domain, $Y := H^1_0(\Omega)$, the operator $B : U \to$

$H^{-1}(\Omega) \equiv Y^*$ denotes the (linear, continuous) control operator, and U_{ad} is assumed to be a closed and convex subset of the Hilbert space U. This problem corresponds to (1.77) with the setting $H = L^2(\Omega)$, $Q : Y \to H$ denoting the injection, $q_d = z$, $g = 0$, $Z = Y^*$, $Y_{\text{ad}} = Y$, $B = -B$, and $A = -\Delta$, compare also problem (1.117).

Let us provide some further examples of control operators and control spaces.

Example 3.1

1. $U := L^2(\Omega)$, $B : L^2(\Omega) \to H^{-1}(\Omega)$ Injection, $U_{\text{ad}} := \{v \in L^2(\Omega); a \le v(x) \le b \text{ a.e. in } \Omega\}$, $a, b \in L^\infty(\Omega)$.
2. $U := \mathbb{R}^m$, $B : \mathbb{R}^m \to H^{-1}(\Omega)$, $Bu := \sum_{j=1}^m u_j F_j$, $F_j \in H^{-1}(\Omega)$ given , $U_{\text{ad}} := \{v \in \mathbb{R}^m; a_j \le v_j \le b_j\}, a < b$.

Due to Theorem 1.43 problem $\mathbb{P}$ admits a unique solution $(y, u) \in Y \times U_{\text{ad}}$. Furthermore, using Remark 1.18 ($\mathbb{P}$) equivalently can be rewritten as the optimization problem

$$\min_{u \in U_{\text{ad}}} \hat{J}(u) \tag{3.7}$$

for the reduced functional

$$\hat{J}(u) := J(y(u), u) \equiv J(SBu, u) \tag{3.8}$$

over the set U_{ad}, where $S : Y^* \to Y$ denotes the solution operator associated with $-\Delta$. The first order necessary (and here also sufficient) optimality conditions here take the form

$$\langle \hat{J}'(u), v - u \rangle_{U^*,U} \ge 0 \quad \text{for all } v \in U_{\text{ad}} \tag{3.9}$$

where $\hat{J}'(u) = \alpha u + B^* S^*(SBu - z) \equiv \alpha u + B^* p \in U^*$, with $p := S^*(SBu - z) \in Y^{**}$ denoting the adjoint variable. Since Y is reflexive here the function p in the present setting satisfies the adjoint equations

$$\begin{aligned} -\Delta p &= y - z \quad \text{in } \Omega, \\ p &= 0 \quad \text{on } \partial\Omega. \end{aligned} \tag{3.10}$$

Remark 3.1 What has been stated so far and what follows also applies to more general elliptic PDEs defined through elliptic operators of the form

$$Ay := -\sum_{i,j=1}^n \partial_{x_j}\left(a_{ij} y_{x_i}\right) + \sum_{i=1}^n b_i y_{x_i} + cy,$$

combined with Dirichlet, Neumann, or Robin-type boundary conditions. Such operators are considered in by Deckelnick and Hinze in [39] together with state constraints.

To discretize ($\mathbb{P}$) we concentrate on finite element approaches and make the following assumptions.

Assumption 3.1 $\Omega \subset \mathbb{R}^n$ denotes a bounded domain (sufficiently smooth, or convex and polygonal, if only this is required), $\bar{\Omega} = \bigcup_{j=1}^{nt} \bar{T}_j$ with admissible quasi-uniform sequences of partitions $\{T_j\}_{j=1}^{nt}$ of Ω, i.e. with $h_{nt} := \max_j \operatorname{diam} T_j$ and $\sigma_{nt} := \min_j\{\sup \operatorname{diam} K; K \subseteq T_j\}$ there holds $c \leq \frac{h_{nt}}{\sigma_{nt}} \leq C$ uniformly in nt with positive constants $0 < c \leq C < \infty$ independent of nt. We abbreviate $\tau_h := \{T_j\}_{j=1}^{nt}$ and set $h = h_{nt}$.

In order to tackle ($\mathbb{P}$) numerically we in the following discuss two different approaches. The first is called *First discretize, then optimize*, the second *First optimize, then discretize*. It will turn out that both approaches under certain circumstances lead to the same numerical results. However, from a structural point of view they are different.

3.2.2 First Discretize, Then Optimize

The *First discretize, then optimize* approach works as follows. All quantities in problem ($\mathbb{P}$) in (3.6) are discretized a-priori, which results in a finite dimensional optimization problem. To discretize we replace the spaces Y and U by finite dimensional spaces Y_h and U_d, the set U_{ad} by some discrete counterpart U_{ad}^d, and the functionals, integrals and dualities by appropriate discrete surrogates. Having in mind Assumption 3.1 we set for $k \in \mathbb{N}$

$$W_h := \{v \in C^0(\bar{\Omega});\, v_{|T_j} \in \mathbb{P}_k(T_j) \text{ for all } 1 \leq j \leq nt\} =: \langle \phi_1, \ldots, \phi_{ng} \rangle, \quad \text{and}$$

$$Y_h := \{v \in W_h, v_{|\partial\Omega} = 0\} =: \langle \phi_1, \ldots, \phi_n \rangle \subseteq Y,$$

with some $0 < n < ng$. The resulting Ansatz for y_h then is of the form $y_h(x) = \sum_{i=1}^n y_i \phi_i$. Further, with $u^1, \ldots, u^m \in U$, we set $U_d := \langle u^1, \ldots, u^m \rangle$ and $U_{\text{ad}}^d := U_{\text{ad}} \cap U_d$. It is convenient to assume that U_{ad}^d may be represented in the form

$$U_{\text{ad}}^d = \left\{ u \in U; u = \sum_{j=1}^m s_j u^j, s \in \mathcal{S} \right\}$$

with $\mathcal{S} \subset \mathbb{R}^m$ denoting a closed, convex set. Finally let $z_h := I_h z = \sum_{i=1}^{ng} z_i \phi_i$, where $I_h : L^2(\Omega) \to W_h$ denotes a continuous interpolation operator. Now we replace problem ($\mathbb{P}$) by

$$(\mathbb{P}_{(h,d)}) \quad \begin{cases} \min_{(y_h,u_d) \in Y_h \times U_d} J_{(h,d)}(y,u) := \frac{1}{2}\|y_h - z_h\|_{L^2(\Omega)}^2 + \frac{\alpha}{2}\|u_d\|_U^2 \\ \text{s.t.} \\ a(y_h, v_h) = \langle B u_d, v_h \rangle_{Y^*,Y} \quad \text{for all } v_h \in Y_h, \\ \text{and} \\ u_d \in U_{\text{ad}}^d. \end{cases} \tag{3.11}$$

Here, we have set $a(y, v) := \int_\Omega \nabla y \nabla v dx$. Introducing the finite element stiffness matrix $A := (a_{ij})_{i,j=1}^n$, $a_{ij} := a(\phi_i, \phi_j)$, the finite element Mass matrix $M := (m_{ij})_{i,j=1}^{ng}$, $m_{ij} := \int_\Omega \phi_i \phi_j dx$, the matrix $E := (e_{ij})_{i=1,\dots,n;\, j=1,\dots,m}$, $e_{ij} = \langle Bu^j, \phi_i \rangle_{Y^*,Y}$, and the control mass matrix $F := (f_{ij})_{i,j=1}^m$, $f_{ij} := (u^i, u^j)_U$, allows us to rewrite $(\mathbb{P}_{(h,d)})$ in the form

$$(\mathbb{P}_{(n,m)}) \qquad \begin{cases} \min_{(y,s)\in\mathbb{R}^n\times\mathbb{R}^m} Q(y,s) := \frac{1}{2}(y-z)^t M(y-z) + \frac{\alpha}{2} s^t F s \\ \text{s.t. } Ay = Es \text{ and } s \in \mathcal{S}. \end{cases} \tag{3.12}$$

This is now a finite dimensional optimization problem with quadratic objective, linear equality constraints, and admissibility characterized by the closed, convex set $\mathcal{S} \subset \mathbb{R}^m$, compare (1.2). Since the matrix A is symmetric positive definite (spd), and thus regular problem $(\mathbb{P}_{(n,m)})$ is equivalent to minimizing the reduced functional $\hat{Q}(s) := Q(A^{-1}Es, s)$ over the set $\mathcal{S}$. It is clear that $(\mathbb{P}_{(n,m)})$ admits a unique solution $(y,s) \in \mathbb{R}^n \times \mathcal{S}$ which is characterized by the finite dimensional variational inequality

$$(\hat{Q}'(s), t - s)_{\mathbb{R}^m} \geq 0 \quad \text{for all } t \in \mathcal{S}, \tag{3.13}$$

with $\hat{Q}'(s) = \alpha F s + E^t A^{-t} M(A^{-1}Es - z) \equiv \alpha F s + E^t p$, where $p := A^{-t}M(A^{-1}Es - z)$ denotes the discrete adjoint vector to whom we associate the discrete adjoint variable $p_h := \sum_{i=1}^n p_i \phi_i$. Comparing this with the expression for $\hat{J}'(u)$ in (3.8), we note that the operator E here takes the role the control operator B there, and the inverse of the stiffness matrix A here that of the solution operator S there.

Problem $(\mathbb{P}_{(n,m)})$ now can be solved numerically with the help of appropriate solution algorithms, which should exploit the structure of the problem. We refer to Chap. 2 for a discussion of this issue.

We fix the following

Note 3.1 In the *First discretize, then optimize* approach the discretization of the adjoint variable p is determined by the Ansatz for the discrete state y_h, more specifically, by the discrete test space used in the variational formulation.

In the *First optimize, then discretize* approach discussed next, this is different.

3.2.3 First Optimize, Then Discretize

The starting point for the present approach is the system of first order necessary optimality conditions for problem $(\mathbb{P})$ stated next, compare (1.114)–(1.116);

$$
(\mathbb{OS}) \quad \begin{cases} -\Delta y = Bu & \text{in } \Omega, \\ y = 0 & \text{on } \partial\Omega, \\ -\Delta p = y - z & \text{in } \Omega, \\ p = 0 & \text{on } \partial\Omega, \\ \langle \alpha u + B^* p, v - u \rangle_{U^*,U} \geq 0 & \text{for all } v \in U_{\text{ad}}. \end{cases} \tag{3.14}
$$

Now we discretize everything related to the state y, the control u, and to functionals, integrals, and dualities as in Sect. 3.2.2. Further, we have the freedom to also select a discretization of the adjoint variable p. Here we choose continuous finite elements of order l on τ, which leads to the Ansatz $p_h(x) = \sum_{i=1}^q p_i \chi_i(x)$, where $\langle \chi_1, \dots, \chi_q \rangle \subset Y$ denotes the Ansatz space for the adjoint variable. Forming the adjoint stiffness matrix $\tilde{A} := (\tilde{a}_{ij})_{i,j=1}^q$, $\tilde{a}_{ij} := a(\chi_i, \chi_j)$, the matrix $\tilde{E} := (\tilde{e}_{ij})_{i=1,\dots,q; j=1,\dots,m}$, $\tilde{e}_{ij} = \langle Bu^j, \chi_i \rangle_{Y^*,Y}$, and $T := (t_{ij})_{i=1,\dots,n; j=1,\dots,q}$, $t_{ij} := \int_\Omega \phi_i \chi_j dx$, the discrete analogon to $(\mathbb{OS})$ reads

$$
(\mathbb{OS})_{(n,q,m)} \quad \begin{cases} Ay = Es, \\ \tilde{A} p = T(y - z), \\ (\alpha F s + \tilde{E}^t p, t - s)_{\mathbb{R}^m} \geq 0 \quad \text{for all } t \in \mathcal{S}. \end{cases} \tag{3.15}
$$

Since the matrices A and $\tilde{A}$ are spd, this system is equivalent to the variational inequality

$$
(\alpha F s + \tilde{E}^t \tilde{A}^{-1} T (A^{-1} E s - z), t - s)_{\mathbb{R}^m} \geq 0 \quad \text{for all } t \in \mathcal{S}. \tag{3.16}
$$

Before we relate the approaches of Sects. 3.2.2 and 3.2.3 let us give some examples, compare also Example 3.1.

Example 3.2

1. $U := L^2(\Omega)$, $B : L^2(\Omega) \to H^{-1}(\Omega)$ Injection, $U_{\text{ad}} := \{v \in L^2(\Omega); a \leq v(x) \leq b \text{ a.e. in } \Omega\}$, $a, b \in L^\infty(\Omega)$. Further let $k = l = 1$ (linear finite elements for y and p), $U_d := \langle u^1, \dots, u^{nt} \rangle$, where $u^k_{|T_i} = \delta_{ki}$ $(k, i = 1, \dots, nt)$ are piecewise constant functions (i.e. $m = nt$), $\mathcal{S} := \prod_{i=1}^{nt} [a_i, b_i]$, where $a_i := a(\text{barycenter}(T_i))$, $b_i := b(\text{barycenter}(T_i))$.
2. As in 1., but $U_d := \langle u^1, \dots, u^{ng} \rangle$, where $u^k_{|D_i} = \delta_{ki}$ $(k, i = 1, \dots, ng)$ are piecewise constant functions (i.e. $m = ng$), with D_i denoting the patch associated to the vertex P_i $(i = 1, \dots, ng)$ of the barycentric dual triangulation of τ, $\mathcal{S} := \prod_{i=1}^{ng} [a_i, b_i]$, where $a_i := a(P_i)$, $b_i := b(P_i)$.
3. As in 1., but $U_d := \langle \phi_1, \dots, \phi_{ng} \rangle$ (i.e. $m = ng$), $\mathcal{S} := \prod_{i=1}^{ng} [a_i, b_i]$, where $a_i := a(P_i)$, $b_i := b(P_i)$, with P_i $(i = 1, \dots, ng)$ denoting the vertices of the triangulation τ.
4. (Compare Example 3.1): As in 1., but $U := \mathbb{R}^m$, $B : \mathbb{R}^m \to H^{-1}(\Omega)$, $Bu := \sum_{j=1}^m u_j F_j$, $F_j \in H^{-1}(\Omega)$ given, $U_{\text{ad}} := \{v \in \mathbb{R}^m; a_j \leq v_j \leq b_j\}$, $a < b$, $U_d := \langle e_1, \dots, e_m \rangle$ with $e_i \in \mathbb{R}^m$ $(i = 1, \dots, m)$ denoting the i-th unit vector, $\mathcal{S} := \prod_{i=1}^{ng} [a_i, b_i] \equiv U_{\text{ad}}$.

3.2.4 Discussion and Implications

Now let us compare the approaches of the two previous sections. It is clear that choosing the same Ansatz spaces for the state y and the adjoint variable p in the *First optimize, then discretize* approach leads to an optimality condition which is identical to that of the *First discretize, then optimize* approach in (3.13), since then $T \equiv M$. However, choosing a different approach for p in general leads to (3.16) with a rectangular, non-quadratic matrix T, with the consequence that the matrix $\alpha F + \tilde{E}^t \tilde{A}^{-1} T A^{-1} E$ not longer represents a symmetric matrix. This is different for the matrix $\hat{Q}''(s) = \alpha F + E^t A^{-1} M A^{-1} E$ of the *First discretize, then optimize* approach. Moreover, the expression $\alpha F s + \tilde{E}^t \tilde{A}^{-1} T (A^{-1} E s - z)$ in general does not represent a gradient, which is different for $\hat{Q}'(s) = \alpha F s + E^t A^{-t} M (A^{-1} E s - z)$ since this in fact is the gradient of the reduced finite dimensional functional $\hat{Q}(s)$.

In many situations of control constrained optimization with PDEs the adjoint variable p admits more regularity than the state y. For example, if z is a smooth function, the domain Ω has smooth boundary and B denotes the injection as in Example 3.1 1., the adjoint variable p admits two more weak derivatives than the state y, whose regularity in the control constrained case is restricted through the regularity of the control u, which in the in the case of e.g. box constraints with constant bounds is not better than $W^{1,r}$ for some $r \leq \infty$, no matter how smooth the boundary of Ω is. So it could be meaningful to use Ansatz functions with higher polynomial degree for p than for y. On the other hand, in the presence of additional state constraints the regularity of the adjoint p is lower than that of the state y, so that it also may be meaningful to consider different Ansatz spaces for the state and the adjoint state.

Up to now there is no general recipe which approach has to be preferred, and it should depend on the application and computational resources which approach to take for tackling the numerical solution of the optimization problem. However, the numerical approach taken should to some extent reflect and preserve the structure which is inherent in the infinite dimensional optimization problem ($\mathbb{P}$). This can be best explained in the case without control constraints, i.e. $U_{\text{ad}} \equiv U$. Then the first order necessary optimality conditions for ($\mathbb{P}$) read

$$\hat{J}'(u) = \alpha u + B^* S^* (S B u - z) \equiv \alpha u + B^* p = 0 \quad \text{in } U^*.$$

Now let us for the moment consider Example 3.1 1., for which this equation becomes

$$\hat{J}'(u) = \alpha u + p = 0 \quad \text{in } L^2(\Omega),$$

since here $U^* = U$ holds. To conserve this identity on also on the discrete level one should relate to each other the discrete Ansatz for the control u and for the adjoint variable p. This argument remains valid also in the presence of control constraints, since then the variational inequality (3.9) can be replaced by the nonsmooth operator equation

$$u = P_{U_{\text{ad}}} \left(-\frac{1}{\alpha} p \right) \quad \text{in } L^2(\Omega), \tag{3.17}$$

where $P_{U_{\mathrm{ad}}}$ denotes the orthogonal projection in U (here $= L^2(\Omega)$) onto the admissible set of controls, compare Lemma 1.10. In any case, optimal control and corresponding adjoint state are related to each other, and this should be reflected by a numerical approach taken for the solution of problem ($\mathbb{P}$).

Note 3.2 Controls should be discretized conservative, i.e. according to the relation between the adjoint state and the control given by the first order optimality condition. This rule should be obeyed in both, the *First discretize, then optimize*, and in the *First optimize, then discretize* approach.

3.2.5 The Variational Discretization Concept

We observe that replacing the function p in (3.17) by its Finite Element approximation p_h, it is possible to compute u if the action of the orthogonal projection $P_{U_{\mathrm{ad}}}$ can be exactly evaluated on a computer. To anticipate discussion this is possible in many practical situations. This motivates to look for a discrete approach to problem (3.7) which leads to (3.17) with p replaced by p_h as optimality condition, and thus avoids explicit discretization of the control u. The following approach is developed by Hinze in [71]. Let us define the discrete reduced functional

$$\hat{J}_h(u) := J(S_h Bu, u), \quad u \in U,$$

and let us consider the following infinite dimensional optimization problem

$$\min_{u \in U_{\mathrm{ad}}} \hat{J}_h(u). \tag{3.18}$$

Similar as (3.7) this problem admits a unique solution $u_h \in U_{\mathrm{ad}}$ which is characterized by the variational inequality

$$\langle \hat{J}_h'(u_h), v - u_h \rangle_{U^*,U} \geq 0 \quad \text{for all } v \in U_{\mathrm{ad}}. \tag{3.19}$$

Using the inverse $R : U^* \to U$ of the Riesz isomorphism between U and U^* this inequality is equivalent to the non-smooth operator equation (see e.g. (1.97))

$$G_h(u) = u - P_{U_{\mathrm{ad}}}(u - \sigma R \hat{J}_h'(u)) \equiv u - P_{U_{\mathrm{ad}}}(u - \sigma(\alpha u + RB^* p_h)) = 0 \quad \text{in } U, \tag{3.20}$$

where we note that $R\hat{J}_h \equiv \nabla \hat{J}_h$. This equation holds for all $\sigma > 0$, and we have

$$J_h'(u) = \alpha u + B^* S_h^*(S_h Bu - z) \equiv \alpha u + B^* p_h(u).$$

We observe that in the setting of (3.17) $G_h(u) = 0$ is exactly (3.17) with p replaced by p_h if we choose $\sigma = \frac{1}{\alpha}$.

So far this is a discrete concept. But is it also possible to compute the solution u_h on a computer? Let us fix $\sigma = \frac{1}{\alpha}$ and consider the following fix point iteration for the numerical solution of

$$0 = G_h(u) = u - P_{U_{\text{ad}}}\left(-\frac{1}{\alpha}RB^* p_h\right) \quad \text{in } U;$$

Algorithm 3.2

- $u \in U$ given
- Do until convergence $u^+ = P_{U_{\text{ad}}}(-\frac{1}{\alpha}RB^* p_h(u))$, $u = u^+$,

where $p_h(u) = S_h^*(S_h Bu - z)$. In this algorithm the variable u (the control) is not discretized. Only state and adjoint are discretized. Two questions immediately arise.

(1) Is Algorithm 3.2 numerically implementable?
(2) Does Algorithm 3.2 converge?

Let us first discuss question (2). Since this algorithm is fix-point iteration, a sufficient condition for convergence is given by the relation $\alpha > \|B^* S_h^* S_h B\|_{\mathcal{L}(U,U^*)}$, since then the mapping $u \mapsto P_{U_{\text{ad}}}(-\frac{1}{\alpha}RB^* p_h(u))$ defines a contraction. This follows from the facts that R is an isometric isomorphism, and $P_{U_{\text{ad}}} : U \to U_{\text{ad}}$ denotes the orthogonal projection, and thus is Lipschitz continuous with Lipschitz constant $L = 1$, see Lemma 1.10(c). Therefore, convergence for Algorithm 3.2 can only be guaranteed if α is large enough. However, (3.20) in many practically relevant situations may also be solved by a semi-smooth Newton algorithm, or a primal-dual active set strategy, see Sect. 2.5.4, and fast local convergence in many practically relevant situations is easy to argue since then the functional $B^* p_h(u) \in U^*$ for given $u \in U$ often is *smoother* than the input control. Furthermore, in these situations, for $\sigma := \frac{1}{\alpha}$ the semi-smooth Newton method, and the primal-dual active set strategy are both numerically implementable in the variational discrete case, see Algorithm 3.9 in Sect. 3.2.7.

Question (1) admits the answer *Yes*, whenever for given u it is possible to numerically evaluate the expression

$$P_{U_{\text{ad}}}\left(-\frac{1}{\alpha}RB^* p_h(u)\right)$$

in the i-th iteration of Algorithm 3.2 with an numerical overhead which is *independent* of the iteration counter of the algorithm. To illustrate this fact let us turn back to Example 3.1 1., i.e. $U = L^2(\Omega)$ and B denoting the injection, with $a \equiv \text{const}_1$, $b \equiv \text{const}_2$. In this case it is easy to verify that

$$P_{U_{\text{ad}}}(v)(x) = P_{[a,b]}(v(x)) = \max\{a, \min\{v(x), b\}\},$$

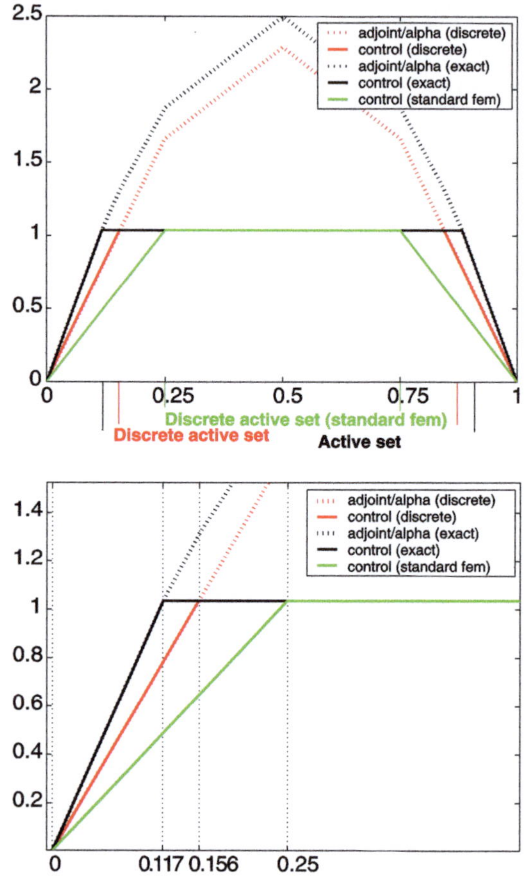

Fig. 3.1 Piecewise linear, continuous interpolation of the continuous control $u = P_{\{u \le \frac{\sqrt{2}-1}{4\alpha}\}}(-\frac{1}{\alpha}p)$ and variational-discrete control $u_h = P_{\{u \le \frac{\sqrt{2}-1}{4\alpha}\}}(-\frac{1}{\alpha}p_h)$ together with their active sets determined by $-\frac{1}{\alpha}p$ and $-\frac{1}{\alpha}p_h$, respectively, for $h = 1/3$ and $\alpha = 0.1$ in the case $n = 1$. Zoom of the same (*bottom*). The decoupling of discrete active set and finite element grid clearly is shown. Results are taken from [71, Sect. 4.2]

see Lemma 1.12, so that in every iteration of Algorithm 3.2 we have to form the control

$$u^+(x) = P_{[a,b]}\left(-\frac{1}{\alpha}p_h(x)\right), \tag{3.21}$$

which for the one-dimensional setting is illustrated in Fig. 3.1.

To construct the function u^+ it is sufficient to characterize the intersection of the bounds a, b (understood as constant functions) and the function $-\frac{1}{\alpha}p_h$ on every simplex T of the triangulation $\tau = \tau_h$. For piecewise linear finite element approximations of p we have the following theorem.

Theorem 3.3 *Let u^+ denote the function of* (3.21), *with p_h denoting a piecewise linear, continuous finite element function, and constant bounds $a < b$. Then there exists a partition $\kappa_h = \{K_1, \dots K_{l(h)}\}$ of Ω such that u^+ restricted to K_j $(j = 1, \dots, l(h))$ is a polynomial either of degree zero or one. For $l(h)$ there holds*

$$l(h) \le Cnt(h),$$

with a positive constant $C \leq 3$ and $nt(h)$ denoting the number of simplexes in τ_h. In particular, the vertices of the discrete active set associated to u^+ need not coincide with finite element nodes.

Proof Abbreviate $\xi_h^a := -\frac{1}{\alpha}p_h^* - a$, $\xi_h^b := b - \frac{1}{\alpha}p_h^*$ and investigate the zero level sets 0_h^a and 0_h^b of ξ_h^a and ξ_h^b, respectively. If 0_h^a or $0_h^b = T_i$ the assertion follows easily.

Case $n = 1$: $0_h^a \cap T_i$ is either empty or a point $S_i^a \in T_i$. Every point S_i^a subdivides T_i into two sub-intervals. Analogously $0_h^b \cap T_i$ is either empty or a point $S_i^b \in T_i$. Further $S_i^a \neq S_i^b$ since $a < b$. The maximum number of sub-intervals of T_i induced by 0_h^a and 0_h^b therefore is equal to three. Therefore, $l(h) \leq 3nt(h)$, i.e. $C = 3$.

Case $n = 2$: $0_h^a \cap T_i$ is either empty or a vertex of τ_h or a line $L_i^a \subset T_i$, analogously $0_h^b \cap T_i$ is either empty or a vertex of τ_h or a line $L_i^b \subset T_i$. Since $a < b$ the lines L_i^a and L_i^b do not intersect. Therefore, similar considerations as in the case $n = 1$ yield $C = 3$.

Case $n \in \mathbb{N}$: $0_h^a \cap T_i$ is either empty or a part of a k-dimensional hyperplane ($k < n$) $L_i^a \subset T_i$, analogously $0_h^b \cap T_i$ is either empty or a part of k-dimensional hyperplane ($k < n$) $L_i^b \subset T_i$. Since $a < b$ the surfaces L_i^a and L_i^b do not intersect. Therefore, similar considerations as in the case $n = 2$ yield $C = 3$. This completes the proof.

It is now clear that the proof of the previous theorem easily extends to functions p_h which are piecewise polynomials of degree $k \in \mathbb{N}$, and bounds a, b which are piecewise polynomials of degree $l \in \mathbb{N}$ and $m \in \mathbb{N}$, respectively, since the difference of a, b and p_h in this case also represents a piecewise polynomial function whose projection on every element can be easily characterized.

We now have that Algorithm 3.2 is numerically implementable in situations like those given in Example 3.1, but only converges for a certain parameter range of α. A locally (super-linear) convergent algorithm for the numerical solution of (3.20) is the semi-smooth Newton method of Sect. 2.5.4, since the function G is semi-smooth in the sense of Sect. 2.1, compare also the work of Hintermüller, Ito, and Kunisch [69], and that of Ulbrich [136, Example 5.6]. We present some details in Sect. 3.2.7.1.

3.2.6 Error Estimates

Next let us investigate the error $\|u - u_h\|_U$ between the solutions u of (3.9) and u_h of (3.18).

Theorem 3.4 *Let u denote the unique solution of* (3.7), *and u_h the unique solution of* (3.18). *Then there holds*

$$\alpha\|u-u_h\|_U^2+\frac{1}{2}\|y(u)-y_h\|^2 \le \langle B^*(p(u)-\tilde{p}_h(u)), u_h-u\rangle_{U*,U}$$
$$+\frac{1}{2}\|y(u)-y_h(u)\|^2, \qquad (3.22)$$

where $\tilde{p}_h(u) := S_h^*(SBu-z)$, $y_h(u) := S_hBu$, *and* $y(u) := SBu$.

Proof Since (3.18) is an optimization problem defined on U_{ad}, the unique solution u of (3.7) is an admissible test function in (3.19). Let us emphasize, that this is different for approaches, where the control space is discretized explicitly. In this case we may only expect that u_h is an admissible test function for the continuous problem (if ever). So let us test (3.9) with u_h, and (3.19) with u, and then add the resulting variational inequalities. This leads to

$$\left\langle\alpha(u-u_h)+B^*S^*(SBu-z)-B^*S_h^*(S_hBu_h-z), u_h-u\right\rangle_{U^*,U} \ge 0.$$

This inequality is equivalent to

$$\alpha\|u-u_h\|_U^2 \le \left\langle B^*(p(u)-\tilde{p}_h(u))+B^*(\tilde{p}_h(u)-p_h(u_h)), u_h-u\right\rangle_{U^*,U}.$$

Let us investigate the second addend on the right hand side of this inequality. By definition of the adjoint variables there holds

$$\begin{aligned}
&\left\langle B^*(\tilde{p}_h(u)-p_h(u), u_h-u\right\rangle_{U^*,U}\\
&=\langle \tilde{p}_h(u)-p_h(u), B(u_h-u)\rangle_{Y,Y^*}\\
&=a(y_h-y_h(u), \tilde{p}_h(u)-p_h(u))=\int_\Omega (y_h(u_h)-y_h(u))(y(u)-y_h(u))dx\\
&=-\|y_h-y\|^2+\int_\Omega (y-y_h)(y-y_h(u))dx \le -\frac{1}{2}\|y_h-y\|^2+\frac{1}{2}\|y-y_h(u)\|^2
\end{aligned}$$

so that the claim of the theorem follows.

What can we learn from Theorem 3.22? It tells us that an error estimate for $\|u-u_h\|_U$ is at hand, if

- An error estimate for $\|RB^*(p(u)-\tilde{p}_h(u)\|_U$ is available, and
- An error estimate for $\|y(u)-y_h(u)\|_{L^2(\Omega)}$ is available.

Note 3.3 The error $\|u-u_h\|_U$ between the solution u of problem (3.7) and u_h of (3.18) is completely determined by the approximation properties of the discrete solution operators S_h and S_h^*.

Let us revisit Example 3.1. Then $U = L^2(\Omega)$ and B denotes the injection. Then $y = SBu \in H^2(\Omega) \cap H_0^1(\Omega)$ (if for example $\Omega \in C^{1,1}$ or Ω convex). Since

$$\begin{aligned}\langle B^*(p(u)-\tilde{p}_h(u)), u-u_h\rangle_{U^*,U} &= \int_\Omega (p(u)-\tilde{p}_h(u))(u-u_h)dx \\ &\le \|p(u)-\tilde{p}_h(u)\|_{L^2(\Omega)}\|u-u_h\|_{L^2(\Omega)} \\ &\le ch^2\|y(u)\|_{L^2(\Omega)}\|u-u_h\|_{L^2(\Omega)},\end{aligned}$$

and

$$\|y-y_h(u)\| \le Ch^2\|u\|_U,$$

we together with the estimate

$$\begin{aligned}\|y-y_h\|_Y^2 &\le Ca(y-y_h, y-y_h(u)) + a(y-y_h, y_h(u)-y_h) \\ &= Ca(y-y_h, y-y_h(u)) + \langle y_h(u)-y_h, B(u-u_h)\rangle_{Y,Y^*} \\ &\le \epsilon\|y-y_h\|_Y^2 + C_\epsilon\{\|y-y_h(u)\|_Y^2 + \|u-u_h\|_U^2\}\end{aligned}$$

for the Y-norm immediately obtain.

Theorem 3.5 *Let u and u_h denote the solutions of problem* (3.7) *and* (3.18), *respectively in the setting of Example* 3.1 1. *Then there holds*

$$\|u-u_h\|_{L^2(\Omega)} + h\|y-y_h\|_Y \le ch^2\left\{\|y(u)\|_{L^2(\Omega)} + \|u\|_{L^2(\Omega)}\right\}.$$

And this theorem is also valid for the setting of Example 3.1 2. if we require $F_j \in L^2(\Omega)$ $(j=1,\dots,m)$. This is an easy consequence of the fact that for a functional $z \in H^{-1}(\Omega)$ there holds $B^*z \in \mathbb{R}^m$ with $(B^*z)_i = \langle F_i, z\rangle_{Y^*,Y}$ for $i=1,\dots,m$.

Theorem 3.6 *Let u and u_h denote the solutions of problem* (3.7) *and* (3.18), *respectively in the setting of Example* 3.1 2. *Then there holds*

$$\|u-u_h\|_{\mathbb{R}^m} + h\|y-y_h\|_Y \le ch^2\left\{\|y(u)\|_{L^2(\Omega)} + \|u\|_{\mathbb{R}^m}\right\},$$

where the positive constant now depends on the functions F_j $(j=1,\dots,m)$.

Proof It suffices to estimate

$$\begin{aligned}&(B^*(p(u)-\tilde{p}_h(u)), u-u_h)_{\mathbb{R}^m} \\ &= \sum_{j=1}^m \left\{\int_\Omega F_j(p(u)-\tilde{p}_h(u))dx(u-u_h)_j\right\} \\ &\le \|p(u)-\tilde{p}_h(u)\|_{L^2(\Omega)}\left(\sum_{j=1}^m\int_\Omega |F_j|^2dx\right)^{\frac{1}{2}}\|u-u_h\|_{\mathbb{R}^m} \\ &\le ch^2\|y(u)\|_{L^2(\Omega)}\|u-u_h\|_{\mathbb{R}^m}.\end{aligned}$$

The reminder terms can be estimated as above.

3.2.6.1 Uniform Estimates

Using Theorem 3.5 in the case of $U = L^2(\Omega)$ with $U_{\rm ad}$ from Example 3.1 1., in combination with discrete Sobolev embeddings, it is also possible to provide error estimates in the L^∞ norm, see Hinze [71] for details. A proof of the following discrete Sobolev embeddings is given by Xu and Zhang in [148], see also the book of Thomée [132] for the case $n = 2$.

Proposition 3.1 *Let τ_h denote a quasi-uniform, regular triangulation of $\Omega \subset \mathbb{R}^n$ ($n = 1, 2, 3$). Then for every piecewise linear, continuous finite element function $v_h \in H_0^1(\Omega)$ there holds*

$$\|v_h\|_\infty \le C \left\{ \begin{array}{c} 1 \\ |\log h|^{\frac{1}{2}} \\ h^{-\frac{1}{2}} \end{array} \right\} |v_h|_1 \quad \text{for} \left\{ \begin{array}{c} n = 1 \\ n = 2 \\ n = 3 \end{array} \right\}, \tag{3.23}$$

where $C > 0$ is a generic constant and $|\cdot|_1$ denotes the H^1 semi-norm.

Theorem 3.7 *Let $z \in L^2(\Omega)$ and let u, u_h denote the solutions of problems* (3.7) *and* (3.18), *respectively. Then there holds*

$$\|u - u_h\|_\infty \le C \left\{ \|(S^* - S_h^*)z\|_\infty + \|(S^* - S_h^*)Su\|_\infty \right.$$

$$\left. + \left\{ \begin{array}{c} h^2 \\ h^2|\log h|^{\frac{1}{2}} \\ h^{\frac{3}{2}} \end{array} \right\} \|u\|_0 \, \text{for} \left\{ \begin{array}{c} n = 1 \\ n = 2 \\ n = 3 \end{array} \right\} \right\}. \tag{3.24}$$

Proof Let $p := S^*(Su^* - z)$ and $p_h := S_h^*(S_h u_h^* - z)$ denote the adjoints associated to u, u_h. Now write $p - p_h = S^*Su - S_h^*S_h u_h + (S_h^* - S^*)z$. Since $U_{\rm ad}$ is defined through box constraints one gets

$$\begin{aligned} \|u - u_h\|_\infty &\le \frac{1}{\alpha}\|p - p_h\|_\infty \\ &\le \frac{1}{\alpha}\{\|(S^* - S_h^*)Su\|_\infty + \|(S^* - S_h^*)z\|_\infty \\ &\quad + \|S_h^*Su - S_h^*S_h u\|_\infty + \|S_h^*S_h u - S_h^*S_h u_h\|_\infty\}. \end{aligned}$$

To estimate the third and fourth addend utilize Proposition 3.1. For the third addend one gets in the case $n = 2$

$$\begin{aligned} \|S_h^*Su - S_h^*S_h u\|_\infty &\le C|\log h|^{\frac{1}{2}}|S_h^*Su - S_h^*S_h u|_1 \\ &\le C|\log h|^{\frac{1}{2}}\|Su - S_h u\|_0 \le C|\log h|^{\frac{1}{2}}h^2\|u\|_0. \end{aligned}$$

Similarly for the fourth addend

$$\|S_h^* S_h u - S_h^* S_h u_h\|_\infty \le C|\log h|^{\frac{1}{2}} |S_h^* S_h u - S_h^* S_h u_h|_1$$
$$\le C|\log h|^{\frac{1}{2}} \|u - u_h\|_0 \le C|\log h|^{\frac{1}{2}} h^2 \{\|z\|_0 + \|u\|_0\},$$

where Theorem 3.5 is used. The exposition for the cases $n = 1, 3$ is similar. This completes the proof.

Remark 3.2 To finalize the L^∞ error estimate for $u - u_h$ it remains to provide estimates for $e_1 := \|(S^* - S_h^*)Su\|_\infty$ and $e_2 := \|(S^* - S_h^*)z\|_\infty$. However, the approximation order for these terms is restricted by 2. In this sense estimate (3.24) is optimal. For example there holds

- $e_i = \mathcal{O}(h^{2-\frac{n}{2}})$, $i = 1, 2$, $n = 1, 2, 3$,
- $e_i = \mathcal{O}(h)$, $i = 1, 2$, $n = 2, 3$, if a discrete maximum principle is satisfied for the finite element spaces, and with the results of Schatz in [119],
- $e_i \le Ch^{2-\frac{n}{q}} |\log h| \{\|Su\|_{W^{2,q}}, \|S^* z\|_{W^{2,q}}\}$, $i = 1, 2$, $n = 1, 2, 3$, if $Su, S^* z \in W^{2,\infty}(\Omega)$,

see [33]. L^∞-error estimates for piecewise linear, continuous approximations of the control for two-dimensional elliptic problems are given Meyer and Rösch in [101], piecewise constant control approximations are considered by Arada, Casas and Tröltzsch in [8], and by Casas, Mateos and Tröltzsch in [30].

3.2.6.2 Numerical Examples for Distributed Control

Now let us present numerical results for variational discretization including a numerical comparisons to other commonly used discrete approaches. Let us begin with the following distributed control problem.

Example 3.3 (Distributed control, generic case) We consider problem (3.6) with Ω denoting the unit circle, $U_{ad} := \{v \in L^2(\Omega); -0.2 \le u \le 0.2\} \subset L^2(\Omega)$ and $B : L^2(\Omega) \to Y^*(\equiv H^{-1}(\Omega))$ the injection. Furthermore we set $z(x) := (1 - |x|^2)x_1$ and $\alpha = 0.1$. The numerical discretization of state and adjoint state is performed with linear, continuous finite elements.

Here we consider the scenario that the exact solution of the problem is not known in advance (although it is easy to construct example problems where exact state, adjoint state and control are known, see the book of Tröltzsch [133]). Instead we use the numerical solutions computed on a grid with $h = \frac{1}{256}$ as reference solutions. To present numerical results it is convenient to introduce the *Experimental Order of Convergence*, brief EOC, which for some positive error functional E is defined by

$$\text{EOC} := \frac{\log E(h_1) - \log E(h_2)}{\log h_1 - \log h_2}. \tag{3.25}$$

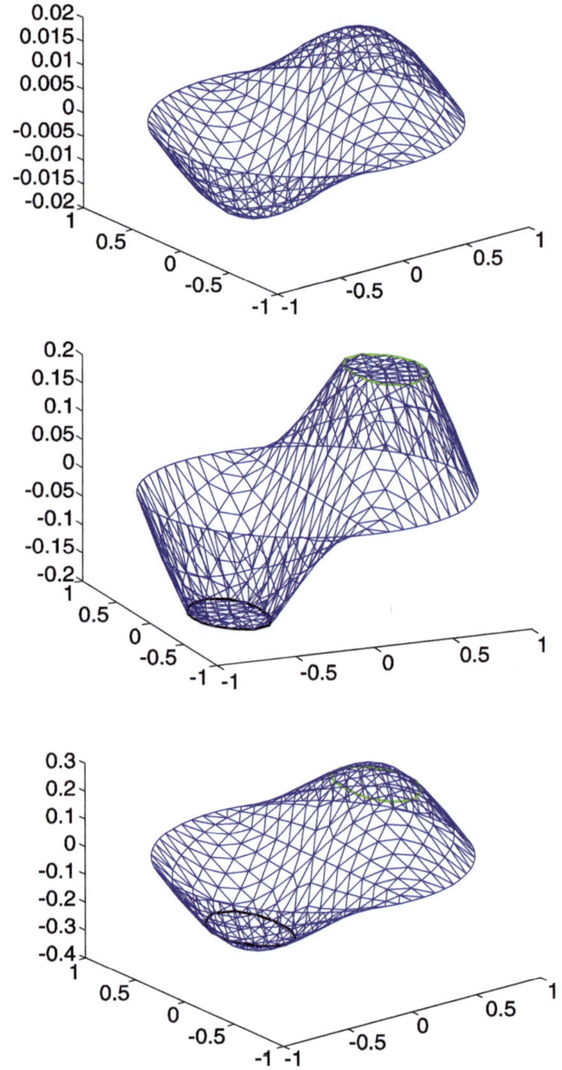

Fig. 3.2 Numerical results of distributed control: Optimal state (*top*), optimal control (*middle*) and corresponding adjoint state (*bottom*). The black and green lines, respectively depict the boarders of the active set

We note that $EOC = \beta$ holds, if $E(h) \sim h^{\beta}$. Figure 3.2 presents the numerical results for $h = \frac{1}{8}$. Figure 3.3 presents a numerical comparison for active sets obtained by variational discretization, and obtained by a conventional approach which uses piecewise linear, continuous finite elements also for the a-priori discretization of controls. We observe a significant better resolution of active sets by the approach presented in the previous subsections. In Tables 3.1–3.3 the experimental order of convergence for different error functionals is presented for the state, adjoint state, and control. We use the abbreviations $E_{y_{L2}}$ for the error in the L^2-norm, $E_{y_{\text{sup}}}$ for the error in the L^{∞}-norm, $E_{y_{\text{sem}}}$ for the error in the H^1-seminorm, and $E_{y_{H_1}}$ for the error in the H^1-norm. Table 3.4 presents the results for the controls of the con-

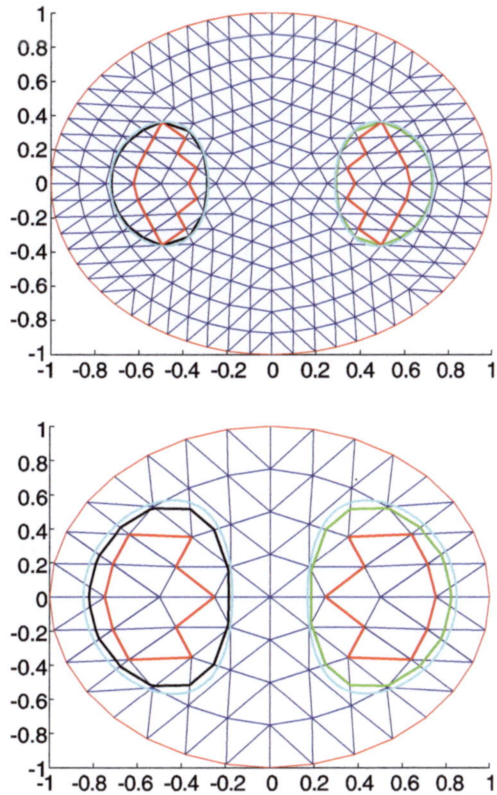

Fig. 3.3 Numerical comparison of active sets obtained by variational discretization, and those obtained by a conventional approach with piecewise linear, continuous controls: $h = \frac{1}{8}$ and $\alpha = 0.1$ (*top*), $h = \frac{1}{4}$ and $\alpha = 0.01$ (*bottom*). The red line depicts the boarder of the active set in the conventional approach, the cyan line the exact boarder, the black and green lines, respectively the boarders of the active set in variational discretization

Table 3.1 Errors (columns left) and EOC (columns right) of state for different error functionals. As reference solution y_h for $h = \frac{1}{256}$ is taken

h	$E_{y_{L2}}$	$E_{y_{\sup}}$	$E_{y_{\text{sem}}}$	$E_{y_{H_1}}$	$\text{EOC}_{y_{L2}}$	$\text{EOC}_{y_{\sup}}$	$\text{EOC}_{y_{\text{sem}}}$	$\text{EOC}_{y_{H_1}}$
1/1	1.47e−2	1.63e−2	5.66e−2	5.85e−2	–	–	–	–
1/2	5.61e−3	6.02e−3	2.86e−2	2.92e−2	1.39	1.44	0.98	1.00
1/4	1.47e−3	1.93e−3	1.38e−2	1.39e−2	1.93	1.64	1.06	1.08
1/8	3.83e−4	5.02e−4	6.89e−3	6.90e−3	1.94	1.95	1.00	1.01
1/16	9.65e−5	1.26e−4	3.44e−3	3.45e−3	1.99	2.00	1.00	1.00
1/32	2.40e−5	3.14e−5	1.71e−3	1.71e−3	2.01	2.00	1.01	1.01
1/64	5.73e−6	7.78e−6	8.37e−4	8.37e−4	2.06	2.01	1.03	1.03
1/128	1.16e−6	1.85e−6	3.74e−4	3.74e−4	2.30	2.07	1.16	1.16

ventional approach which should be compared to the numbers of Table 3.3. Table 3.5 presents the order of convergence of the active sets for variational discretization, and for the conventional approach. As error functional we use in this case the

Table 3.2 Errors (columns left) and EOC (columns right) of adjoint state for different error functionals. As reference solution p_h for $h = \frac{1}{256}$ is taken

h	$E_{p_{L2}}$	$E_{p_{\sup}}$	$E_{p_{\text{sem}}}$	$E_{p_{H_1}}$	$\text{EOC}_{p_{L2}}$	$\text{EOC}_{p_{\sup}}$	$\text{EOC}_{p_{\text{sem}}}$	$\text{EOC}_{p_{H_1}}$
1/1	2.33e−2	2.62e−2	8.96e−2	9.26e−2	–	–	–	–
1/2	6.14e−3	7.75e−3	4.36e−2	4.40e−2	1.92	1.76	1.04	1.07
1/4	1.59e−3	2.50e−3	2.17e−2	2.18e−2	1.95	1.64	1.00	1.02
1/8	4.08e−4	6.52e−4	1.09e−2	1.09e−2	1.97	1.94	0.99	0.99
1/16	1.03e−4	1.64e−4	5.48e−3	5.48e−3	1.99	1.99	1.00	1.00
1/32	2.54e−5	4.14e−5	2.73e−3	2.73e−3	2.01	1.99	1.01	1.01
1/64	6.11e−6	1.04e−5	1.33e−3	1.33e−3	2.06	1.99	1.03	1.03
1/128	1.27e−6	2.61e−6	5.96e−4	5.96e−4	2.27	1.99	1.16	1.16

Table 3.3 Errors (columns left) and EOC (columns right) of control for different error functionals. As reference solution u_h for $h = \frac{1}{256}$ is taken

h	$E_{u_{L2}}$	$E_{u_{\sup}}$	$E_{u_{\text{sem}}}$	$E_{u_{H_1}}$	$\text{EOC}_{u_{L2}}$	$\text{EOC}_{u_{\sup}}$	$\text{EOC}_{u_{\text{sem}}}$	$\text{EOC}_{u_{H_1}}$
1/1	2.18e−1	2.00e−1	8.66e−1	8.93e−1	–	–	–	–
1/2	5.54e−2	7.75e−2	4.78e−1	4.81e−1	1.97	1.37	0.86	0.89
1/4	1.16e−2	2.30e−2	2.21e−1	2.22e−1	2.25	1.75	1.11	1.12
1/8	3.02e−3	5.79e−3	1.15e−1	1.15e−1	1.94	1.99	0.94	0.95
1/16	7.66e−4	1.47e−3	6.09e−2	6.09e−2	1.98	1.98	0.92	0.92
1/32	1.93e−4	3.67e−4	2.97e−2	2.97e−2	1.99	2.00	1.03	1.03
1/64	4.82e−5	9.38e−5	1.41e−2	1.41e−2	2.00	1.97	1.07	1.07
1/128	1.17e−5	2.37e−5	6.40e−3	6.40e−3	2.04	1.98	1.14	1.14

Table 3.4 Conventional approach: Errors (columns left) and EOC (columns right) of control for different error functionals. As reference solution u_h for $h = \frac{1}{256}$ is taken

h	$E_{u_{L2}}$	$E_{u_{\sup}}$	$E_{u_{\text{sem}}}$	$E_{u_{H_1}}$	$\text{EOC}_{u_{L2}}$	$\text{EOC}_{u_{\sup}}$	$\text{EOC}_{u_{\text{sem}}}$	$\text{EOC}_{u_{H_1}}$
1/1	2.18e−1	2.00e−1	8.66e−1	8.93e−1	–	–	–	–
1/2	6.97e−2	9.57e−2	5.10e−1	5.15e−1	1.64	1.06	0.76	0.79
1/4	1.46e−2	3.44e−2	2.39e−1	2.40e−1	2.26	1.48	1.09	1.10
1/8	4.66e−3	1.65e−2	1.53e−1	1.54e−1	1.65	1.06	0.64	0.64
1/16	1.57e−3	8.47e−3	9.94e−2	9.94e−2	1.57	0.96	0.63	0.63
1/32	5.51e−4	4.33e−3	6.70e−2	6.70e−2	1.51	0.97	0.57	0.57
1/64	1.58e−4	2.09e−3	4.05e−2	4.05e−2	1.80	1.05	0.73	0.73
1/128	4.91e−5	1.07e−3	2.50e−2	2.50e−2	1.68	0.96	0.69	0.69

area

$$E_a := |(A \setminus A_h) \cup (A_h \setminus A)|$$

Table 3.5 Errors (columns left) and EOC (columns right) of active sets. As reference set that corresponding to the control u_h for $h = \frac{1}{256}$ is taken. The order of convergence seems to tend to 1.5 in the classical approach. The order of convergence of variational discretization is clearly 2, and its errors are two orders of magnitude smaller than those produced by the conventional approach

h	Conventional E_a	Approach EOC_a	Variational E_a	Discretization EOC_a
1/1	5.05e−1	–	5.11e−1	–
1/2	5.05e−1	0.00	3.38e−1	0.60
1/4	5.05e−1	0.00	1.25e−1	1.43
1/8	2.60e−1	0.96	2.92e−2	2.10
1/16	1.16e−1	1.16	7.30e−3	2.00
1/32	4.98e−2	1.22	1.81e−3	2.01
1/64	1.88e−2	1.41	4.08e−4	2.15
1/128	6.98e−3	1.43	8.51e−5	2.26

of the symmetric difference of discrete and continuous active sets. EOC with the corresponding subscripts denotes the associated experimental order of convergence.

As a result we obtain, that variational discretization provides a much better approximation of the controls and active sets than the conventional approach. In particular the errors in the L^2- and L^∞-norm are much smaller than the corresponding ones in the conventional approach. Let us also note that the results in the conventional approach would become even more worse if we would use piecewise constants as Ansatz for the controls. For theoretical and numerical results of conventional approaches let us refer to the work of Arada, Casas and Tröltzsch [8].

Let us note that similar numerical results can be obtained by an approach of Meyer and Rösch presented in [100]. The authors in a preliminary step compute a piecewise constant optimal control $\bar{u}$ and with its help compute in a post-processing step a projected control u^P through

$$u^P = P_{U_{\mathrm{ad}}}\left(-\frac{1}{\alpha}B^* p_h(\bar{u})\right)$$

which satisfies

$$\|u - u^P\| = \mathcal{O}(h^2)$$

in the setting of Example 3.1 1. However, the numerical analysis of their approach requires some sort of strict complementarity of the continuous solution u, which is not necessary to impose for obtaining the result of Theorem 3.5 for variational discretization. In particular Meyer and Rösch have to require that the $(d-1)$-dimensional Hausdorff measure of the discrete active set induced by the optimal control only intersects with a certain number of simplexes of the triangulation. This requirement for example can be satisfied, if the gradient of the adjoint in the solution does not vanish on the boarder of the active set and L^∞-estimates are available for the finite element approximation of the adjoint in the solution.

The next example considers control of an elliptic equation on an L-shape domain. In this situation the solution does not admit integrable second derivatives,

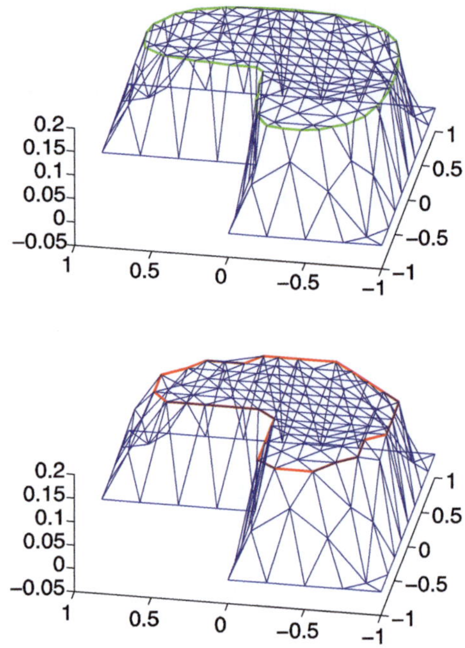

Fig. 3.4 Optimal state with variational discretization (*top*), and classical discretization

so that the approximation properties of finite element approximations are limited. To obtain appropriate finite element approximations graded meshes should be used. This technique combined with post processing of Meyer and Rösch [100] is investigated by Apel, Rösch and Winkler in [7], where also a numerical investigation can be found. For appropriately graded meshes they prove in [7, Theorem 1] the estimate

$$\|\bar{u} - \bar{u}_h\| = \mathcal{O}(h^2).$$

Let us note that this estimate for variational discretization is an immediate consequence of [7, Lemma 4] combined with (3.22), where an assumption like [7, (21)] on strict complementarity of in the continuous solution is not necessary.

Example 3.4 (L-shape) We consider the minimization problem (3.6) with $\Omega = (-1,1)^2 \setminus ([-1,0] \times [0,1])$ denoting an L-shape domain, $U_{\text{ad}} := \{v \in L^2(\Omega); -0.2 \le u \le 0.2\} \subset L^2(\Omega)$ and $B : L^2(\Omega) \to Y^*(\equiv H^{-1}(\Omega))$ the injection. Further we set $z(x) := (1 - |x|^2)$ and $\alpha = 0.1$. Figures 3.4–3.5 show the numerical solutions of the variational approach and the classical approach with piecewise linear, continuous control approximations, both obtained with Algorithm 3.2. For the presentation of the active sets a coarse grid is used.

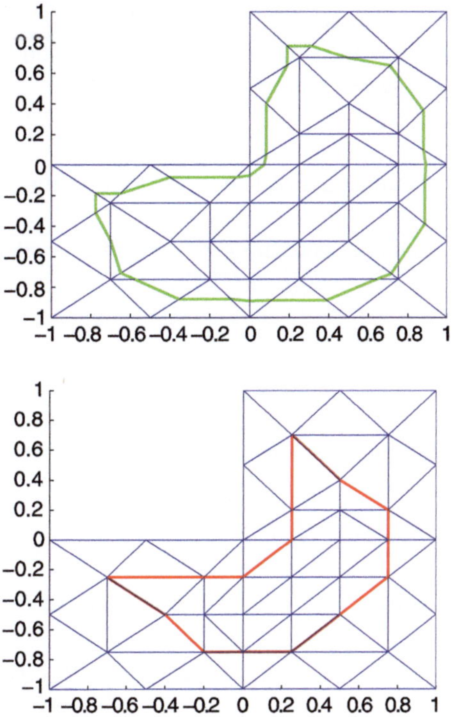

Fig. 3.5 Numerical comparison of active sets obtained with variational discretization (*top*), and those obtained by a conventional approach with piecewise linear, continuous controls (*bottom*): $h = \frac{1}{2}$ and $\alpha = 0.1$. The red line depicts the boarder of the active set in the conventional approach, the green line the boarder of the active set of variational discretization

3.2.7 Boundary Control

Concerning their structure most of the considerations of the previous subsections remain valid also for inhomogeneous Neumann and Dirichlet boundary control problems. Let us consider the model problems

$$(\mathbb{NC}) \quad \begin{cases} \min_{(y,u)\in Y\times U} J(y,u) := \frac{1}{2}\|y-z\|^2_{L^2(\Omega)} + \frac{\alpha}{2}\|u\|^2_U \\ \text{s.t.} \\ -\Delta y = 0 \qquad \text{in } \Omega, \\ \partial_\eta y = Bu - \gamma y \quad \text{on } \partial\Omega, \\ \text{and} \\ u \in U_{\text{ad}} \subseteq U, \end{cases} \tag{3.26}$$

and

$$(\mathbb{DC}) \quad \begin{cases} \min_{(y,u)\in Y\times U} J(y,u) := \frac{1}{2}\|y-z\|^2_{L^2(\Omega)} + \frac{\alpha}{2}\|u\|^2_U \\ \text{s.t.} \\ -\Delta y = 0 \quad \text{in } \Omega, \\ y = Bu \qquad \text{on } \partial\Omega, \\ \text{and} \\ u \in U_{\text{ad}} \subseteq U, \end{cases} \tag{3.27}$$

where in both cases $B : U \to L^2(\Gamma)$ with $\Gamma := \partial\Omega$. Let us note that the Dirichlet problem for y in ($\mathbb{DC}$) for $Bu \in L^2(\Gamma)$ is understood in the very weak sense, see (3.43) for the associated bilinear form.

3.2.7.1 Neumann and Robin-Type Boundary Control

We first consider problem ($\mathbb{NC}$) which equivalently can be rewritten in the form

$$\min_{u\in U_{\text{ad}}} \hat{J}(u) \tag{3.28}$$

for the reduced functional $\hat{J}(u) := J(y(u), u) \equiv J(SBu, u)$ over the set U_{ad}, where $S : Y^* \to Y$ for $Y := H^1(\Omega)$ denotes the weak solution operator of the Neumann boundary value problem for $-\Delta$, i.e. $y = Sf$ iff

$$a(y, v) := \int_\Omega \nabla y \nabla v dx + \int_\Gamma \gamma y v d\Gamma = \langle f, v\rangle_{Y^*,Y} \quad \text{for all } v \in Y,$$

and the action of $Bu \in L^2(\Gamma)$ as an element $EBu \in Y^*$ is defined by

$$\langle EBu, v\rangle_{Y^*,Y} := \int_\Gamma Buv d\Gamma \quad \text{for all } v \in Y. \tag{3.29}$$

Problem (3.28) admits a unique solution u which satisfies the first order necessary (and here also sufficient) optimality conditions

$$\langle \hat{J}'(u), v - u\rangle_{U^*,U} \geq 0 \quad \text{for all } v \in U_{\text{ad}}, \tag{3.30}$$

where $\hat{J}'(u) = \alpha u + B^* E^* S^*(SEBu - z) \equiv \alpha u + B^* E^* p$, with $p := S^*(SEBu - z)$ denoting the adjoint variable. Here $E^* : Y \to L^2(\Gamma)$ denotes the trace operator. From here onwards let us not longer distinguish between B and EB. The function p in our setting satisfies the following Poisson problem with Neumann (Robin-type) boundary conditions;

$$\begin{aligned} -\Delta p &= y - z \quad \text{in } \Omega, \\ \partial_\eta p + \gamma p &= 0 \quad \text{on } \partial\Omega. \end{aligned}$$

We now define the variational-discrete analogon to problem (3.28) as in the previous subsection;

$$\min_{u\in U_{\text{ad}}} \hat{J}_h(u), \tag{3.31}$$

where for $u \in U$ we set $\hat{J}_h(u) := J(S_h Bu, u)$ with S_h denoting the discrete analogon of S. According to (3.28) this problem admits a unique solution $u_h \in U_{\text{ad}}$ which is characterized by the variational inequality

$$\langle \hat{J}_h'(u_h), v - u_h\rangle_{U^*,U} \geq 0 \quad \text{for all } v \in U_{\text{ad}}, \tag{3.32}$$

where similar as above

$$J_h'(u) = \alpha u + B^* S_h^*(S_h Bu - z) \equiv \alpha u + B^* p_h(u)$$

for $u \in U$. We notice that the whole exposition can be done by *copy and paste* from Sect. 3.2.5, and the structure of the optimization problem as well as its discretization do not depend on where control is applied. It is completely characterized by the operators S, S_h, and B (as well as by E). For Neumann boundary control the analogon to Theorem 3.4 reads

Theorem 3.8 *Let u denote the unique solution of* (3.28), *and u_h the unique solution of* (3.31). *Then there holds*

$$\alpha\|u - u_h\|_U^2 + \frac{1}{2}\|y - y_h\|^2 \le \langle B^*(p(u) - \tilde{p}_h(u)), u_h - u\rangle_{U^*,U} + \frac{1}{2}\|y - y_h(u)\|^2, \tag{3.33}$$

where $\tilde{p}_h(u) := S_h^*(SBu - z)$ and $y_h(u) := S_h Bu$.

The proof of this theorem is analogous to that of Theorem 3.4. Now let us formulate some direct consequences of this theorem. The following corollary immediately follows from (3.33).

Corollary 3.1 *Let u denote the solution of* (3.28) *with associated state $y = y(u)$, and adjoint state $p = p(u)$, and let u_h denote the solution to* (3.31) *with associated discrete state $y_h = y_h(u_h)$. Then*

$$\alpha\|u - u_h\|_U^2 + \|y - y_h\|^2 \le C_\alpha\|p - p^h\|_{L^2(\Gamma)}^2 + \|y - y^h\|^2, \tag{3.34}$$

where y^h, p^h denote the unique solutions to $a(y^h, v_h) = \langle Bu, v_h\rangle_{U^*,U}$, and $a(v_h, p^h) = \int_\Omega (y - z)v_h$ for all $v_h \in W_h$.

We observe that finite element estimates in L^2 for the Galerkin approximations y^h, p^h to y and p, respectively imply an estimate for $\|u - u_h\|_U$. Next we prove L^∞ error estimates for optimal Neumann boundary controls in the case $U = L^2(\Gamma)$ (i.e. $B = Id$) and $U_{\text{ad}} = \{v \in L^2(\Gamma), a \le v \le b\}$ with $a < b$ denoting constants.

Corollary 3.2 *Let u denote the solution of* (3.51) *with associated state y, and u_h the solution to* (3.59) *with associated discrete state y_h. Then*

$$\|u - u_h\|_{L^\infty(\Gamma)} \le C\left\{\|p - p^h\|_{L^\infty(\Gamma)} + \gamma(h)\|y - y^h\|\right\}, \tag{3.35}$$

where $\gamma(h) = |\log h|$ for $d = 2$, and $\gamma(h) = h^{-1/2}$ for $d = 3$.

Proof Using

$$u = P_{U_{\text{ad}}}\left(-\frac{1}{\alpha}p\right), \text{ and } u_h = P_{U_{\text{ad}}}\left(-\frac{1}{\alpha}p_h\right),$$

we obtain

$$\|u-u_h\|_{L^\infty(\Gamma)} = \left\| P_{U_{\text{ad}}}\left(-\frac{1}{\alpha}p\right) - P_{U_{\text{ad}}}\left(-\frac{1}{\alpha}p_h\right)\right\|_{L^\infty(\Gamma)} \le \frac{1}{\alpha}\|p-p_h\|_{L^\infty(\Gamma)}$$
$$\le \frac{1}{\alpha}\|p-p^h\|_{L^\infty(\Omega)} + \frac{1}{\alpha}\|p^h-p_h\|_{L^\infty(\Omega)}$$
$$\le \frac{1}{\alpha}\|p-p^h\|_{L^\infty(\Omega)} + \gamma(h)\|p^h-p_h\|_{H^1(\Omega)},$$

where $\gamma(h)=|\log h|$ for $d=2$, and $\gamma(h)=h^{-1/2}$ for $d=3$, see the paper of Xu and Zhang [148]. We proceed with estimating $\|p^h-p_h\|_{H^1(\Omega)}$ according to

$$\|p^h-p_h\|^2_{H^1(\Omega)} \le Ca(p^h-p_h, p^h-p_h) \le C\|p^h-p_h\|\|y-y_h\|.$$

This completes the proof.

From the estimates (3.34) and (3.35) we again deduce that the approximation quality of the control is steered by the approximation quality of finite element solutions y^h to the state y, and by the finite element approximation p^h of the adjoint p.

Let us give some examples.

Example 3.5 (Specific settings for finite element error estimates in Neumann control)

1. Let us consider the situation in the paper [27, Sect. 5,6] of Casas and Mateos, where Ω is a two-dimensional convex polygonal domain, i.e. $d=2$, $B=Id$, $U=L^2(\Gamma)$ and $U_{\text{ad}}=\{v\in U, a\le u\le b\}$. Further let $z\in L^2(\Omega)$. Then $y,p\in H^2(\Omega)$, so that by [27, Theorem 4.1] we have $\|y-y^h\|\le Ch^2$ and $\|p-p^h\|_{L^2(\Gamma)}\le h^{3/2}$. Thus, (3.34) directly yields
$$\|u-u_h\|_{L^2(\Gamma)} \le Ch^{3/2}.$$
2. Let us consider a smooth, bounded two- or three-dimensional domain Ω and let the approximation properties A1–A4 in the work [119] of Schatz be satisfied. Bootstrapping yields at least $y\in H^2(\Omega)$ and $p\in H^4(\Omega)\hookrightarrow W^{2,\infty}(\Omega)$ for $d<4$. Thus we deduce from [119, Theorem 2.2]
$$\|p-p^h\|_\infty \le Ch^{2-\frac{d}{q}}|\log h|\|p\|_{W^{2,q}} \quad \text{for all } d\le q\le\infty,$$
compare also [39, Lemma 3.4], and again $\|y-y^h\|\le Ch^2$. Thus, (3.35) directly delivers
$$\|u-u_h\|_{L^\infty(\Gamma)} \le C\left\{h^{2-\frac{d}{q}}|\log h| + \gamma(h)h^2\right\} \quad \text{for all } d\le q\le\infty.$$
We should note that when using finite element approximations defined over partitions formed of simplexes one has to consider also an error induced by boundary

approximations. However, locally, for small enough gridsizes the smooth boundary may be parameterized as graph over the faces of the corresponding simplex. For smooth boundaries the difference of the areas of the face and the corresponding graph is bounded by the square of the gridsize, so that error estimates of the same quality as in this example also hold for the accordingly transformed continuous solution, see [43]. Furthermore, appropriate quadrature could be used to evaluate the quantities living on boundary simplexes or boundary faces.

Now let us briefly describe the application of the semi-smooth Newton algorithm of Sect. 2.5.4 to the numerical solution of problem (3.31) in the case $B = E$ with E denoting the extension defined in (3.29), and $U_{\mathrm{ad}} = \{v \in L^2\Gamma), a \le v \le b\}$. Starting point is the non-smooth operator equation

$$G(u) := u - P_{U_{\mathrm{ad}}}\left(-\frac{1}{\alpha}B^* p_h(u)\right) = 0 \quad \text{in } U. \tag{3.36}$$

Let us recall here that for given $u \in U$ with associated discrete state $y_h(u)$ the function $p_h(u)$ solves $a(v_h, p_h(u)) = \int_\Omega (y_h(u) - z)v_h$ for all $v_h \in W_h$. It follows from (3.32) and the fact that $P_{U_{\mathrm{ad}}}$ denotes the orthogonal projection onto U_{ad} that this equation admits the unique solution $u_h \in U_{\mathrm{ad}}$ of problem (3.31). Moreover, it follows with Theorem 2.14 that G in the present setting is semi-smooth in the sense that

$$\sup_{M \in \partial G(u+s)} \|G(u+s) - G(u) - Ms\|_U = o(\|s\|_U) \quad \text{as } \|s\|_U \to 0,$$

where

$$\partial G(u) := \left\{I + D(u)\left(\frac{1}{\alpha}B^* p_h'(u)\right)\right\}$$

$$\text{with } D(u)(x) = \begin{cases} 0, & \text{if } -\frac{1}{\alpha}B^* p_h(u)(x) \notin [a,b], \\ \in [0,1], & \text{if } -\frac{1}{\alpha}B^* p_h(u)(x) \in \{a,b\}, \\ 1, & \text{if } -\frac{1}{\alpha}B^* p_h(u)(x) \in (a,b), \end{cases}$$

denotes the generalized differential. Here we also refer to the papers [69] of Hintermüller, Ito, and Kunisch, and [136] of Michael Ulbrich. With $g \equiv g(u)$ denoting the indicator function of the inactive set $\mathcal{I}(u) := \{x \in \Gamma; -\frac{1}{\alpha}B^* p_h(u)(x) \in (a,b)\}$ we set

$$G'(u) := I + \frac{1}{\alpha} g B^* p_h'(u) \in \partial G(u).$$

It follows from the considerations below related to (3.40) that $G'(u)$ is bounded invertible, since $p_h'(u) = S_h^* S_h B$ with S_h denoting the finite element solution operator. Thus, $B^* p_h'(u) = B^* S_h^* S_h B$ is positive semi-definite on U.

We are now in the position to formulate the semi-smooth Newton algorithm of Sect. 2.5.

Algorithm 3.9 Choose $u \in U$

While $G(u) \neq 0$ solve

$$G'(u)u^{\text{new}} = G'(u)u - G(u) \tag{3.37}$$

for u^{new} and set $u = u^{\text{new}}$.

We emphasize that this algorithm works in the infinite-dimensional space U so that it is not obvious that this algorithm is numerically implementable. For a related discussion we refer to [71].

Using

$$\beta := (I - g)\text{bounds} \equiv \begin{cases} a, & \text{if } -\frac{1}{\alpha} B^* p_h(u) < a, \\ b, & \text{if } -\frac{1}{\alpha} B^* p_h(u) > b, \\ 0, & \text{else}, \end{cases}$$

a short calculation shows, that the Newton equation (3.37) can be rewritten in the form

$$u^{\text{new}} = \text{bounds on } \mathcal{A}(u) := \Gamma \setminus \mathcal{I}(u), \quad \text{and} \tag{3.38}$$

$$(\alpha g I + g B^* S_h^* S_h B g) u^{\text{new}} = -g B^* (S_h^* y_0 - S_h^* S_h B \beta). \tag{3.39}$$

Here, bounds stands either for a or for b. We solve the equation (compare (2.26))

$$(\alpha g I + g B^* S_h^* S_h B g) u^{\text{new}} = -g B^* (S_h^* y_0 - S_h^* S_h B \beta)$$

with a conjugate gradient method. This is feasible since for given $u \in U$ the operator $\mathcal{E}_I^* (\alpha I + B^* S_h^* S_h B) \mathcal{E}_I$ is positive definite on $L^2(\mathcal{I}(u))$, where the function $\mathcal{E}_I f \in L^2(\Gamma)$ denotes the extension-by-zero to Γ of functions $f \in L^2(\mathcal{I}(u))$, and $\mathcal{E}_I^*$ denotes its adjoint whose action for $s \in L^2(\Gamma)$ is given by $\mathcal{E}_I^* s = (gs)_{|\mathcal{I}(u)}$. Thus, formally solving (3.39), (3.38) corresponds to solving

$$\mathcal{E}_I^* (\alpha I + B^* S_h^* S_h B) \mathcal{E}_I u_I^{\text{new}} = -\mathcal{E}_I^* B^* (S_h^* y_0 - S_h^* S_h B \beta) \tag{3.40}$$

and then setting $u^{\text{new}} = u_I^{\text{new}}$ on $\mathcal{I}(u)$, and $u^{\text{new}} = \text{bounds on } \mathcal{A}(u)$, compare also [69, (4.7)] in the paper of Hintermüller, Ito and Kunisch.

It is now clear from these considerations that the Newton iterates may develop kinks or even jumps along the border of the active set, see the numerical results of the next section. However, it follows from the definition of the active set $\mathcal{A}(u)$ that its border consists of polygons, since we use continuous, piecewise linear ansatz functions for the state. We note that this border in general consists of piecewise polynomials of the same degree as that of the finite element ansatz functions, if higher order finite elements are used. Therefore, Algorithm 3.9 is numerically implementable, since in every of its iterations only a finite number of degrees of freedom has to be managed, which in the present case of linear finite elements can not exceed $3nv + 2ne$, where nv denotes the number of finite element nodes, and *ne* the number of finite element edges, compare also with the arguments in the proof of Theorem 3.3, and see [71] for details. Moreover, the main ingredient of the cg algorithm applied to solve the Newton equation (3.40) consists in evaluating $\mathcal{E}_I^*(\alpha I + B^* S_h^* S_h B)\mathcal{E}_I f$ for functions $f \in L^2(\mathcal{I}(u))$. From the definitions of B and S_h it is then clear which actions have to be taken for this evaluation.

It is also clear, that only local convergence of the semi-smooth Newton algorithm can be expected, where the convergence radius at the solution depends on the penalization parameter α. For the numerical examples presented in the next section and the considered values of α it is sufficient to use a cascade approach where linear interpolations of numerical solutions on coarse grids are used as starting values on the next finer grid. Further details on the semi-smooth Newton methods applied to variationally discretized optimal control problems can be found in the paper [80] of Hinze and Vierling, where, among other things, also time-dependent problems are considered and globalization strategies are proposed.

3.2.7.2 Numerical Examples for Robin-Type Boundary Control

Now we consider numerical examples for Robin-type boundary control and in particular compare the results obtained with variational discretization to that obtained by Casas, Mateos and Tröltzsch in [30] with the conventional approach. In order to compare our numerical results to exact solutions we consider an optimal control problem which slightly differs from that formulated in (3.26). The following example is taken from the paper [30] of Casas, Mateos and Tröltzsch. The computational domain is the unit square $\Omega := (0,1)^2 \subset \mathbb{R}^2$. The optimization problem reads

$$\min J(y,u) = \frac{1}{2}\int_\Omega (y(x) - y_\Omega)^2 dx + \frac{\alpha}{2}\int_\Gamma u(x)^2 d\sigma(x) + \int_\Gamma e_u(x)u(x)d\sigma(x) + \int_\Gamma e_y(x)y(x)d\sigma(x)$$

s.t. $(y,u) \in H^1(\Omega) \times L^2(\Gamma)$, $u \in U_{\text{ad}} = \{u \in L^2(\Gamma) : 0 \le u(x) \le 1 \text{ a.e. on } \Gamma\}$, and (y,u) satisfying the linear state equation

$$\begin{aligned} -\Delta y(x) + c(x)y(x) &= e_1(x) \quad \text{in } \Omega \\ \partial_\nu y(x) + y(x) &= e_2(x) + u(x) \quad \text{on } \Gamma, \end{aligned}$$

where $\alpha = 1$, $c(x_1,x_2) = 1 + x_1^2 - x_2^2$, $e_y(x_1,x_2) = 1$, $y_\Omega(x_1,x_2) = x_1^2 + x_1x_2$, $e_1(x_1,x_2) = -2 + (1 + x_1^2 - x_2^2)(1 + 2x_1^2 + x_1x_2 - x_2^2)$,

$$e_u(x_1,x_2) = \begin{cases} -1 - x_1^3 & \text{on } \Gamma_1 \\ -1 - \min(8(x_2-0.5)^2 + 0.5, \\ \qquad 1 - 15x_2(x_2-0.25)(x_2-0.75)(x_2-1)) & \text{on } \Gamma_2 \\ -1 - x_1^2 & \text{on } \Gamma_3 \\ -1 - x_2(1-x_2) & \text{on } \Gamma_4, \end{cases}$$

and

$$e_2(x_1,x_2)=\begin{cases}1-x_1+2x_1^2-x_1^3 & \text{on } \Gamma_1\\ 7+2x_2-x_2^2-\min(8(x_2-0.5)^2+0.5,1) & \text{on } \Gamma_2\\ -2+2x_1+x_1^2 & \text{on } \Gamma_3\\ 1-x_2-x_2^2 & \text{on } \Gamma_4.\end{cases}$$

Here $\Gamma_1,\dots,\Gamma_4$ denote the boundary parts of the unit square numbered counterclockwise beginning at bottom. The adjoint equation for this example is given by

$$\begin{aligned}-\Delta p+c(x)p &= y(x)-y_\Omega(x) \quad \text{in } \Omega\\ \partial_\nu p+p &= e_y(x) \quad \text{on } \Gamma,\end{aligned}$$

and the optimal control is given by

$$u=\mathrm{Proj}_{U_{\mathrm{ad}}}\left(-\frac{1}{\alpha}(p+e_u)\right) \quad \text{on } \Gamma. \tag{3.41}$$

To compute the variational control $u_h\in U_{\mathrm{ad}}$ we in the present example iterate (3.41), i.e. we apply the fix-point iteration of Algorithm 3.2. The corresponding numerical results can be found in Tables 3.6–3.7 and Figs. 3.6–3.7. The reported EOC in Table 3.7 confirms the findings in Example 3.5. Moreover, should it be noted, that the error in the controls of the variational discrete approach on the initial grid are smaller than those produced by the conventional approach on refinement level 4, see Table 3.6.

Now let us consider an example for a semilinear state equation. It is taken from the paper [27, Sect. 7.1] of Casas and Mateos. For details of the numerical results we refer to the paper [73] of Hinze and Matthes. The optimization problem reads

$$\begin{aligned}\min \hat{J}(u) &= \frac{1}{2}\int_\Omega (y_u(x)-y_\Omega)^2dx+\frac{\alpha}{2}\int_\Gamma u(x)^2dx\\ &\quad +\int_\Gamma e_u(x)u(x)dx+\int_\Gamma e_y(x)y_u(x)dx\end{aligned}$$

subject to $u\in U_{\mathrm{ad}}=\{u\in L^2; 0\le u(x)\le 1 \text{ a.e. } x\in\Gamma\}$, where y_u satisfies the semilinear equation

$$\begin{aligned}-\Delta y_u(x)+c(x)y_u(x) &= e_1(x) && \text{in } \Omega\\ \partial_\nu y_u(x)+y_u(x) &= e_2(x)+u(x)-y(x)^2 && \text{on } \Gamma.\end{aligned}$$

Table 3.6 Errors in variational discretization (top part) and in the approach of [30] (bottom part). We observe that the error in the controls in variational discretization on the initial grid already is smaller than the error produced by the approach of [30] on a grid with mesh size $h = 2^{-7}$

h	δy_{L^2}	δy_{L^∞}	δp_{L^2}	δp_{L^∞}	δu_{L^2}	δu_{L^∞}
2^{-0}	0.21922165	0.16660113	0.00981870	0.01171528	0.01293312	0.00975880
2^{-1}	0.05490636	0.05592789	0.00283817	0.00375928	0.00412034	0.00375928
2^{-2}	0.01379774	0.01802888	0.00077525	0.00108642	0.00111801	0.00099280
2^{-3}	0.00345809	0.00554111	0.00019969	0.00028092	0.00028729	0.00025594
2^{-4}	0.00086531	0.00165357	0.00005038	0.00007065	0.00007250	0.00006447
2^{-5}	0.00021639	0.00048246	0.00001263	0.00001769	0.00001819	0.00001615
2^{-6}	0.00005410	0.00013819	0.00000316	0.00000443	0.00000455	0.00000404
2^{-7}	0.00001353	0.00003899	0.00000079	0.00000111	0.00000114	0.00000101
2^{-8}	0.00000338	0.00001086	0.00000020	0.00000028	0.00000028	0.00000025
2^{-4}	0.00056188				0.04330776	0.11460900
2^{-5}	0.00014240				0.02170775	0.05990258
2^{-6}	0.00003500				0.01086060	0.03060061
2^{-7}	0.00000897				0.00543114	0.01546116

Table 3.7 EOC for the variational discrete approach in the case of Robin-type boundary control. For a comparison to the approach taken by Casas, Mateos and Tröltzsch in [30] see also Fig. 3.7

h	y_{L^2}	y_{L^∞}	p_{L^2}	p_{L^∞}	u_{L^2}	u_{L^∞}
2^{-1}	1.997345	1.574758	1.790572	1.639862	1.650235	1.376247
2^{-2}	1.992541	1.633258	1.872222	1.790877	1.881837	1.920876
2^{-3}	1.996386	1.702064	1.956905	1.951362	1.960359	1.955685
2^{-4}	1.998688	1.744588	1.986941	1.991434	1.986431	1.989070
2^{-5}	1.999575	1.777112	1.996193	1.997494	1.995161	1.997047
2^{-6}	1.999873	1.803728	1.998912	1.999222	1.998106	1.999024
2^{-7}	1.999964	1.825616	1.999700	1.999725	1.999174	1.999834
2^{-8}	1.999991	1.843640	1.999932	1.999950	1.999609	1.999918

Here, $\Omega = (0,1)^2$, $\alpha = 1$, $c(x) = x_2^2 + x_1 x_2$, $e_y(x) = -3 - 2x_1^2 - 2x_1 x_2$, $y_\Omega(x) = 1 + (x_1 + x_2)^2$, $e_1(x) = -2 + (1 + x_1^2 + x_1 x_2)(x_2^2 + x_1 x_2)$,

$$e_u(x) = \begin{cases} 1 - x_1^3 & \text{on } \Gamma_1 \\ 1 - \min\left\{ \begin{matrix} 8(x_2 - 0.5)^2 + 0.58 \\ 1 - 16x_2(x_2 - y_1^*)(x_2 - y_2^*)(x_2 - 1) \end{matrix} \right\} & \text{on } \Gamma_2 \\ 1 - x_1^2 & \text{on } \Gamma_3 \\ 1 + x_2(1 - x_2) & \text{on } \Gamma_4, \end{cases}$$

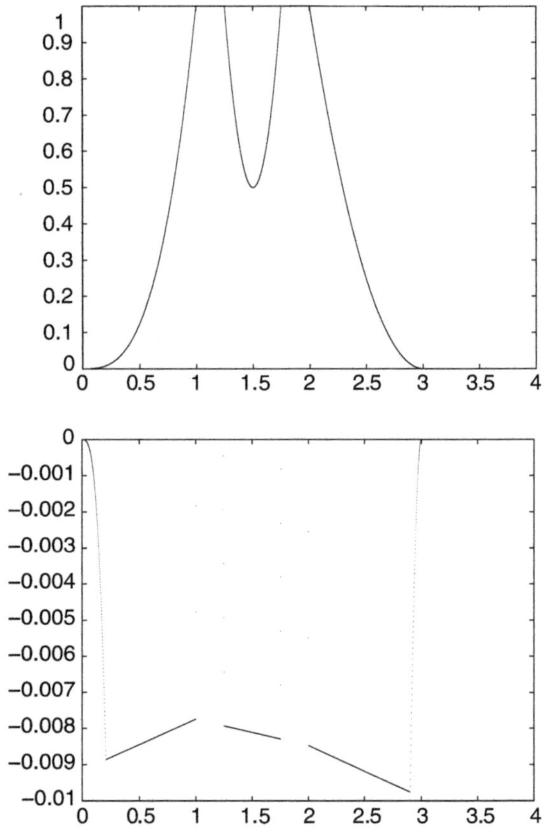

Fig. 3.6 Exact control (*top*) and error of exact and numerically computed control on the initial grid containing 4 triangles, i.e. $h = \frac{1}{2}$ (*bottom*)

with $y_1^* = \frac{1}{2} - \frac{\sqrt{21}}{20}$ and $y_2^* = \frac{1}{2} + \frac{\sqrt{21}}{20}$. Furthermore,

$$e_2(x) = \begin{cases} 2 - x_1 + 3x_1^2 - x_1^3 + x_1^4 & \text{on } \Gamma_1 \\ 8 + 6x_2 + x_2^2 - \min\{8(x_2 - 0.5)^2 + 0.58, 1\} & \text{on } \Gamma_2 \\ 2 + 4x_1 + 3x_1^2 + 2x_1^3 + x_1^4 & \text{on } \Gamma_3 \\ 2 - x_2 & \text{on } \Gamma_4. \end{cases}$$

The adjoint equation is given by

$$\begin{aligned} -\Delta\phi(x) + c(x)\phi(x) &= y_u(x) - y_\Omega(x) && \text{in } \Omega \\ \partial_\nu\phi(x) + \phi(x) &= e_y(x) - 2y(x)\phi(x) && \text{on } \Gamma. \end{aligned}$$

Again a short calculation shows that

$$\bar{u}(x) = \begin{cases} x_1^3 & \text{on } \Gamma_1 \\ \min\{8(x_2 - 0.5)^2 + 0.58, 1\} & \text{on } \Gamma_2 \\ x_1^2 & \text{on } \Gamma_3 \\ 0 & \text{on } \Gamma_4 \end{cases}$$

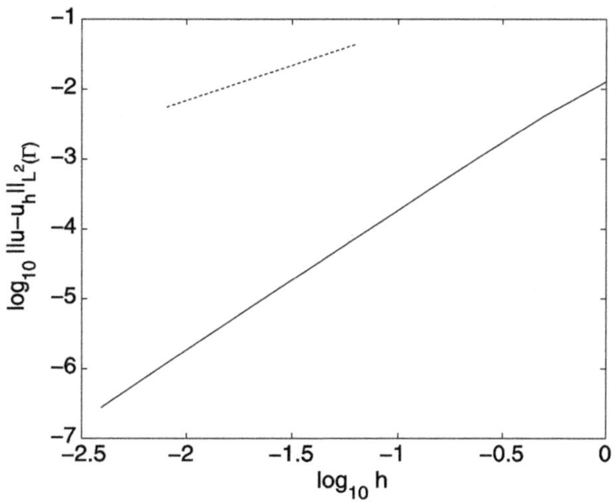

Fig. 3.7 Numerical comparison of EOC of controls for $E(h) := \|u - u_h\|_{L^2(\Gamma)}$: Approach of Casas, Mateos and Tröltzsch in [30] (*dashed*) and the variational approach (*solid*). The latter yields quadratic convergence, whereas the approach of [30] only shows linear convergence

denotes the optimal control with corresponding optimal state $\bar{y}(x) = 1 + x_1^2 + x_1 x_2$ and adjoint $\bar{\phi}(x) = -1$.

For the numerical solution of the present example again a semi-smooth Newton method is applied. Since we are dealing with nonlinear state equations the determination of u^{new} in (3.39) has to be replaced by

$$(\alpha g I + g B^* p_h'(u) g) u^{\text{new}} = -g B^*(p_h(u) - p_h'(u)(u - \beta)), \quad \text{and}$$
$$u^{\text{new}} = \text{bounds on } \Omega \setminus \mathcal{I}(u).$$

The numerical results are very similar to that of the previous example. This is due to the fact that the nonlinearity in the state equation is monotone.

The errors and EOCs for the present example are shown in Table 3.8 for the Casas-Mateos-Ansatz and the variational discretization, respectively. The EOC of the numerical experiments of Casas and Mateos is calculated from tables of Casas and Mateos in [27]. The EOC of the numerical experiments of Casas and Mateos is 1.5 and about 1.0 for the L^2 and L^∞ norm, respectively. The EOC is 2 for variational discretization. We note that also for this example already the errors on the coarsest mesh for $h = 1$ are smaller in our approach than those for $h = 2^{-4}$ in the conventional Casas-Mateos-ansatz.

The Newton iteration is terminated if with G of (3.36) $\|G(u^i)\|/\|G(u^0)\| \le 10^{-5}$ and $\|u^i - u^{i-1}\|/\max(\|u^i\|, \|u^{i-1}\|) \le 10^{-5}$ holds. The inner cg iteration is terminated if $\|r\| \le \frac{10^{-4}}{i} \min\{1, \|G(u^i)\|/, \|G(u^i)\|/\|G(u^0)\|\}$ holds with r denoting the current residuum of the Newton system.

In Fig. 3.8 the optimal control together with the error for $h = 0.5$ and the finite element grid is shown.

Table 3.8 Errors in u for the semilinear example

h	Casas $E_{u_{L^2}}$	Mateos $E_{u_{L^\infty}}$	This $E_{u_{L^2}}$	Paper $E_{u_{L^\infty}}$	Casas $\mathrm{EOC}_{u_{L^2}}$	Mateos $\mathrm{EOC}_{u_{L^\infty}}$	This $\mathrm{EOC}_{u_{L^2}}$	Paper $\mathrm{EOC}_{u_{L^\infty}}$
2^{-0}	–	–	1.13e−2	1.83e−2	–	–	–	–
2^{-1}	–	–	4.72e−3	6.43e−3	–	–	1.26	1.51
2^{-2}	–	–	1.33e−3	2.19e−3	–	–	1.82	1.55
2^{-3}	–	–	3.45e−4	6.69e−4	–	–	1.95	1.71
2^{-4}	8.5e−3	4.1e−2	8.75e−5	1.89e−4	–	–	1.98	1.82
2^{-5}	3.0e−3	1.5e−2	2.20e−5	5.11e−5	1.5	1.5	1.99	1.89
2^{-6}	1.1e−3	1.1e−2	5.50e−6	1.33e−5	1.4	0.4	2.00	1.94
2^{-7}	3.8e−4	3.8e−3	1.38e−6	3.42e−6	1.5	1.5	2.00	1.96
2^{-8}	1.4e−4	2.7e−3	3.44e−7	8.66e−7	1.4	0.5	2.00	1.98
2^{-9}	–	–	8.61e−8	2.18e−7	–	–	2.00	1.99
2^{-10}	–	–	2.15e−8	5.47e−8	–	–	2.00	1.99

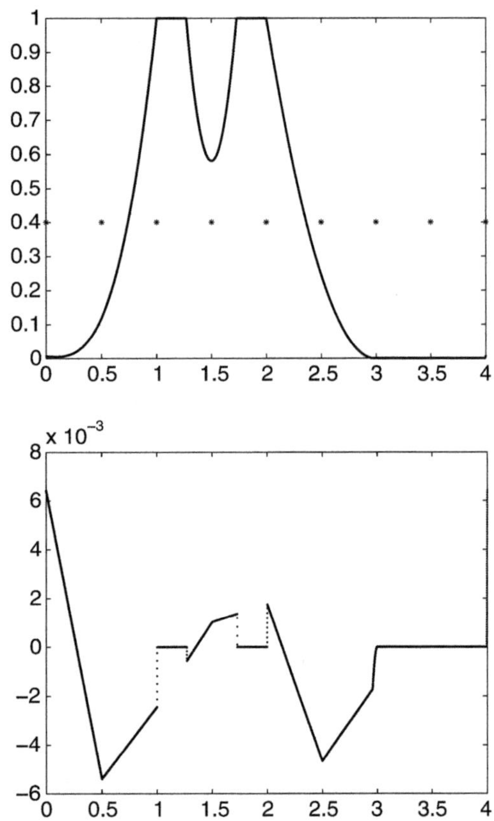

Fig. 3.8 Optimal control u *top*, error in u *bottom*, both for $h = 0.5$. *Bold dots* depict the finite element grid on the boundary

3.2.7.3 Dirichlet Boundary Control

Now we switch to problem ($\mathbb{DC}$). The numerical analysis of this problem is carried out by Casas and Raymond in [28]. There it is shown that problem ($\mathbb{DC}$) equivalently can be rewritten in the form

$$\min_{u \in U_{\text{ad}}} \hat{J}(u) \tag{3.42}$$

for the reduced functional $\hat{J}(u) := J(y(u), u) \equiv J(SBu, u)$ over the set U_{ad}, where $S : Y^* \to L^2(\Omega)$ for $Y := H^2(\Omega) \cap H_0^1(\Omega)$ denotes the very-weak solution operator of the Dirichlet boundary value problem for $-\Delta$, i.e. for $f \in Y^*$ and $u \in U$ there holds $y = S(f + EBu)$ iff

$$a(y, v) := \int_\Omega y(-\Delta v)dx = \langle f, v\rangle_{Y^*,Y} - \int_\Gamma Bu\partial_\eta v d\Gamma \quad \text{for all } v \in Y. \tag{3.43}$$

Here, the action of $Bu \in L^2(\Gamma)$ as an element $EBu \in Y^*$ is defined by

$$\langle EBu, v\rangle_{Y^*,Y} := \int_\Gamma Bu\partial_\eta v d\Gamma \quad \text{for all } v \in Y.$$

The first order necessary (and here also sufficient) optimality conditions here again take the form

$$\langle \hat{J}'(u), v - u\rangle_{U^*,U} \geq 0 \quad \text{for all } v \in U_{\text{ad}}, \tag{3.44}$$

where $\hat{J}'(u) = \alpha u - B^* E^* S^*(SEBu - z) \equiv \alpha u - B^* E^* p$, with $p := S^*(SEBu - z)$ denoting the adjoint variable. Here $E^* : Y \to L^2(\Gamma)$ denotes the trace operator of first order, i.e. for $v \in Y$ there holds $E^* v = (\partial_\eta v)_{|\Gamma}$. From here onwards let us not longer distinguish between B and EB, so that $\hat{J}'(u) = \alpha u - B^* \partial_\eta p$. The function p in our setting satisfies the following Poisson problem with homogeneous Dirichlet boundary conditions;

$$\begin{aligned} -\Delta p &= y - z \quad \text{in } \Omega, \\ p &= 0 \quad \text{on } \partial\Omega. \end{aligned}$$

To define an appropriate discrete approach for (3.42) in the present situation is a little bit more involved due to the following fact.

Note 3.4 We intend to approximate the solution y of the Dirichlet boundary value problem in (3.42) and the adjoint variable p by piecewise polynomials y_h and p_h of order k greater or equal to one, say. Then it is clear that it might not be meaningful to prescribe boundary values for y_h represented by (restrictions of) piecewise polynomials of order $k - 1$. However, the discrete analogon of the variational inequality (3.44) exactly proposes this, since $\partial_\eta p_h$ is a piecewise polynomial of order $k - 1$ on Γ.

We now introduce the common discrete concept for the approximation of very weak solutions to elliptic Dirichlet boundary value problems, compare the paper [11] of Berggren. For this purpose we use the L^2 projection Π_h onto boundary functions which are piecewise polynomials of degree $k \geq 1$ and are continuous on the boundary grid induced by triangulation of Ω on the boundary Γ. For $v \in L^2(\Gamma)$ we define $\Pi_h v$ to be the continuous, piecewise polynomial of degree k defined by the relation

$$\int_\Gamma \Pi_h v w_h d\Gamma = \int_\Gamma v w_h d\Gamma \quad \text{for all } w_h \in \text{trace}(W_h),$$

where W_h is defined in Sect. 3.2.2. The numerical approximation $S_h Bu := y_h \in W_h$ of the very weak solution y of the state equation with boundary values Bu is defined by the relation

$$\int_\Omega \nabla y_h \nabla v_h dx = 0 \quad \text{for all } v_h \in Y_h, \text{ and } y_h = \Pi_h(Bu) \text{ on } \Gamma,$$

and the numerical approximation p_h of the adjoint variable p as the usual finite element approximation $p_h := S_h^*(S_h E(Bu) - z)$, i.e.

$$\int_\Omega \nabla p_h \nabla v_h dx = \int_\Omega (y_h - z) v_h dx \quad \text{for all } v_h \in Y_h.$$

The variational discrete analogon of the optimization problem (3.42) reads

$$\min_{u \in U_{\text{ad}}} \hat{J}_h(u), \tag{3.45}$$

where for $u \in U$ we set $\hat{J}_h(u) := J(S_h Bu, u)$ with S_h denoting the discrete analogon to S. It admits a unique solution $u_h \in U_{\text{ad}}$. To derive the first order optimality conditions we use the Lagrange approach of Sect. 1.6.4. The Lagrangian of problem (3.45) is defined as

$$\begin{aligned} L(y_h, u, p_h, \kappa_h) = {} & \frac{1}{2}\|y_h - z\|^2 + \frac{\alpha}{2}\|u\|_U^2 - \int_\Omega \nabla y_h \nabla p_h dx \\ & - \int_\Gamma y_h \kappa_h d\Gamma + \int_\Gamma Bu \kappa_h d\Gamma, \end{aligned}$$

so that a short calculation yields for $u \in U_{\text{ad}}$

$$\hat{J}_h'(u) = \alpha u + B^* \kappa_h(u),$$

where $\kappa_h(u)$ in the latter equation is a continuous, piecewise polynomial function of degree k on the boundary grid defined through the relation

$$\int_\Gamma \kappa_h(u) w_h d\Gamma := -\int_\Omega \nabla p_h \nabla w_h dx + \int_\Omega (y_h(u) - z) w_h dx \quad \text{for all } w_h \in W_h.$$

Here we have used the fact that the derivative of the reduced cost functional $\hat{J}$ is given by the derivative of the Lagrangian w.r.t. the control u, i.e. $\hat{J}' = L_u$, see (1.89). The discrete numerical flux $\partial_\eta p_h(u)$ of the discrete adjoint $p_h(u)$ is a continuous, piecewise polynomial function of degree k on the boundary grid, and then is given by

$$\partial_\eta p_h(u) = -\kappa_h(u).$$

With this we obtain the following representation of the derivative of the reduced cost functional

$$\hat{J}_h'(u) = \alpha u - B^* \partial_\eta p_h(u), \tag{3.46}$$

which also is given by Casas and Raymond in [28].

The unique solution $u_h \in U_{\mathrm{ad}}$ of problem (3.45) satisfies the variational inequality

$$\langle J_h'(u_h), v - u_h \rangle_{U^*,U} \geq 0 \text{ for all } v \in U_{\mathrm{ad}}, \tag{3.47}$$

which also represents a sufficient condition for u_h to solve problem (3.45). For Dirichlet boundary control the analogon to Theorem 3.4 reads

Theorem 3.10 *Let u, u_h denote the unique solutions to* (3.42), *and* (3.47), *respectively, and y, y_h the corresponding optimal states. Then there holds*

$$\alpha \|u - u_h\|_U^2 + \frac{1}{2}\|y - y_h\|^2 \leq -\big\langle B^*(\partial_\eta p(u) - \partial_\eta p_h(u)), u_h - u\big\rangle_{U^*,U}$$
$$+ \frac{1}{2}\|y(u) - y_h(u)\|^2, \tag{3.48}$$

where $\partial_\eta p_h(u)$ denotes the discrete flux associated to $y(u) = SBu$, and $y_h(u) := S_h Bu$.

Proof We test (3.44) with u_h, (3.47) with the solution u of problem (3.42), and add the variational inequalities (3.44) and (3.47). This leads to

$$\alpha \|u - u_h\|_U \leq -\langle B^*(\partial_\eta p(u) - \partial_\eta p_h(u)), u_h - u\rangle_{U^*,U}$$
$$- \langle B^*(\partial_\eta p_h(u) - \partial_\eta p_h), u_h - u\rangle_{U^*,U}.$$

From the definition of B, Π_h and of S_h it follows that

$$-\big\langle B^*(\partial_\eta p_h(u) - \partial_\eta p_h), u_h - u\big\rangle_{U^*,U}$$
$$= \int_\Gamma (y_h(u) - y_h)(\partial_\eta p_h(u) - \partial_\eta p_h) d\Gamma$$
$$= \underbrace{\int_\Omega \nabla(y_h(u) - y_h)\nabla(p - p_h)dx}_{=0} - \int_\Omega (y_h(u) - y_h)(y - y_h)dx$$

$$\leq -\frac{1}{2}\|y - y_h\|^2 + \frac{1}{2}\|y_h(u) - y\|^2,$$

which together with the first estimate gives the desired result.

To provide estimates for the error in the controls it suffices to estimate the norms

$$\|\partial_\eta p - \partial_\eta p_h(u)\|_{L^2(\Gamma)}, \quad \text{and} \quad \|y - y_h(u)\|.$$

Now let us assume $B = Id$ and U_{ad} is defined by box constraints $a \leq u \leq b$, so that we deal with the setting presented by Casas and Raymond in [28]. The domain Ω considered in their work is two-dimensional and polygonal, so that $p \in W^{2,q}(\Omega)$ for some $q \geq 2$. This in turn implies $\partial_\eta p \in W^{1-1/q,q}(\Gamma)$. From estimates of the projection error for Π_h we expect that

$$\|\partial_\eta p - \partial_\eta p_h(u)\|_{L^2(\Gamma)} \sim h^{1-1/q},$$

and, since $u = P_{[a,b]}\partial_\eta p \in W^{1-1/q,q}(\Gamma)$,

$$\|y - y_h(u)\| \sim h,$$

so that $\|u - u_h\|_{L^2(\Gamma)} \sim h^{1-1/q}$ should be expected. In fact this is what Casas and Raymond prove for Dirichlet boundary control with box constraints on two-dimensional convex polygonal domains in [28]. Their main result there reads

$$\|u - u_h\|_{L^2(\Gamma)} \leq Ch^{1-1/q}, \tag{3.49}$$

where u_h denotes the optimal discrete boundary control which they sought in the space of piecewise linear, continuous finite elements on Γ. Here $q \geq 2$ depends on the smallest angle of the boundary polygon. May, Rannacher and Vexler study Dirichlet boundary control without control constraints in [94]. They also consider two dimensional convex polygonal domains and among other things provide optimal error estimates in weaker norms. In particular they address

$$\|u - u_h\|_{H^{-1}(\Gamma)} + \|y - y_h\|_{H^{-1/2}(\Omega)} \sim h^{2-2/q}.$$

Let us finally note that Vexler in [142] for $U_{\text{ad}} = \{u \in \mathbb{R}^n; a \leq u \leq b\}$ and

$$Bu := \sum_{i=1}^{n} u_i f_i$$

with $f_i \in H^{5/2}(\Gamma)$ provides finite element analysis for problem (3.42) with bounded, two-dimensional polygonal domains. Among other things he in [142, Theorem 3.4] shows that

$$|u - u_h| \leq Ch^2. \tag{3.50}$$

In Sect. 3.2.7.4 we present a numerical example for two-dimensional polygonal domains which shows that the result obtained by Casas and Raymond in fact is optimal for their setting.

3.2.7.4 Numerical Example for Dirichlet Boundary Control

Here we consider problem (3.27) with $U = L^2(\Gamma)$, $\alpha = 1$ and $U_{\text{ad}} = \{u \in U; 0 \le u \le 0.9\}$, i.e. $B \equiv Id$. Again we choose $\Omega = (0,1)^2$. The desired state is given by $z = -\text{sign}(x - 0.5 - \frac{0.1}{\pi})$. State and adjoint state are discretized with piecewise linear, continuous Ansatz functions as described in Sect. 3.2.7.3. The variational inequality (3.47) motivates as solution algorithm the iteration

$$u_h^+ = P_{U_{\text{ad}}}\left(\frac{1}{\alpha}\partial_\eta p_h(u_h)\right).$$

We investigate two different approaches; approach 1 in this algorithm uses $\partial_\eta p_h(u_h)$, which represents a piecewise constant (on the boundary grid) L^2 function. Let us emphasize that we not yet have available theory for this approach (which in fact seems to be the natural one if we would replace the continuous quantities in (3.44) by their discrete counterparts). The second approach in this algorithm uses the piecewise linear, continuous discrete flux $\partial_\eta p_h(u_h)$ defined by (3.46). For $h = 2^{-6}$ the value of the cost functional in the optimal solution for the second approach is $J = 0.47473792124624$. The numerical results are summarized in Table 3.9 and are better than those predicted by the theoretical investigations of Berggren in [11] for the state equation, and are in accordance with the predictions of Casas and Raymond in [28] for the control problem.

Let us present a numerical example that shows that the estimate (3.49) in fact is optimal for two-dimensional polygonal domains. In particular we consider problem (3.42) without constraints on the control in the form

$$\min_{u \in L^2(\Gamma)} J(u) = \frac{1}{2}\|y - y_0\| + \frac{\alpha}{2}\|u\|_{L^2(\Gamma)}, \quad \text{s.t.} \quad -\Delta y = f \quad \text{in } \Omega,\, y = u \text{ on } \Gamma$$

with

$$\bar{\Omega} = \text{conv}\left\{\left(\cos\frac{2\pi(i-1)}{12}, \sin\frac{2\pi(i-1)}{12}\right) : i = 1\ldots, 12\right\}, \quad \alpha = 1, \quad \text{and}$$

$$y_0(x_1, x_2) = 4(x_1 - 0.4)^2 - 4(x_2 - 0.6)^2, \quad f = 0.$$

The triangulation of the domain is depicted in Fig. 3.9. The maximum inner angle of the polygon is given by $\omega_{\max} = \frac{5}{6}\pi$, so that the critical exponent in estimate (3.49) is given by

$$p^* = \omega_{\max}\left(\omega_{\max} - \frac{\pi}{2}\right)^{-1} = \frac{5}{2}.$$

The experimental order of convergence reported in Table 3.10 confirms the estimate (3.49) of Casas and Raymond. EOC for two different finite element approaches to problem (3.42) are presented. (I1) presents the results for the approach of Casas and Raymond, whereas (I2) presents the results for variational discrretization combined with a mixed finite element approximation of the state equation based on the lowest

Table 3.9 EOC for Dirichlet boundary control: Approach 1 (top part), for which theory is not yet available, Approach 2 (bottom part), for which the theory of Sect. 3.2.7.3 applies. In both cases we observe linear convergence of the states and controls. The adjoint state also converges linear for approach 1, but seems to converge quadratically in approach 2

h	y_{L^2}	y_{L^∞}	p_{L^2}	p_{L^∞}	u_{L^2}	u_{L^∞}
1–2	−44.315839	−45.874172	2.252319	1.449921	−Inf	−Inf
2–3	−2.658752	−2.692762	0.890090	0.631871	−2.710238	−2.947286
3–4	0.513148	0.230017	1.605929	1.322948	0.559113	0.709528
4–5	0.864432	0.633565	1.641025	1.616581	0.867286	0.687088
5–6	0.955413	0.898523	1.474113	1.599350	0.937568	0.794933
6–7	0.969762	0.711332	1.239616	1.497993	0.936822	0.878459
7–8	0.992879	0.987835	1.106146	1.342300	0.986749	0.960009
8–9	0.990927	0.858741	1.035620	1.177092	0.982189	0.976724
1–2	−0.015094	−0.950093	2.273887	1.599015	−0.464738	−0.950093
2–3	1.479164	1.040787	0.909048	0.498459	1.194508	1.040787
3–4	1.484622	0.855688	1.720355	1.540523	0.979140	0.855688
4–5	1.647971	0.701102	1.873278	1.835947	1.360098	0.701102
5–6	1.545075	0.764482	1.910160	1.895133	1.253975	0.764482
6–7	1.424251	0.798198	1.955067	1.875618	1.227700	0.798198
7–8	1.163258	0.825129	1.915486	1.819988	1.173902	0.825129
8–9	1.020300	0.845442	1.742227	1.722124	1.099603	0.845442

Table 3.10 EOC for Dirichlet boundary control on a polygonal domain, no constraints on the control

i	np	h	(I1)		(I2)	
			$\|u-u_h\|_{L^2(\Gamma)}$	EOC	$\|u-u_h\|_{L^2(\Gamma)}$	EOC
1	21	0.61966	0.372593	–	0.488032	–
2	69	0.30983	0.330050	0.175	0.325708	0.583
3	249	0.15491	0.214437	0.622	0.222048	0.553
4	945	0.07746	0.144640	0.568	0.145601	0.609
5	3681	0.03873	0.095540	0.598	0.089347	0.705
6	14529	0.01936	0.057251	0.739	0.047105	0.924
7	57729	0.00968	–	–	–	–

order Raviart Thomas element. As exact solution the finite element approximation obtained with refinement level 7 is taken. In Fig. 3.10 the optimal control together with the optimal state for approach (I1) is shown. One clearly observes a loss of regularity in the corners of the polygon.

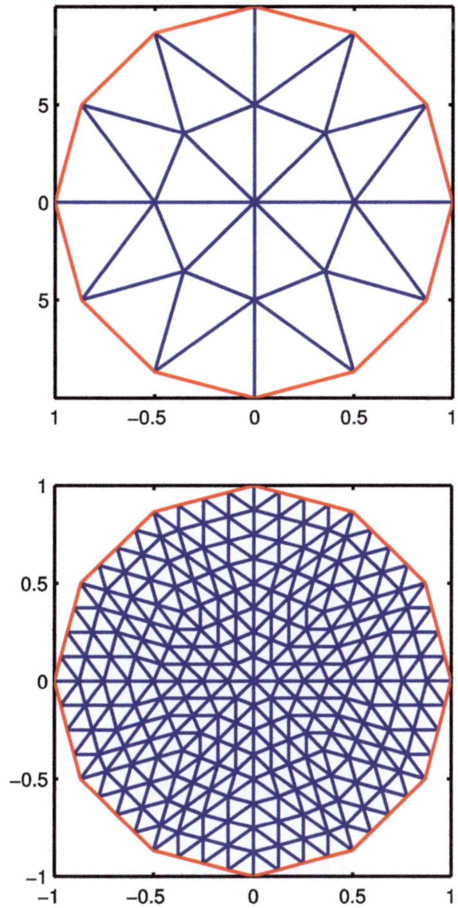

Fig. 3.9 Triangulations of the polygonal domain for 2 different refinement levels

Note 3.5 We note that in some numerical examples presented in the previous subsections, (variants) of the fix-point iteration of Algorithm 3.2 are used. Convergence of this algorithm can only be guaranteed for parameter values $\alpha > 0$ large enough. For small parameters $\alpha > 0$ primal-dual active set strategies as proposed by Hintermüller, Ito and Kunisch in [69], or semi-smooth Newton methods from the paper [136] of M. Ulbrich could be applied to the numerical solution of the discrete problems, see Sect. 2.5 and compare the discussion associated to (3.20). Finally we note that our solution algorithms perform independent of the finite element mesh, i.e. is mesh-independent, compare the discussion in Sect. 2.8.1, and in the work of Hintermüller and Ulbrich [68]. This may easily be explained by the fact that the iteration of Algorithm 3.2 is defined on the infinite dimensional space U of controls, rather than on a finite dimensional subspace of U. Thus, the finite element discretization from the viewpoint of the control problem has more of the flavor of a parametrization than of a discretization.

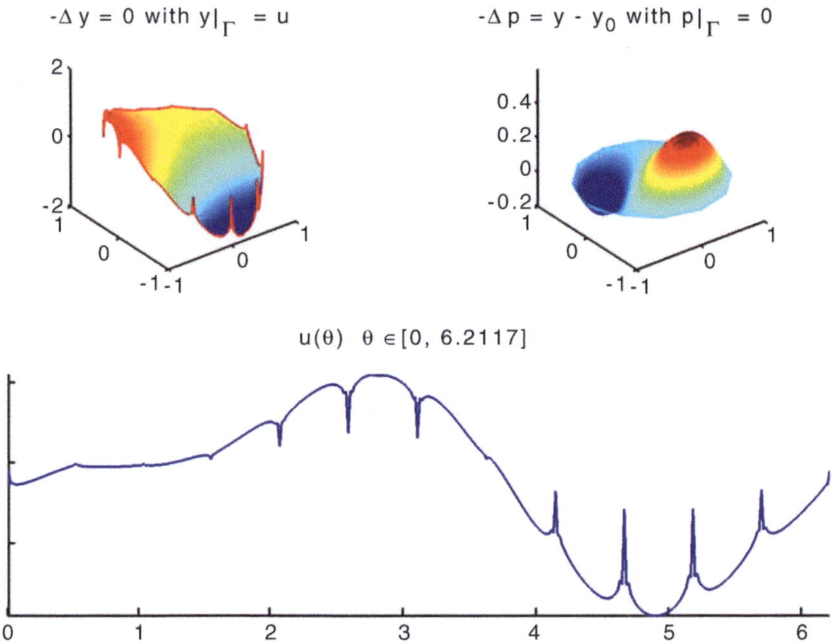

Fig. 3.10 Top: State (*left*) and adjoint state, bottom: optimal control for control on polygonal domain

3.2.8 Some Literature Related to Control Constraints

There are many contributions to finite element analysis for elliptic control problems with constraints on the controls. For an introduction to the basic techniques we refer to the book [133] of Tröltzsch. Falk [48], and Geveci [54] present finite element analysis for piecewise constant approximations of the controls. For semilinear state equations Arada, Casas, and Tröltzsch in [8] present a finite element analysis for piecewise constant discrete controls. Among other things they prove that the sequence $(u_h)_h$ of discrete controls contains a subsequence converging to a solution u of the continuous optimal control problem. Assuming certain second order sufficient conditions for u they are also able to prove optimal error estimates of the form

$$\|u - u_h\| = \mathcal{O}(h) \quad \text{and} \quad \|u - u_h\|_\infty = \mathcal{O}(h^\lambda),$$

with $\lambda = 1$ for triangulations of non-negative type, and $\lambda = 1/2$ in the general case. In [29] these results are extended in that Casas and Tröltzsch prove that every nonsingular local solution u (i.e. a solution satisfying a second order sufficient condition) locally can be approximated by a sequence $(u_h)_h$ of discrete controls, also satisfying these error estimates. There are only few results considering uniform estimates. For piecewise linear controls in the presence of control constraints are Meyer and Rösch in [101] for two-dimensional bounded domains with $C^{1,1}$-boundary prove the estimate

$$\|u - u_h\|_\infty = \mathcal{O}(h),$$

which seems to be optimal in regard of Table 3.4, and is one order less than the approximation order obtained with variational discretization, compare Remark 3.2.

Casas, Mateos and Tröltzsch in [30] present numerical analysis for Neumann boundary control of semilinear elliptic equations and prove the estimate

$$\|u - u_h\|_{L^2(\Gamma)} = \mathcal{O}(h)$$

for piecewise constant control approximations. In [27] Casas and Mateos extend these investigations to piecewise linear, continuous control approximations, and also to variational discrete controls. Requiring some second order sufficient conditions at the continuous solution u they are able to prove the estimates

$$\|u - u_h\|_{L^2(\Gamma)} = o(h), \quad \text{and} \quad \|u - u_h\|_{L^\infty(\Gamma)} = o(h^{\frac{1}{2}}),$$

for a general class of control problems, where u_h denotes the piecewise linear, continuous approximation to u. For variational discrete controls u_h^v they show the better estimate

$$\|u - u_h^v\|_{L^2(\Gamma)} = \mathcal{O}(h^{\frac{3}{2}-\epsilon}) \quad (\epsilon > 0).$$

Furthermore, they improve their results for objectives which are quadratic w.r.t. the control and obtain

$$\|u - u_h\|_{L^2(\Gamma)} = \mathcal{O}(h^{\frac{3}{2}}), \quad \text{and} \quad \|u - u_h\|_{L^\infty(\Gamma)} = \mathcal{O}(h).$$

These results are in accordance with those presented in Table 3.8.

Let us finally recall the contribution [28] of Casas and Raymond to numerical analysis of Dirichlet boundary control, who prove the optimal estimate (3.49), and the contribution of Vexler [142], who for a control in $\mathbb{R}^n$ proves the estimate (3.50).

Let us also briefly mention some contributions to a posteriori adaptive concepts in PDE constrained optimization. Residual based estimators for problems with control constraints are investigated by Liu and Yan in e.g. [91], by Hintermüller and Hoppe in [65], and by Gaevskaya, Hoppe, and S. Repin in [51]. For an excellent overview of the dual weighted residual method applied to optimal control problems we refer to the work [9] of Becker and Rannacher. An application of this method in the presence of control constraints is provided by Vexler and Wollner in [143], where also a recent survey of the literature in the field is given.

3.3 Constraints on the State

Next we also consider constraints on the state. The numerical analysis in this situation becomes more involved since the multipliers associated to constraints on the state in general appear to be Borel measures or derivatives of Borel measures.

3.3.1 Pointwise Bounds on the State

As model problem with pointwise bounds on the state we take the Neumann problem

$$(\mathbb{S})\quad \begin{cases} \min_{(y,u)\in Y\times U_{\mathrm{ad}}} J(y,u) := \frac{1}{2}\int_\Omega |y-y_0|^2 + \frac{\alpha}{2}\|u-u_0\|_U^2 \\ \text{s.t.} \\ \left.\begin{array}{ll} Ay = Bu & \text{in } \Omega, \\ \partial_\eta y = 0 & \text{on } \Gamma, \end{array}\right\} :\Longleftrightarrow y = \mathcal{G}(Bu) \\ \text{and} \\ y \in Y_{\mathrm{ad}} := \{y \in L^\infty(\Omega),\, y(x) \le b(x) \text{ a.e. in } \Omega\}. \end{cases} \tag{3.51}$$

Here, $Ay := -\Delta y + y$, and $\Omega \subset \mathbb{R}^d$ $(d = 2, 3)$ denotes an open, bounded sufficiently smooth (or polyhedral) domain. Furthermore, we again suppose that $\alpha > 0$ and that $y_0 \in H^1(\Omega)$, $u_0 \in U$ and $b \in W^{2,\infty}(\Omega)$ are given. $(U, (\cdot,\cdot)_U)$ denotes a Hilbert space and $B : U \to L^2(\Omega) \subset H^1(\Omega)^*$ a linear, continuous operator. By $R : U^* \to U$ we again denote the inverse of the Riesz isomorphism. In the special case $U \equiv L^2(\Omega)$ without control constraints, i.e. $U_{\mathrm{ad}} \equiv L^2(\Omega)$ the finite element analysis of problem (3.51) is carried out by Deckelnick and Hinze in [39]. Here we extend the analysis to the case of control and pointwise state constraints, where we use techniques which are applicable to a wider class of control problems. The exposition is closely related to the work [40] of Deckelnick and Hinze, where more general elliptic state equations are considered, and contains the results of [39] as a special case.

Problem (3.51) admits the form of problem (1.138), and more specifically that of problem (1.144). To ensure Robinson's regularity condition for our optimization problem it is due to Lemma 1.14 sufficient to impose the so called *Slater condition* or interior point condition.

Assumption 3.11

$$\exists \tilde{u} \in U_{\mathrm{ad}} \quad \mathcal{G}(B\tilde{u}) < b \quad \text{in } \bar{\Omega}.$$

Since the state constraints form a convex set and the set of admissible controls is closed and convex it is not difficult to establish the existence of a unique solution $u \in U_{\mathrm{ad}}$ to this problem, compare the analysis of problem (1.144). In order to characterize this solution we introduce the space $\mathcal{M}(\bar{\Omega})$ of Radon measures which is defined as the dual space of $C^0(\bar{\Omega})$ and endowed with the norm

$$\|\mu\|_{\mathcal{M}(\bar{\Omega})} = \sup_{f\in C^0(\bar{\Omega}),|f|\le 1} \int_{\bar{\Omega}} f\, d\mu.$$

For the problem under consideration we now have the following theorem, which specifies the KKT system (1.140)–(1.143) for the present setting, compare also the considerations related to (1.144).

Theorem 3.12 *Let $u \in U_{\mathrm{ad}}$ denote the unique solution to* (3.51). *Then there exist $\mu \in \mathcal{M}(\bar{\Omega})$ and $p \in L^2(\Omega)$ such that with $y = \mathcal{G}(Bu)$ there holds*

$$\int_\Omega pAv = \int_\Omega (y - y_0)v + \int_{\bar{\Omega}} v d\mu \quad \forall v \in H^2(\Omega) \quad \text{with } \partial_\eta v = 0 \quad \text{on } \partial\Omega, \tag{3.52}$$

$$(RB^* p + \alpha(u - u_0), v - u_0)_U \geq 0 \quad \forall v \in U_{\mathrm{ad}}, \tag{3.53}$$

$$\mu \geq 0, \quad y(x) \leq b(x) \quad \text{in } \Omega \quad \text{and} \quad \int_{\bar{\Omega}} (b - y) d\mu = 0. \tag{3.54}$$

The proof of this theorem in the presented form is given by Casas in [22, Theorem 5.2], compare also [21, Theorem 2].

We now develop and analyze a finite element approximation of problem (3.51). We start by approximating the cost functional J by a sequence of functionals J_h where h is a mesh parameter related to a sequence of triangulations. The definition of J_h involves only the approximation of the state equation by linear finite elements and enforces constraints on the state in the nodes of the triangulation, whereas the controls are still sought in U_{ad}. We shall prove that the minima of J_h converge in L^2 to the minimum of J as $h \to 0$ and that the states convergence strongly in H^1 with corresponding error bounds. We thereby extend the variational discretization approach developed in Sect. 3.2.5 to problems with control and state constraints. We prove the following error bounds

$$\|u - u_h\|_U, \ \|y - y_h\|_{H^1} = \begin{cases} O(h^{\frac{1}{2}}), & \text{if } d = 2, \\ O(h^{\frac{1}{4}}), & \text{if } d = 3, \end{cases}$$

where u_h and y_h are the discrete control and state respectively. If in addition $Bu \in W^{1,s}(\Omega)$ we obtain

$$\|u - u_h\|_U, \ \|y - y_h\|_{H^1} \leq C h^{\frac{3}{2} - \frac{d}{2s}} \sqrt{|\log h|},$$

and if $Bu \in L^\infty(\Omega)$ also

$$\|u - u_h\|_U, \ \|y - y_h\|_{H^1} \leq C h |\log h|,$$

where the latter estimate is valid for $d = 2, 3$.

Roughly speaking, the idea is to test (3.53) with u_h and (3.62), the discrete counterpart of (3.53), with the continuous solution u. This is feasible since controls are not discretized explicitly. An important tool in the analysis is the use of L^∞-error estimates for finite element approximations of the Neumann problem developed by Schatz in [119]. The need for uniform estimates is due to the presence of the measure μ in (3.52).

3.3.1.1 Finite Element Discretization

For the convenience of the reader we recall the finite element setting. To begin with let $\mathcal{T}_h$ be a triangulation of Ω with maximum mesh size $h := \max_{T \in \mathcal{T}_h} \mathrm{diam}(T)$

and vertices $x_1, \ldots, x_m$. We suppose that $\bar{\Omega}$ is the union of the elements of $\mathcal{T}_h$ so that element edges lying on the boundary are curved. In addition, we assume that the triangulation is quasi-uniform in the sense that there exists a constant $\kappa > 0$ (independent of h) such that each $T \in \mathcal{T}_h$ is contained in a ball of radius $\kappa^{-1}h$ and contains a ball of radius κh. Let us define the space of linear finite elements,

$$X_h := \{v_h \in C^0(\bar{\Omega}) \mid v_h \text{ is a linear polynomial on each } T \in \mathcal{T}_h\}$$

with the appropriate modification for boundary elements. In what follows it is convenient to introduce a discrete approximation of the operator $\mathcal{G}$. For a given function $v \in L^2(\Omega)$ we denote by $z_h = \mathcal{G}_h(v) \in X_h$ the solution of the discrete Neumann problem

$$a(z_h, v_h) = \int_\Omega v v_h \quad \text{for all } v_h \in X_h.$$

It is well-known that for all $v \in L^2(\Omega)$

$$\|\mathcal{G}(v) - \mathcal{G}_h(v)\| \le Ch^2 \|v\|, \tag{3.55}$$

$$\|\mathcal{G}(v) - \mathcal{G}_h(v)\|_{L^\infty} \le Ch^{2-\frac{d}{2}} \|v\|. \tag{3.56}$$

The estimate (3.56) can be improved provided one strengthens the assumption on v.

Lemma 3.1

(a) *Suppose that* $v \in W^{1,s}(\Omega)$ *for some* $1 < s < \frac{d}{d-1}$. *Then*

$$\|\mathcal{G}(v) - \mathcal{G}_h(v)\|_{L^\infty} \le Ch^{3-\frac{d}{s}} |\log h| \|v\|_{W^{1,s}}.$$

(b) *Suppose that* $v \in L^\infty(\Omega)$. *Then*

$$\|\mathcal{G}(v) - \mathcal{G}_h(v)\|_{L^\infty} \le Ch^2 |\log h|^2 \|v\|_{L^\infty}.$$

Proof (a) Let $z = \mathcal{G}(v)$, $z_h = \mathcal{G}_h(v)$. Elliptic regularity theory implies that $z \in W^{3,s}(\Omega)$ from which we infer that $z \in W^{2,q}(\Omega)$ with $q = \frac{ds}{d-s}$ using a well–known embedding theorem. Furthermore, we have

$$\|z\|_{W^{2,q}} \le c\|z\|_{W^{3,s}} \le c\|v\|_{W^{1,s}}. \tag{3.57}$$

Using Theorem 2.2 and the following Remark in [119] we have

$$\|z - z_h\|_{L^\infty} \le c|\log h| \inf_{\chi \in X_h} \|z - \chi\|_{L^\infty}, \tag{3.58}$$

which, combined with a well-known interpolation estimate, yields

$$\|z - z_h\|_{L^\infty} \le ch^{2-\frac{d}{q}} |\log h| \|z\|_{W^{2,q}} \le ch^{3-\frac{d}{s}} |\log h| \|v\|_{W^{1,s}}$$

in view (3.57) and the relation between s and q.

(b) Elliptic regularity theory in the present case implies that $z \in W^{2,q}(\Omega)$ for all $1 \leq q < \infty$ with

$$\|z\|_{W^{2,q}} \leq Cq\|v\|_{L^q}$$

where the constant C is independent of q. For the dependence on q in this estimate we refer to the work of Agmon, Douglis and Nirenberg [2], see also [53] and [55, Chap. 9]. Proceeding as in (a) we have

$$\begin{aligned}\|z - z_h\|_{L^\infty} &\leq Ch^{2-\frac{d}{q}}|\log h|\|z\|_{W^{2,q}} \leq Cqh^{2-\frac{d}{q}}|\log h|\|v\|_{L^q} \\ &\leq Cqh^{2-\frac{d}{q}}|\log h|\|v\|_{L^\infty},\end{aligned}$$

so that choosing $q = |\log h|$ gives the result.

Problem (3.51) is now approximated by the variational discretization concept of [71]. This delivers the following sequence of control problems depending on the mesh parameter h:

$$\begin{aligned}&\min_{u \in U_{\text{ad}}} J_h(u) := \frac{1}{2}\int_\Omega |y_h - y_0|^2 + \frac{\alpha}{2}\|u - u_{0,h}\|_U^2 \\ &\text{subject to} \quad y_h = \mathcal{G}_h(Bu) \quad \text{and} \quad y_h(x_j) \leq b(x_j) \quad \text{for } j = 1, \ldots, m.\end{aligned} \tag{3.59}$$

Here, $u_{0,h}$ denotes an approximation to u_0 which is assumed to satisfy

$$\|u_0 - u_{0,h}\|_U \leq Ch. \tag{3.60}$$

Problem (3.59) represents a convex infinite-dimensional optimization problem of similar structure as problem (3.51), but with only finitely many equality and inequality constraints for the state, which form a convex admissible set. So we are again in the setting of (1.138) with Y replaced by the finite element space X_h (compare also the analysis of Casas presented in [24])

Lemma 3.2 *Problem* (3.59) *has a unique solution* $u_h \in U_{\text{ad}}$. *There exist* $\mu_1, \ldots, \mu_m \in \mathbb{R}$ *and* $p_h \in X_h$ *such that with* $y_h = \mathcal{G}_h(Bu_h)$ *and* $\mu_h = \sum_{j=1}^m \mu_j \delta_{x_j}$ *we have*

$$a(v_h, p_h) = \int_\Omega (y_h - y_0)v_h + \int_{\bar{\Omega}} v_h d\mu_h \quad \forall v_h \in X_h, \tag{3.61}$$

$$\left(RB^* p_h + \alpha(u_h - u_{0,h}), v - u_h\right)_U \geq 0 \quad \forall v \in U_{\text{ad}}, \tag{3.62}$$

$$\mu_j \geq 0, \quad y_h(x_j) \leq b(x_j), \quad j = 1, \ldots, m \quad \textit{and} \quad \int_{\bar{\Omega}} (I_h b - y_h) d\mu_h = 0. \tag{3.63}$$

Here, δ_x denotes the Dirac measure concentrated at x and I_h is the usual Lagrange interpolation operator.

Remark 3.3 Problem (3.59) is still an infinite-dimensional optimization problem, but with finitely many state constraints. This is reflected by the well known fact that the variational inequalities (3.53) and (3.62) can be rewritten in the form

$$u = P_{U_{\text{ad}}}\left(-\frac{1}{\alpha}RB^*p + u_0\right) \quad \text{and} \quad u_h = P_{U_{\text{ad}}}\left(-\frac{1}{\alpha}RB^*p_h + u_{0,h}\right), \tag{3.64}$$

respectively, where $P_{U_{\text{ad}}} : U \to U_{\text{ad}}$ denotes the orthogonal projection onto U_{ad}, and $R : U^* \to U$ the inverse of the Riesz isomorphism. Due to the presence of $P_{U_{\text{ad}}}$ in variational discretization the function u_h will in general not belong to X_h even in the case $U = L^2(\Omega)$, $B = Id$. This is different for the purely state constrained problem, for which $P_{U_{\text{ad}}} \equiv Id$, so that in this specific setting $u_h = -\frac{1}{\alpha}p_h + u_{0,h} \in X_h$ by (3.64). In that case the space $U = L^2(\Omega)$ in (3.59) may be replaced by X_h to obtain the same discrete solution u_h, which results in a finite-dimensional discrete optimization problem instead. However, we emphasize, that the infinite-dimensional formulation of (3.59) is very useful for our numerical analysis in the Sect. 3.3.1.2.

As a first result for (3.59) we prove that the sequence of optimal controls, states and the measures μ_h are uniformly bounded.

Lemma 3.3 *Let $u_h \in U_{\text{ad}}$ be the optimal solution of (3.59) with corresponding state $y_h \in X_h$ and adjoint variables $p_h \in X_h$ and $\mu_h \in \mathcal{M}(\bar{\Omega})$. Then there exists $\bar{h} > 0$ so that*

$$\|y_h\|, \|u_h\|_U, \|\mu_h\|_{\mathcal{M}(\bar{\Omega})} \le C \quad \textit{for all } 0 < h \le \bar{h}.$$

Proof Let $\tilde{u}$ denote an element satisfying Assumption 3.11. Since $\mathcal{G}(B\tilde{u})$ is continuous, Assumption 3.11 implies that there exists $\delta > 0$ such that

$$\mathcal{G}(B\tilde{u}) \le b - \delta \quad \text{in } \bar{\Omega}. \tag{3.65}$$

It follows from (3.56) that there is $h_0 > 0$ with

$$\mathcal{G}_h(B\tilde{u}) \le b \quad \text{in } \bar{\Omega} \text{ for all } 0 < h \le h_0$$

so that $J_h(u_h) \le J_h(\tilde{u}) \le C$ uniformly in h giving

$$\|u_h\|_U, \|y_h\| \le C \quad \text{for all } h \le h_0. \tag{3.66}$$

Next, let u denote the unique solution to problem (3.51). We infer from (3.65) and (3.56) that $v := \frac{1}{2}u + \frac{1}{2}\tilde{u}$ satisfies

$$\begin{aligned} \mathcal{G}_h(Bv) &\le \frac{1}{2}\mathcal{G}(Bu) + \frac{1}{2}\mathcal{G}(B\tilde{u}) + Ch^{2-\frac{d}{2}}(\|Bu\| + \|B\tilde{u}\|) \\ &\le b - \frac{\delta}{2} + Ch^{2-\frac{d}{2}}(\|u\|_U + \|\tilde{u}\|_U) \le b - \frac{\delta}{4} \quad \text{in } \bar{\Omega} \end{aligned} \tag{3.67}$$

provided that $h \le \bar{h}, \bar{h} \le h_0$. Since $v \in U_{\rm ad}$, (3.62), (3.61), (3.66) and (3.67) imply

$$\begin{aligned} 0 &\le \left(RB^* p_h + \alpha(u_h - u_{0,h}), v - u_h\right)_U \\ &= \int_\Omega B(v-u_h)p_h + \alpha(u_h - u_{0,h}, v - u_h)_U \\ &= a(\mathcal{G}_h(Bv) - y_h, p_h) + \alpha(u_h - u_{0,h}, v - u_h)_U \\ &= \int_\Omega (\mathcal{G}_h(Bv) - y_h)(y_h - y_0) + \int_{\bar\Omega} (\mathcal{G}_h(Bv) - y_h) d\mu_h + \alpha(u_h - u_{0,h}, v - u_h)_U \\ &\le C + \sum_{j=1}^m \mu_j \left(b(x_j) - \frac{\delta}{4} - y_h(x_j)\right) = C - \frac{\delta}{4}\sum_{j=1}^m \mu_j, \end{aligned}$$

where the last equality is a consequence of (3.63). It follows that

$$\|\mu_h\|_{\mathcal{M}(\bar\Omega)} \le C$$

and the lemma is proved.

3.3.1.2 Error Analysis

An important ingredient in our analysis is an error bound for a solution of a Neumann problem with a measure valued right hand side. Let A be defined as above and consider

$$A^* q = \tilde\mu_{|\Omega} \quad \text{in } \Omega, \qquad \partial_\eta q = \tilde\mu_{|\partial\Omega} \quad \text{on } \partial\Omega. \tag{3.68}$$

Theorem 3.13 *Let $\tilde\mu \in \mathcal{M}(\bar\Omega)$. Then there exists a unique weak solution $q \in L^2(\Omega)$ of* (3.68), *i.e.*

$$\int_\Omega qAv = \int_{\bar\Omega} v d\tilde\mu \quad \forall v \in H^2(\Omega) \quad \textit{with} \sum_{i,j=1}^d a_{ij} v_{x_i} \nu_j = 0 \quad \textit{on } \partial\Omega.$$

Furthermore, q belongs to $W^{1,s}(\Omega)$ for all $s \in (1, \frac{d}{d-1})$. For the finite element approximation $q_h \in X_h$ of q defined by

$$a(v_h, q_h) = \int_{\bar\Omega} v_h d\tilde\mu \quad \textit{for all } v_h \in X_h,$$

the following error estimate holds;

$$\|q - q_h\| \le Ch^{2-\frac{d}{2}} \|\tilde\mu\|_{\mathcal{M}(\bar\Omega)}. \tag{3.69}$$

Proof A corresponding result is proved by Casas in [20] for the case of an operator A subject to Dirichlet conditions, but the arguments can be adapted to the present situation. We omit the details.

Clearly, A in our setting is self adjoint, so that $A \equiv A^*$. However we note, that all considerations in this Sections also apply to more general elliptic operators containing e.g. transport terms, see [40] for details. We are now prepared to prove the main theorem for the optimal controls in the present section.

Theorem 3.14 *Let u and u_h be the solutions of* (3.51) *and* (3.59) *respectively. Then*

$$\|u - u_h\|_U + \|y - y_h\|_{H^1} \le Ch^{1-\frac{d}{4}}.$$

If in addition $Bu \in W^{1,s}(\Omega)$ for some $s \in (1, \frac{d}{d-1})$ then

$$\|u - u_h\|_U + \|y - y_h\|_{H^1} \le Ch^{\frac{3}{2}-\frac{d}{2s}}\sqrt{|\log h|}.$$

Proof We test (3.53) with u_h, (3.62) with u and add the resulting inequalities. This gives

$$\big(RB^*(p - p_h) - \alpha(u_0 - u_{0,h}) + \alpha(u - u_h), u_h - u\big)_U \ge 0,$$

which in turn yields

$$\alpha\|u - u_h\|_U^2 \le \int_\Omega B(u_h - u)(p - p_h) - \alpha\left(u_0 - u_{0,h}, u_h - u\right)_U. \tag{3.70}$$

Let $y^h := \mathcal{G}_h(Bu) \in X_h$ and denote by $p^h \in X_h$ the unique solution of

$$a(w_h, p^h) = \int_\Omega (y - y_0)w_h + \int_{\bar{\Omega}} w_h d\mu \quad \text{for all } w_h \in X_h.$$

Applying Theorem 3.13 with $\tilde{\mu} = (y - y_0) + \mu$ we infer

$$\|p - p^h\| \le Ch^{2-\frac{d}{2}}\left(\|y - y_0\| + \|\mu\|_{\mathcal{M}(\bar{\Omega})}\right). \tag{3.71}$$

Recalling that $y_h = \mathcal{G}_h(Bu_h)$, $y^h = \mathcal{G}_h(Bu)$ and observing (3.61) as well as the definition of p^h we can rewrite the first term on the right-hand-side of (3.70)

$$\begin{aligned}\int_\Omega B(u_h - u)(p - p_h) &= \int_\Omega B(u_h - u)(p - p^h) + \int_\Omega B(u_h - u)(p^h - p_h)\\ &= \int_\Omega B(u_h - u)(p - p^h) + a(y_h - y^h, p^h - p_h)\\ &= \int_\Omega B(u_h - u)(p - p^h) + \int_\Omega (y - y_h)(y_h - y^h)\end{aligned}$$

$$
\begin{aligned}
&+\int_{\bar\Omega}(y_h-y^h)d\mu-\int_{\bar\Omega}(y_h-y^h)d\mu_h\\
&=\int_\Omega B(u_h-u)(p-p^h)-\|y-y_h\|^2\\
&+\int_\Omega(y-y_h)(y-y^h)+\int_{\bar\Omega}(y_h-y^h)d\mu\\
&+\int_{\bar\Omega}(y^h-y_h)d\mu_h. \qquad (3.72)
\end{aligned}
$$

After inserting (3.72) into (3.70) and using Young's inequality we obtain in view of (3.71), (3.55) and (3.60)

$$
\begin{aligned}
&\frac{\alpha}{2}\|u-u_h\|_U^2+\frac{1}{2}\|y-y_h\|^2\\
&\le C\big(\|p-p^h\|^2+\|y-y^h\|^2+\|u_0-u_{0,h}\|_U^2\big)+\int_{\bar\Omega}(y_h-y^h)d\mu\\
&+\int_{\bar\Omega}(y^h-y_h)d\mu_h\\
&\le Ch^{4-d}+\int_{\bar\Omega}(y_h-y^h)d\mu+\int_{\bar\Omega}(y^h-y_h)d\mu_h. \qquad (3.73)
\end{aligned}
$$

It remains to estimate the integrals involving the measures μ and μ_h. Since

$$y_h-y^h\le(I_hb-b)+(b-y)+(y-y^h)\quad\text{in }\bar\Omega$$

we deduce with the help of (3.54)

$$\int_{\bar\Omega}(y_h-y^h)d\mu\le\|\mu\|_{\mathcal{M}(\bar\Omega)}\Big(\|I_hb-b\|_{L^\infty}+\|y-y^h\|_{L^\infty}\Big).$$

Similarly, (3.63) implies

$$\int_{\bar\Omega}(y^h-y_h)d\mu_h\le\|\mu_h\|_{\mathcal{M}(\bar\Omega)}\Big(\|b-I_hb\|_{L^\infty}+\|y-y^h\|_{L^\infty}\Big).$$

Inserting the above estimates into (3.73) and using Lemma 3.3 as well as an interpolation estimate we infer

$$\|u-u_h\|_U^2+\|y-y_h\|^2\le Ch^{4-d}+C\|y-y^h\|_{L^\infty}. \qquad (3.74)$$

The estimates on $\|u-u_h\|_U$ now follow from (3.56) and Lemma 3.1 respectively. Finally, in order to bound $\|y-y_h\|_{H^1}$ we note that

$$a(y-y_h,v_h)=\int_\Omega B(u-u_h)v_h$$

for all $v_h \in X_h$, from which one derives the desired estimates using standard finite element techniques and the bounds on $\|u - u_h\|_U$.

For controls $u, u_h \in L^\infty$ uniformly we also have for $d = 2, 3$

Corollary 3.3 *Let u and u_h be the solutions of* (3.51) *and* (3.59) *respectively. Let us assume that $u, u_h \in L^\infty(\Omega)$ with $\|u_h\|_\infty \le C$ uniformly in h. Then, for h small enough*

$$\|u - u_h\|_U + \|y - y_h\|_{H^1} \le Ch|\log h|$$

with some positive constant C independent of h.

Proof In order to avoid the dependence on the dimension we should avoid finite element approximations of the adjoint variable p, which due to its low regularity only allows error estimates in the L^2 norm. We therefore provide a proof technique which completely avoids the use of finite element approximations of the adjoint variable. To begin with we start with the basic estimate (3.70)

$$\alpha\|u - u_h\|_U^2 \le \int_\Omega B(u_h - u)(p - p_h) - \alpha(u_0 - u_{0,h}, u_h - u)_U$$

and write

$$\begin{aligned}
&\int_\Omega B(u_h - u)(p - p_h) \\
&= \int_\Omega pA(\tilde{y} - y) - a(y_h - y^h, p_h) \\
&= \int_\Omega (y - y_0)(\tilde{y} - y) + \int_{\bar{\Omega}} \tilde{y} - y\, d\mu - \int_\Omega (y_h - y_0)(y_h - y^h) \\
&\quad + \int_{\bar{\Omega}} y_h - y^h d\mu_h,
\end{aligned}$$

where $\tilde{y} := \mathcal{G}(Bu_h)$. Proceeding similar as in the proof of the previous theorem we obtain

$$\begin{aligned}
&\int_\Omega (y - y_0)(\tilde{y} - y) + \int_{\bar{\Omega}} \tilde{y} - y\, d\mu - \int_\Omega (y_h - y_0)(y_h - y^h) + \int_{\bar{\Omega}} y_h - y^h d\mu_h \\
&\le C\{\|\mu\|_{\mathcal{M}(\bar{\Omega})} + \|\mu_h\|_{\mathcal{M}(\bar{\Omega})}\}\{\|b - I_h b\|_{L^\infty} + \|y - y^h\|_{L^\infty} + \|\tilde{y} - y_h\|_{L^\infty}\} \\
&\quad - \|y - y_h\|^2 + C\{\|y - y^h\| + \|\tilde{y} - y_h\|\}.
\end{aligned}$$

Using Lemma 3.3 together with Lemma 3.1 then yields

$$\alpha\|u - u_h\|_U^2 + \|y - y_h\|^2 \le C\{h^2 + h^2|\log h|^2\},$$

so that the claim follows as in the proof of the previous theorem.

Remark 3.4 Let us note that the approximation order of the controls and states in the presence of control and state constraints is the same as in the purely state constrained case, if $Bu \in W^{1,s}(\Omega)$. This assumption holds for the important example $U = L^2(\Omega)$, $B = Id$ and $u_{0_h} = P_h u_0$, with $u_0 \in H^1(\Omega)$ and $P_h : L^2(\Omega) \to X_h$ denoting the L^2-projection, and subsets of the form

$$U_{\text{ad}} = \{v \in L^2(\Omega), a_l \leq v \leq a_u \text{ a.e. in } \Omega\},$$

with bounds $a_l, a_u \in W^{1,s}(\Omega)$, since $u_0 \in H^1(\Omega)$, and $p \in W^{1,s}(\Omega)$. Moreover, $u, u_h \in L^\infty(\Omega)$ with $\|u_h\|_\infty \leq C$ uniformly in h holds if for example $a_l, a_u \in L^\infty(\Omega)$.

Remark 3.5 We mention here a second approach that differs from the one discussed above in the way in which the inequality constraints are realized. Denote by $D_1, \ldots, D_m$ the cells of the dual mesh. Each cell D_i is associated with a vertex x_i of $\mathcal{T}_h$ and we have

$$\bar{\Omega} = \bigcup_{i=1}^{m} D_i, \qquad \text{int}(D_i) \cap \text{int}(D_j) = \emptyset, \quad i \neq j.$$

In (3.59), we now impose the constraints

$$⨍_{D_j} (y_h - I_h b) \leq 0 \quad \text{for } j = 1, \ldots, m \tag{3.75}$$

on the discrete solution $y_h = \mathcal{G}_h(Bu)$. Here, we have abbreviated $⨍_{D_j} f = \frac{1}{|D_j|}\int_{D_j} f$. The measure μ_h that appears in Lemma 3.2 now has the form $\mu_h = \sum_{j=1}^m \mu_j ⨍_{D_j} \cdot dx$, and the pointwise constraints in (3.63) are replaced by those of (3.75). The error analysis for the resulting numerical method can be carried out in the same way as shown above with the exception of Theorem 3.14, where the bounds on $\tilde{y} - b$ and $\tilde{y}_h - I_h b$ require a different argument. In this case, additional terms of the form

$$\left\| f - ⨍_{D_j} f \right\|_{L^\infty(D_j)}$$

have to be estimated. Since these will in general only be of order $O(h)$, this analysis would only give $\|u - u_h\|, \|y - y_h\|_{H^1} = O(\sqrt{h})$. The numerical test example in Sect. 3.3.1.4 suggests that at least $\|u - u_h\| = O(h)$, but we are presently unable to prove such an estimate.

3.3.1.3 Piecewise Constant Controls

In the presence of state constraints a result similar to that of Corollary 3.3 can also be shown for piecewise constant control approximations with box constraints on the control. Let now B denote the identity and let $U_{\text{ad}} = \{v \in L^2(\Omega); a_l \leq v \leq a_u$

a.e. in $\Omega\}$, where $a_l < a_u$ are given constants. We present the corresponding result which is taken from the paper [41] of Deckelnick and Hinze. For this purpose we define the space of piecewise constant functions

$$Y_h := \{v_h \in L^2(\Omega) \mid v_h \text{ is constant on each } T \in \mathcal{T}_h\}.$$

and denote by $Q_h : L^2(\Omega) \to Y_h$ the orthogonal projection onto Y_h so that

$$(Q_h v)(x) := \fint_T v, \quad x \in T, T \in \mathcal{T}_h,$$

where $\fint_T v$ denotes the average of v over T. In order to approximate (3.51) we introduce a discrete counterpart of U_{ad},

$$U_{\text{ad}}^h := \{v_h \in Y_h \mid a_l \le v_h \le a_u \text{ in } \Omega\}.$$

Note that $U_{\text{ad}}^h \subset U_{\text{ad}}$ and that $Q_h v \in U_{\text{ad}}^h$ for $v \in U_{\text{ad}}$. Since $Q_h v \to v$ in $L^2(\Omega)$ as $h \to 0$ we infer from the continuous embedding $H^2(\Omega) \hookrightarrow C^0(\bar{\Omega})$ and Lemma 3.1 that

$$\mathcal{G}_h(Q_h v) \to \mathcal{G}(v) \quad \text{in } L^\infty(\Omega) \text{ for all } v \in U_{\text{ad}}. \tag{3.76}$$

Problem (3.51) is now approximated by the following sequence of control problems depending on the mesh parameter h:

$$\begin{aligned} &\min_{u \in U_{\text{ad}}^h} J_h(u) := \frac{1}{2}\int_\Omega |y_h - y_0|^2 + \frac{\alpha}{2}\int_\Omega |u|^2 \\ &\text{subject to} \quad y_h = \mathcal{G}_h(u) \quad \text{and} \quad y_h(x_j) \le b(x_j) \quad \text{for } j = 1, \dots, m. \end{aligned} \tag{3.77}$$

Problem (3.77), as problem (3.59), represents a convex finite-dimensional optimization problem of similar structure as problem (3.51), but with only finitely many equality and inequality constraints for state and control, which form a convex admissible set. The following optimality conditions can be argued as those given in (3.2) for problem (3.59).

Lemma 3.4 *Problem* (3.77) *has a unique solution* $u_h \in U_{\text{ad}}^h$. *There exist* $\mu_1, \dots, \mu_m \in \mathbb{R}$ *and* $p_h \in X_h$ *such that with* $y_h = \mathcal{G}_h(u_h)$ *and* $\mu_h = \sum_{j=1}^m \mu_j \delta_{x_j}$ *we have*

$$a(v_h, p_h) = \int_\Omega (y_h - y_0)v_h + \int_{\bar{\Omega}} v_h d\mu_h \quad \forall v_h \in X_h, \tag{3.78}$$

$$\int_\Omega (p_h + \alpha u_h)(v_h - u_h) \ge 0 \quad \forall v_h \in U_{\text{ad}}^h, \tag{3.79}$$

$$\mu_j \ge 0, \quad y_h(x_j) \le b(x_j), \quad j = 1, \dots, m \quad \textit{and} \quad \int_{\bar{\Omega}} (I_h b - y_h) d\mu_h = 0. \tag{3.80}$$

Here, δ_x denotes the Dirac measure concentrated at x and I_h is the usual Lagrange interpolation operator.

For (3.77) we now prove bounds on the discrete states and the discrete multipliers. Similar to Lemma 3.3 we have

Lemma 3.5 *Let $u_h \in U_{\text{ad}}^h$ be the optimal solution of (3.77) with corresponding state $y_h \in X_h$ and adjoint variables $p_h \in X_h$ and $\mu_h \in \mathcal{M}(\bar{\Omega})$. Then there exists $\bar{h} > 0$ such that*

$$\|y_h\|, \|\mu_h\|_{\mathcal{M}(\bar{\Omega})} \le C, \qquad \|p_h\|_{H^1} \le C\gamma(d,h) \quad \textit{for all } 0 < h \le \bar{h},$$

where $\gamma(2,h) = \sqrt{|\log h|}$ and $\gamma(3,h) = h^{-\frac{1}{2}}$.

Proof Since $\mathcal{G}(\tilde{u}) \in C^0(\bar{\Omega})$, Assumption 3.11 implies that there exists $\delta > 0$ such that

$$\mathcal{G}(\tilde{u}) \le b - \delta \quad \text{in } \bar{\Omega}. \tag{3.81}$$

It follows from (3.76) that there is $\bar{h} > 0$ with

$$\mathcal{G}_h(Q_h\tilde{u}) \le b - \frac{\delta}{2} \quad \text{in } \bar{\Omega} \text{ for all } 0 < h \le \bar{h}. \tag{3.82}$$

Since $Q_h\tilde{u} \in U_{\text{ad}}^h$, (3.80), (3.79) and (3.82) imply

$$\begin{aligned}
0 \le \int_\Omega (p_h + \alpha u_h)(Q_h\tilde{u} - u_h) &= \int_\Omega p_h(Q_h\tilde{u} - u_h) + \alpha\int_\Omega u_h(Q_h\tilde{u} - u_h) \\
&= a(\mathcal{G}_h(Q_h\tilde{u}) - y_h, p_h) + \alpha\int_\Omega u_h(Q_h\tilde{u} - u_h) \\
&= \int_\Omega (\mathcal{G}_h(Q_h\tilde{u}) - y_h)(y_h - y_0) + \int_{\bar{\Omega}} (\mathcal{G}_h(Q_h\tilde{u}) - y_h)d\mu_h + \alpha\int_\Omega u_h(Q_h\tilde{u} - u_h) \\
&\le C - \frac{1}{2}\|y_h\|^2 + \sum_{j=1}^m \mu_j\left(b(x_j) - \frac{\delta}{2} - y_h(x_j)\right) = C - \frac{1}{2}\|y_h\|^2 - \frac{\delta}{2}\sum_{j=1}^m \mu_j
\end{aligned}$$

where the last equality is a consequence of (3.80). It follows that $\|y_h\|, \|\mu_h\|_{\mathcal{M}(\bar{\Omega})} \le C$. In order to bound $\|p_h\|_{H^1}$ we insert $v_h = p_h$ into (3.79) and deduce with the help of the coercivity of A, a well-known inverse estimate and the bounds we have already obtained that

$$\begin{aligned}
c_1\|p_h\|_{H^1}^2 \le a(p_h, p_h) &= \int_\Omega (y_h - y_0)p_h + \int_{\bar{\Omega}} p_h d\mu_h \\
&\le \|y_h - y_0\|\,\|p_h\| + \|p_h\|_{L^\infty}\|\mu_h\|_{\mathcal{M}(\bar{\Omega})} \le C\|p_h\| + C\gamma(d,h)\|p_h\|_{H^1}.
\end{aligned}$$

Hence $\|p_h\|_{H^1} \le C\gamma(d,h)$ and the lemma is proved.

We are now prepared to prove the analogue to Theorem 3.14 for piecewise constant control approximations.

Theorem 3.15 *Let u and u_h be the solutions of* (3.51) *and* (3.77) *respectively. Then we have for* $0 < h \le \bar{h}$

$$\|u - u_h\| + \|y - y_h\|_{H^1} \le \begin{cases} Ch|\log h|, & \text{if } d = 2 \\ C\sqrt{h}, & \text{if } d = 3. \end{cases}$$

Proof We test (3.53) with u_h, (3.80) with $Q_h u$ and add the resulting inequalities. Keeping in mind that $u - Q_h u \perp Y_h$ we obtain

$$\int_\Omega (p - p_h + \alpha(u - u_h))(u_h - u)$$
$$\ge \int_\Omega (p_h + \alpha u_h)(u - Q_h u) = \int_\Omega (p_h - Q_h p_h)(u - Q_h u).$$

As a consequence,

$$\alpha\|u - u_h\|^2 \le \int_\Omega (u_h - u)(p - p_h) - \int_\Omega (p_h - Q_h p_h)(u - Q_h u) \equiv I + II. \quad (3.83)$$

Let $y^h := \mathcal{G}_h(u) \in X_h$ and denote by $p^h \in X_h$ the unique solution of

$$a(w_h, p^h) = \int_\Omega (y - y_0)w_h + \int_{\bar\Omega} w_h d\mu \quad \text{for all } w_h \in X_h.$$

Applying Theorem 3.13 with $\tilde\mu = (y - y_0) + \mu$ we infer

$$\|p - p^h\| \le Ch^{2-\frac{d}{2}}\left(\|y - y_0\| + \|\mu\|_{\mathcal{M}(\bar\Omega)}\right). \quad (3.84)$$

Recalling that $y_h = \mathcal{G}_h(u_h)$, $y^h = \mathcal{G}_h(u)$ and observing (3.79) as well as the definition of p^h we can rewrite the first term in (3.83)

$$\begin{aligned} I &= \int_\Omega (u_h - u)(p - p^h) + \int_\Omega (u_h - u)(p^h - p_h) \\ &= \int_\Omega (u_h - u)(p - p^h) + a(y_h - y^h, p^h - p_h) \\ &= \int_\Omega (u_h - u)(p - p^h) + \int_\Omega (y - y_h)(y_h - y^h) + \int_{\bar\Omega} (y_h - y^h)d\mu \\ &\quad - \int_{\bar\Omega} (y_h - y^h)d\mu_h \\ &= \int_\Omega (u_h - u)(p - p^h) - \|y - y_h\|^2 + \int_\Omega (y - y_h)(y - y^h) \\ &\quad + \int_{\bar\Omega} (y_h - y^h)d\mu + \int_{\bar\Omega} (y^h - y_h)d\mu_h. \end{aligned} \quad (3.85)$$

Applying Young's inequality we deduce

$$|I| \leq \frac{\alpha}{4}\|u-u_h\|^2 - \frac{1}{2}\|y-y_h\|^2 + C\big(\|p-p^h\|^2 + \|y-y^h\|^2\big) \\ + \int_{\bar{\Omega}} (y_h - y^h)d\mu + \int_{\bar{\Omega}} (y^h - y_h)d\mu_h. \tag{3.86}$$

Let us estimate the integrals involving the measures μ and μ_h. Since $y_h - y^h \leq (I_h b - b) + (b - y) + (y - y^h)$ in $\bar{\Omega}$ we deduce with the help of (3.54), Lemma 3.1 and an interpolation estimate

$$\int_{\bar{\Omega}} (y_h - y^h)d\mu \leq \|\mu\|_{\mathcal{M}(\bar{\Omega})}\big(\|I_h b - b\|_{L^\infty} + \|y - y^h\|_{L^\infty}\big) \leq Ch^2|\log h|^2.$$

On the other hand $y^h - y_h \leq (y^h - y) + (b - I_h b) + (I_h b - y_h)$, so that (3.80), Lemma 3.1 and Lemma 3.5 yield

$$\int_{\bar{\Omega}} (y^h - y_h)d\mu_h \leq \|\mu_h\|_{\mathcal{M}(\bar{\Omega})}\big(\|b - I_h b\|_{L^\infty} + \|y - y^h\|_{L^\infty}\big) \leq Ch^2|\log h|^2.$$

Inserting these estimates into (3.86) and recalling (3.55) as well as (3.69) we obtain

$$|I| \leq \frac{\alpha}{4}\|u-u_h\|^2 - \frac{1}{2}\|y-y_h\|^2 + Ch^{4-d} + Ch^2|\log h|^2. \tag{3.87}$$

Let us next examine the second term in (3.83). Since $u_h = Q_h u_h$ and Q_h is stable in $L^2(\Omega)$ we have

$$|II| \leq 2\|u-u_h\|\,\|p_h - Q_h p_h\| \leq \frac{\alpha}{4}\|u-u_h\|^2 + Ch^2\|p_h\|^2_{H^1} \\ \leq \frac{\alpha}{4}\|u-u_h\|^2 + Ch^2\gamma(d,h)^2$$

using an interpolation estimate for Q_h and Lemma 3.5. Combining this estimate with (3.87) and (3.83) we finally obtain

$$\|u-u_h\|^2 + \|y-y_h\|^2 \leq Ch^{4-d} + Ch^2|\log h|^2 + Ch^2\gamma(d,h)^2$$

which implies the estimate on $\|u-u_h\|$. In order to bound $\|y-y_h\|_{H^1}$ we note that

$$a(y-y_h, v_h) = \int_\Omega (u-u_h)v_h$$

for all $v_h \in X_h$, from which one derives the desired estimate using standard finite element techniques and the bound on $\|u-u_h\|$.

Remark 3.6 An inspection of the proof of Theorem 3.15 shows that we also could avoid to use error estimates for the auxiliary function p^h if we would use a technique for the term I similar to that used in the proof of Corollary 3.3. However, our approach to estimate II is based on inverse estimates which finally lead to the dimension dependent error estimate presented in Theorem 3.15.

3.3.1.4 Numerical Examples for Pointwise Constraints on the State

Example 3.6 The following test problem is taken—in a slightly modified form—from the paper [104, Example 6.2] of Meyer, Prüfert and Tröltzsch. Let $\Omega := B_1(0)$, $\alpha > 0$,

$$y_0(x) := 4 + \frac{1}{\pi} - \frac{1}{4\pi}|x|^2 + \frac{1}{2\pi}\log|x|, \qquad u_0(x) := 4 + \frac{1}{4\alpha\pi}|x|^2 - \frac{1}{2\alpha\pi}\log|x|$$

and $b(x) := |x|^2 + 4$. We consider the cost functional

$$J(u) := \frac{1}{2}\int_\Omega |y - y_0|^2 + \frac{\alpha}{2}\int_\Omega |u - u_0|^2,$$

where $y = \mathcal{G}(u)$. By checking the optimality conditions of first order one verifies that $u \equiv 4$ is the unique solution of (3.51) with corresponding state $y \equiv 4$ and adjoint states

$$p(x) = \frac{1}{4\pi}|x|^2 - \frac{1}{2\pi}\log|x| \quad \text{and} \quad \mu = \delta_0.$$

The finite element counterparts of y, u, p and μ are denoted by y_h, u_h, p_h and μ_h.

To investigate the experimental order of convergence (see (3.25) for its definition) for our model problem we choose a sequence of uniform partitions of Ω containing five refinement levels, starting with eight triangles forming a uniform octagon as initial triangulation of the unit disc. The corresponding grid sizes are $h_i = 2^{-i}$ for $i = 1, \ldots, 5$. As error functionals we take $E(h) = \|(u, y) - (u_h, y_h)\|$ and $E(h) = \|(u, y) - (u_h, y_h)\|_{H^1}$ and note, that the error $p - p_h$ is related to $u - u_h$ via (3.62). We solve problems (3.59) using the QUADPROG routine of the MATLAB OPTIMIZATION TOOLBOX. The required finite element matrices for the discrete state and adjoint systems are generated with the help of the MATLAB PDE TOOLBOX. Furthermore, for discontinuous functions f we use the quadrature rule

$$\int_\Omega f(x)dx \approx \sum_{T \in \mathcal{T}_h} f\left(x_{s(T)}\right)|T|,$$

where $x_{s(T)}$ denotes the barycenter of T. In all computations we set $\alpha = 1$.

In Table 3.11, we present EOCs for problem (3.59) (case $S = D$) and the approach sketched in Remark 3.5 (case $S = M$). As one can see, the error $\|u - u_h\|$ behaves in the case $S = D$ as predicted by Theorem 3.14, whereas the errors $\|y - y_h\|$ and $\|y - y_h\|_{H^1}$ show a better convergence behaviour. On the finest level we have $\|u - u_h\| = 0.003117033$, $\|y - y_h\| = 0.000123186$ and $|y - y_h|_{H^1} = 0.000083757$. Furthermore, all coefficients of μ_h are equal to zero, except the one in front of δ_0 whose value is 0.99946494. The errors $\|u - u_h\|$, $\|y - y_h\|$ and $\|y - y_h\|_{H^1}$ in the case $S = M$ show a better EOC than in the case $S = D$. This can be explained by the fact that the exact solutions y and u are contained in the finite element space, and that the relaxed form of the state constraints introduce a smearing effect on the numerical solutions at the origin. On the finest level we have

Table 3.11 Experimental order of convergence

Level	$(S=D)$ $\|u-u_h\|$	$(S=M)$ $\|u-u_h\|$	$(S=D)$ $\|y-y_h\|$	$(S=M)$ $\|y-y_h\|$	$(S=D)$ $\|y-y_h\|_{H^1}$	$(S=M)$ $\|y-y_h\|_{H^1}$
1	0.788985	0.654037	0.536461	0.690302	0.860516	0.688531
2	0.759556	1.972784	1.147861	2.017836	1.272400	2.015602
3	0.919917	1.962191	1.389378	2.004383	1.457095	2.004286
4	0.966078	1.856687	1.518381	1.989727	1.564204	1.990566
5	0.986686	1.588722	1.598421	1.979082	1.632772	1.979945

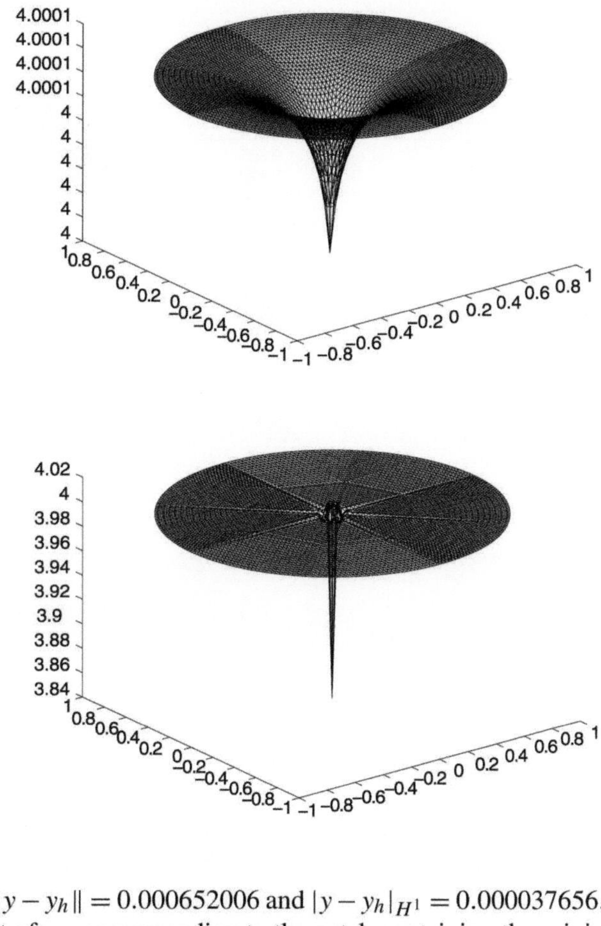

Fig. 3.11 Numerically computed state y_h (*top*) and control u_h (*bottom*) for $h = 2^{-5}$ in the case $S = D$

$\|u-u_h\| = 0.001020918$, $\|y-y_h\| = 0.000652006$ and $|y-y_h|_{H^1} = 0.000037656$. Furthermore, the coefficient of μ_h corresponding to the patch containing the origin has the value 1.0640946.

Figures 3.11 and 3.12 present the numerical solutions y_h and u_h for $h = 2^{-5}$ in the case $S = D$ and $S = M$, respectively. We note that using equal scales on all axes would give completely flat graphs in all four figures.

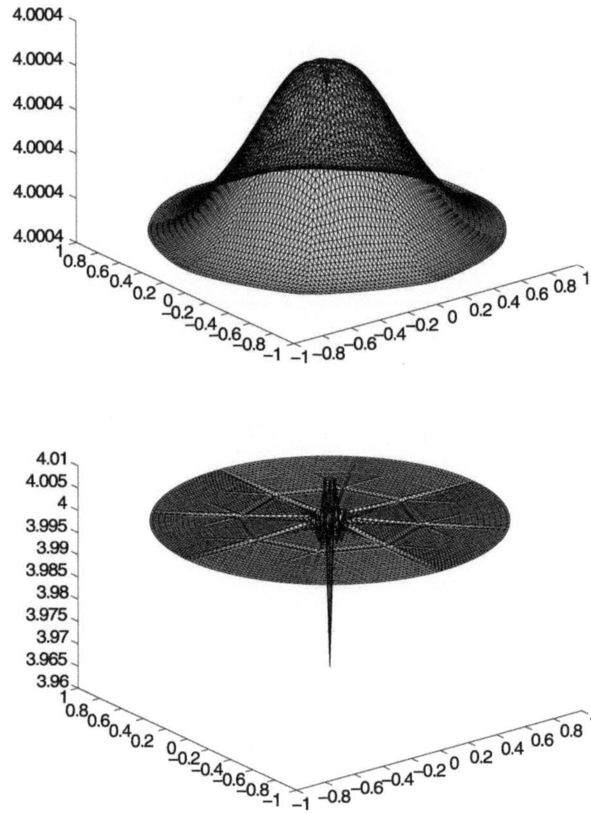

Fig. 3.12 Numerically computed state y_h (*top*) and control u_h (*bottom*) for $h = 2^{-5}$ in the case $S = M$

Example 3.7 The second test problem is taken from the work [102], Example 2 of Meyer, Rösch, and Tröltzsch. It reads

$$\min_{u \in L^2(\Omega)} J(u) = \frac{1}{2} \int_\Omega |y - y_0|^2 + \frac{1}{2} \int_\Omega |u - u_0|^2$$

$$\text{subject to} \quad y = \mathcal{G}(u) \quad \text{and} \quad y(x) \geq b(x) \quad \text{in } \Omega.$$

Here, Ω denotes the unit square,

$$b(x) = \begin{cases} 2x_1 + 1, & x_1 < \frac{1}{2}, \\ 2, & x_1 \geq \frac{1}{2}, \end{cases} \qquad y_0(x) = \begin{cases} x_1^2 - \frac{1}{2}, & x_1 < \frac{1}{2}, \\ \frac{1}{4}, & x_1 = \frac{1}{2}, \\ \frac{3}{4}, & x_1 > \frac{1}{2}, \end{cases}$$

and

$$u_0(x) = \begin{cases} \frac{5}{2} - x_1^2, & x_1 < \frac{1}{2}, \\ \frac{9}{4}, & x_1 \geq \frac{1}{2}. \end{cases}$$

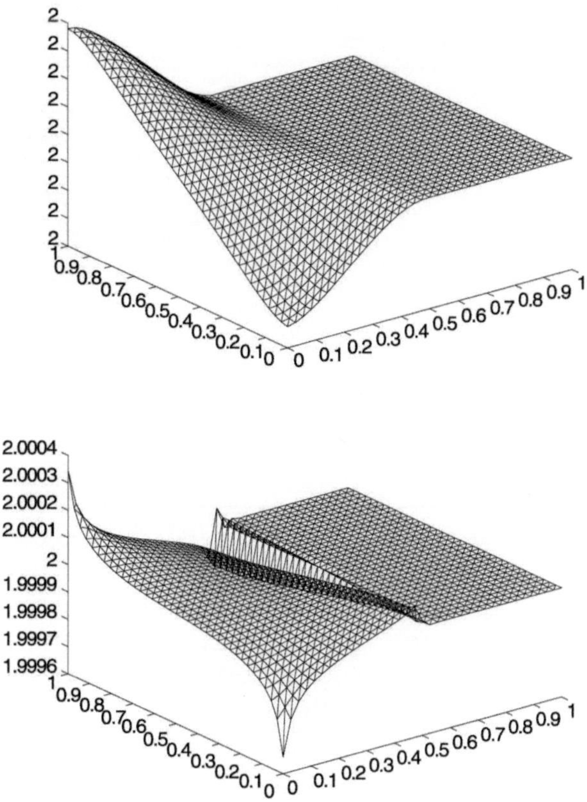

Fig. 3.13 Numerically computed state y_h (*top*) and control u_h (*bottom*) for $h = \frac{\sqrt{2}}{36}$ in the case $S = D$

The exact solution is given by $y \equiv 2$ and $u \equiv 2$ in Ω. The corresponding Lagrange multiplier $p \in H^1(\Omega)$ is given by

$$p(x) = \begin{cases} \frac{1}{2} - x_1^2, & x_1 < \frac{1}{2}, \\ \frac{1}{4}, & x_1 \geq \frac{1}{2}. \end{cases}$$

The multiplier μ has the form

$$\int_{\bar{\Omega}} f d\mu = \int_{\{x_1 = \frac{1}{2}\}} f ds + \int_{\{x_1 > \frac{1}{2}\}} f dx, \quad f \in C^0(\bar{\Omega}). \tag{3.88}$$

In our numerical computations we use uniform grids generated with the POIMESH function of the MATLAB PDE TOOLBOX. Integrals containing y_0, u_0 are numerically evaluated by substituting y_0, u_0 by their piecewise linear, continuous finite element interpolations $I_h y_0, I_h u_0$. The grid size of a grid containing l horizontal and l vertical lines is given by $h_l = \frac{\sqrt{2}}{l+1}$. Figure 3.13 presents the numerical results for a grid with $h = \frac{\sqrt{2}}{36}$ in the case $(S = D)$. The corresponding values of μ_h on the same grid are presented in Fig. 3.14. They reflect the fact that the measure consists

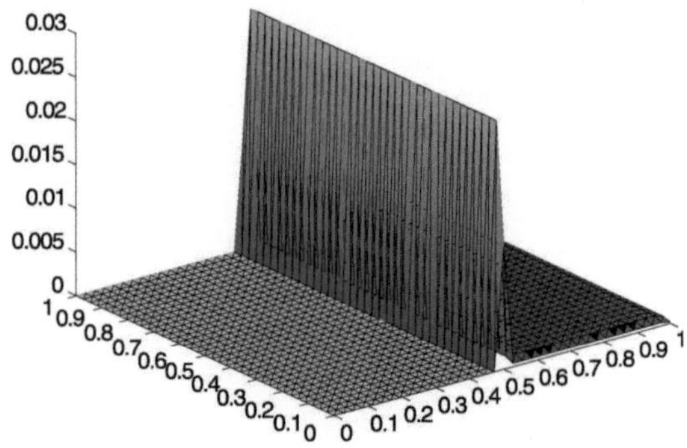

Fig. 3.14 Numerically computed multiplier μ_h for $h = \frac{\sqrt{2}}{36}$ in the case $S = D$

Table 3.12 Experimental order of convergence, $x_1 = \frac{1}{2}$ grid line

Level	$(S = D)$ $\|u - u_h\|$	$(S = M)$ $\|u - u_h\|$	$(S = D)$ $\|y - y_h\|$	$(S = M)$ $\|y - y_h\|$	$(S = D)$ $\|y - y_h\|_{H^1}$	$(S = M)$ $\|y - y_h\|_{H^1}$
1	1.669586	0.448124	1.417368	0.544284	1.594104	0.384950
2	1.922925	1.184104	1.990906	1.473143	1.992097	1.239771
3	2.000250	1.456908	2.101633	1.871948	2.080739	1.745422
4	2.029556	1.530303	2.125168	2.427634	2.108241	2.348036
5	2.041913	1.260744	2.124773	2.743918	2.116684	2.563363
6	2.047106	1.142668	2.117184	1.430239	2.117739	1.318617
7	2.048926	1.177724	2.107828	1.503463	2.115633	1.409563
8	2.049055	1.194893	2.098597	1.578342	2.112152	1.497715
9	2.048312	1.194802	2.090123	1.622459	2.108124	1.549495

of a lower dimensional part which is concentrated on the line $\{x \in \Omega \mid x_1 = \frac{1}{2}\}$ and a regular part with a density $\chi_{|\{x_1 > \frac{1}{2}\}}$. We again note that using equal scales on all axes would give completely flat graphs for y_h as well as for u_h.

We compute EOCs for the two different sequences of grid-sizes $s_o = \{h_1, h_3, \ldots, h_{19}\}$ and $s_e = \{h_0, h_2, \ldots, h_{18}\}$. We note that the grids corresponding to s_o contain the line $x_1 = \frac{1}{2}$. Table 3.12 presents EOCs for s_o, and Table 3.13 presents EOCs for s_e. For the sequence s_o we observe super-convergence in the case $(S = D)$, although the discontinuous function y_0 for the quadrature is replaced by its piecewise linear, continuous finite element interpolant $I_h y_0$. Let us note that further numerical experiments show that the use of the quadrature rule (3.6) for integrals containing the function y_0 decreases the EOC for $\|u - u_h\|$ to $\frac{3}{2}$, whereas EOCs remain close to 2 for the other two errors $\|y - y_h\|$ and $\|y - y_h\|_{H^1}$. For this sequence also the case $(S = M)$ behaves twice as good as expected by our arguments in Remark 3.5. For the sequence s_e the error $\|u - u_h\|$ in the case $(S = D)$ approximately behaves as predicted by our theory, in the case $(S = M)$ it behaves as for the sequence s_o. The errors $\|y - y_h\|$ and $\|y - y_h\|_{H^1}$ behave that well, since the exact solutions y and u are contained in the finite element space. For h_{19} we have in the case $(S = D)$

Table 3.13 Experimental order of convergence, $x_1 = \frac{1}{2}$ not a grid line

Level	$(S=D)$ $\|u-u_h\|$	$(S=M)$ $\|u-u_h\|$	$(S=D)$ $\|y-y_h\|$	$(S=M)$ $\|y-y_h\|$	$(S=D)$ $\|y-y_h\|_{H^1}$	$(S=M)$ $\|y-y_h\|_{H^1}$
1	0.812598	0.460528	1.160789	2.154570	0.885731	1.473561
2	1.361946	0.406917	2.042731	0.597846	1.918942	0.405390
3	1.228268	1.031763	1.832573	1.392796	1.700124	1.088595
4	1.245030	1.262257	1.678233	1.621110	1.570580	1.392408
5	1.252221	1.416990	1.646124	1.844165	1.554434	1.686808
6	1.256861	1.505759	1.696309	2.128776	1.620231	2.021210
7	1.264456	1.489061	1.627539	2.507863	1.559065	2.415552
8	1.260157	1.316627	1.640964	2.989867	1.580113	2.818148
9	1.265599	1.169109	1.686579	1.601263	1.635084	1.460153

Table 3.14 Approximation of the multiplier in the case $(S = D)$, $x_1 = \frac{1}{2}$ grid line

Level	$\sum_{x_i \in \{x_1=1/2\}} \mu_i$	$\sum_{x_i \in \{x_1>1/2\}} \mu_i$
1	1.13331662624081	0.36552954225441
2	1.06315278164899	0.43644163287114
3	1.03989323182608	0.45990635060758
4	1.02893022155910	0.47095098878247
5	1.02265064139378	0.47727091447291
6	1.01855129775903	0.48139306499280
7	1.01569011772403	0.48426838085822
8	1.01359012331610	0.48637773715316
9	1.01198410389649	0.48799027450619

$\|u-u_h\| = 0.000103428$, $\|y-y_h\| = 0.000003233$ and $|y-y_h|_{H^1} = 0.000015155$, and in the case $(S = M)$ $\|u-u_h\| = 0.011177577$, $\|y-y_h\| = 0.000504815$ and $|y-y_h|_{H^1} = 0.001547907$. We observe that the errors in the case $S = M$ are two magnitudes larger than in the case $(S = D)$. This can be explained by the fact that an Ansatz for the multiplier μ with a linear combination of Dirac measures is better suited to approximate measures concentrated on singular sets than a piecewise constant Ansatz as in the case $(S = M)$. Finally, Table 3.14 presents $\sum_{x_i \in \{x_1=1/2\}} \mu_i$ and $\sum_{x_i \in \{x_1>1/2\}} \mu_i$ for s_o in the case $(S = D)$. As one can see $\sum_{x_i \in \{x_1=1/2\}} \mu_i$ tends to 1, the length of $\{x_1 = 1/2\}$, and $\sum_{x_i \in \{x_1>1/2\}} \mu_i$ tends to $1/2$, the area of $\{x_1 > 1/2\}$. These numerical findings indicate that $\mu_h = \sum_{i=1}^m \mu_i \delta_{x_i}$ well approximates μ, since $\int_{\bar{\Omega}} d\mu_h = \sum_{i=1}^m \mu_i$, and that μ_h also well resolves the structure of μ, see (3.88). For all numerical computations of this example we have $\mu_i = 0$ for $x_i \in \{x_1 < 1/2\}$.

3.3.1.5 Some Literature for (Control and) State Constraints

To the authors knowledge only few attempts have been made to develop a finite element analysis for elliptic control problems in the presence of control and state constraints. In [24] Casas proves convergence of finite element approximations to optimal control problems for semi-linear elliptic equations with finitely many state constraints. Casas and Mateos extend these results in [26] to a less regular setting for the states and prove convergence of finite element approximations to semi-linear distributed and boundary control problems. In [99] Meyer considers a fully discrete strategy to approximate an elliptic control problem with pointwise state and control constraints. He obtains the approximation order

$$\|\bar{u} - \bar{u}_h\| + \|\bar{y} - \bar{y}_h\|_{H^1} = \mathcal{O}(h^{2-d/2-\epsilon}) \quad (\epsilon > 0),$$

where d denotes the spatial dimension. His results confirm those obtained by the Deckelnick and Hinze in [39] for the purely state constrained case, and are in accordance with Theorem 3.14. Meyer also considers variational discretization and in the presence of L^∞ bounds on the controls shows

$$\|\bar{u} - \bar{u}_h\| + \|\bar{y} - \bar{y}_h\|_{H^1} = \mathcal{O}(h^{1-\epsilon}|\log h|) \quad (\epsilon > 0),$$

which is a result of a similar quality as that given in Corollary 3.3.

Let us comment also on further approaches that tackle optimization problems for PDEs with control and state constraints. A *Lavrentiev-type regularization* of problem (3.51) is investigated by Meyer, Rösch and Tröltzsch in [102]. In this approach the state constraint $y \le b$ in (3.51) is replaced by the mixed constraint $\epsilon u + y \le b$, with $\epsilon > 0$ denoting a regularization parameter, see problem (2.70). It turns out that the associated Lagrange multiplier μ_ϵ belongs to $L^2(\Omega)$. Numerical analysis for this approach with emphasis on the coupling of gridsize and regularization parameter ϵ is presented by Hinze and Meyer in [74]. The resulting optimization problems are solved either by interior-point methods or primal-dual active set strategies, compare the work [104] Meyer, Prüfert and Tröltzsch.

Hintermüller and Kunisch in [66, 67] consider the Moreau-Yosida relaxation approach to problem classes containing (3.51). In this approach the state constraint is relaxed in that it is dropped and a L^2 regularization term of the form $\frac{1}{2\gamma}\int_\Omega |\max(0, \gamma\, \mathcal{G}(Bu))|^2$ is added to the cost functional instead, where γ denotes the relaxation parameter, see problem (2.56). Numerical analysis for this approach with emphasis on the coupling of gridsize and relaxation parameter γ is presented by Hintermüller and Hinze in [65].

Schiela in [121] chooses a different way to relax state constraints in considering barrier functionals of the form $-\mu \int_\Omega \log(-\mathcal{G}(Bu))dx$ which penalize the state constraints. In [79] he together with Hinze presents numerical analysis for this approach with emphasis on the coupling of gridsize and barrier parameter μ.

Adaptive approaches to state constrained optimal control problems are only very recently reported. Hoppe and Kieweg present an residual based approach in e.g. [81]. Günther and Hinze in [58] apply the dual weighted residual method to elliptic optimal control problems with state constraints. A related approach is presented by Bendix and Vexler in [10]. Wollner in [147] presents an adaptive approach using interior point methods with applications to elliptic problems with state constraints, and he also considers problems with constraints on the gradient of the state.

3.3.2 Pointwise Bounds on the Gradient of the State

We now consider constraints on the gradient of the state. These kind of constraints play an important role in practical applications where cooling of melts forms a critical process. In order to accelerate such production processes it is highly desirable to speed up the cooling processes while avoiding damage of the products caused by large material stresses. Cooling processes as those considered in Sect. 4.2 frequently are described by systems of partial differential equations involving the temperature as a system variable, so that large (Von Mises) stresses in the optimization process can be avoided by imposing pointwise bounds on the gradient of the temperature. Pointwise bounds on the gradient in optimization in general deliver adjoint variables admitting low regularity only. This fact then necessitates the development of tailored discrete concepts which take into account the low regularity of adjoint variables and multipliers involved in the optimality conditions of the underlying optimization problem.

We again consider open bounded domains $\Omega \subset \mathbb{R}^d$ $(d = 2, 3)$ with a smooth boundary $\partial\Omega$ together with the differential operator $A := -\Delta + Id$. It then follows that for a given $f \in L^r(\Omega)$ $(1 < r < \infty)$ the elliptic boundary value problem

$$\begin{aligned} Ay &= f \quad \text{in } \Omega \\ y &= 0 \quad \text{on } \partial\Omega \end{aligned} \tag{3.89}$$

has a unique solution $y \in W^{2,r}(\Omega) \cap W_0^{1,r}(\Omega)$ which we denote by $y = \mathcal{G}(f)$. Furthermore,

$$\|y\|_{W^{2,r}} \le C\|f\|_{L^r},$$

where $\|\cdot\|_{L^r}$ and $\|\cdot\|_{W^{k,r}}$ denote the usual Lebesgue and Sobolev norms. Let $r > d$, $\alpha > 0$ and $y_0 \in L^2(\Omega)$ be given. We now consider the control problem

$$\begin{aligned} &\min_{u \in L^r(\Omega)} J(u) = \frac{1}{2}\int_\Omega |y - y_0|^2 + \frac{\alpha}{r}\int_\Omega |u|^r. \\ &\text{subject to} \quad y = \mathcal{G}(u) \quad \text{and} \quad \nabla y \in \mathcal{K}. \end{aligned} \tag{3.90}$$

Here,

$$\mathcal{K} = \{z \in C^0(\bar{\Omega})^d \mid |z(x)| \le \delta, x \in \bar{\Omega}\}, \tag{3.91}$$

so that we are in the setting of problem (1.138) with $U = U_{\text{ad}} = L^r(\Omega)$. Since $r > d$ we have $y \in W^{2,r}(\Omega)$ and hence $\nabla y \in C^0(\bar{\Omega})^d$ by a well-known embedding result. To ensure Robinsons regularity condition (1.139) it is sufficient to impose the following Slater condition:

$$\exists \hat{u} \in K \quad |\nabla \hat{y}(x)| < \delta,\ x \in \bar{\Omega} \quad \text{where } \hat{y} \text{ solves (3.89) with } f = \hat{u}. \tag{3.92}$$

Since $\hat{u}$ is feasible for (3.90) we deduce from the work [25, Theorem 3] of Casas and Fernandéz, that the above control problem has a unique solution $u \in L^r(\Omega)$.

For the KKT system of problem (3.90) we obtain with the help of (1.140)–(1.143) (compare also [25, Corollary 1])

Theorem 3.16 *An element $u \in L^r(\Omega)$ is a solution of* (3.90) *if and only if there exist $\boldsymbol{\mu} \in \mathcal{M}(\bar{\Omega})^d$ and $p \in L^t(\Omega)$ ($t < \frac{d}{d-1}$) such that*

$$\int_\Omega pAz - \int_\Omega (y - y_0)z - \int_{\bar{\Omega}} \nabla z \cdot d\boldsymbol{\mu} = 0 \quad \forall z \in W^{2,t'}(\Omega) \cap W_0^{1,t'}(\Omega) \tag{3.93}$$

$$p + \alpha |u|^{r-2} u = 0 \quad \text{in } \Omega \tag{3.94}$$

$$\int_{\bar{\Omega}} (\mathbf{z} - \nabla y) \cdot d\boldsymbol{\mu} \le 0 \quad \forall \mathbf{z} \in \mathcal{K}. \tag{3.95}$$

Here, y is the solution of (3.89) *and* $\frac{1}{t} + \frac{1}{t'} = 1$. *Further we recall that $\mathcal{M}(\bar{\Omega})$ denotes the space of regular Borel measures.*

Remark 3.7 Lemma 1 in the paper [25] of Casas and Fernandéz shows that the vector valued measure $\boldsymbol{\mu}$ appearing in Theorem 3.16 can be written in the form

$$\boldsymbol{\mu} = \frac{1}{\delta} \nabla y \mu,$$

where $\mu \in \mathcal{M}(\bar{\Omega})$ is a nonnegative measure that is concentrated on the set $\{x \in \bar{\Omega} \mid |\nabla y(x)| = \delta\}$.

Remark 3.8 Let us present an example which shows that an optimal control u, and thus the associated adjoint variable p, in general does not admit weak derivatives. To begin with we consider (3.90) with the choices $\Omega = B_2(0) \subset \mathbb{R}^2$, $\alpha = 1$,

$$\mathcal{K} = \left\{ \mathbf{z} \in C^0(\bar{\Omega})^2 \mid |\mathbf{z}(x)| \le \frac{1}{2}, x \in \bar{\Omega} \right\}$$

as well as

$$y_0(x) := \begin{cases} \frac{1}{4} + \frac{1}{2}\log 2 - \frac{1}{4}|x|^2, & 0 \le |x| \le 1, \\ \frac{1}{2}\log 2 - \frac{1}{2}\log|x|, & 1 < |x| \le 2. \end{cases}$$

In order to construct a test example we allow an additional right hand side f in the state equation and replace (3.89) by

$$\begin{aligned} -\Delta y &= f + u \quad \text{in } \Omega \\ y &= 0 \quad \text{on } \partial\Omega, \end{aligned}$$

where

$$f(x) := \begin{cases} 2, & 0 \le |x| \le 1, \\ 0, & 1 < |x| \le 2. \end{cases}$$

The optimization problem then has the unique solution

$$u(x) = \begin{cases} -1, & 0 \le |x| \le 1 \\ 0, & 1 < |x| \le 2 \end{cases}$$

with corresponding state $y \equiv y_0$. We note that we obtain equality in (3.94), i.e. $p = -u$. Furthermore, the action of the measure $\boldsymbol{\mu}$ applied to a vectorfield $\phi \in C^0(\bar{\Omega})^2$ is given by

$$\int_{\bar{\Omega}} \phi \cdot d\boldsymbol{\mu} = -\int_{\partial B_1(0)} x \cdot \phi dS.$$

Having in mind to consider finite element approximations of problem (3.90) discrete concepts for the control u should be considered which reflect the low regularity of the control.

3.3.2.1 Finite Element Discretization

We sketch an approach which uses classical piecewise linear, continuous approximations of the states in the setting of Sect. 3.3.1.1. In [42] Deckelnick, Günther and Hinze present a finite element approximation to problem (3.90) which uses mixed finite element approximations for the states.

Let us recall the definition of the space of linear finite elements,

$$X_h := \{v_h \in C^0(\bar{\Omega}) \mid v_h \text{ is a linear polynomial on each } T \in \mathcal{T}_h\}$$

with the appropriate modification for boundary elements, and let $X_{h0} := X_h \cap H_0^1(\Omega)$. Here $\mathcal{T}_h$ again denotes a quasi-uniform triangulation of Ω with maximum mesh size $h := \max_{T \in \mathcal{T}_h} \operatorname{diam}(T)$. We suppose that $\bar{\Omega}$ is the union of the elements of $\mathcal{T}_h$ so that element edges lying on the boundary are curved. Furthermore let us recall the definition of the discrete approximation of the operator $\mathcal{G}$. For a given function $v \in L^2(\Omega)$ we denote by $z_h = \mathcal{G}_h(v) \in X_{h0}$ the solution of

$$a(z_h, v_h) = \int_\Omega v v_h \quad \text{for all } v_h \in X_{h0}.$$

It is well-known that for all $v \in L^r(\Omega)$

$$\begin{aligned} \|\mathcal{G}(v) - \mathcal{G}_h(v)\|_{W^{1,\infty}} &\le C \inf_{z_h \in S_{h0}} \|\mathcal{G}(v) - z_h\|_{W^{1,\infty}} \\ &\le Ch^{1-\frac{d}{r}} \|\mathcal{G}(v)\|_{W^{2,r}} \le Ch^{1-\frac{d}{r}} \|v\|_{L^r}. \end{aligned} \tag{3.96}$$

For each $T \in \mathcal{T}_h$ let $z_T \in \mathbb{R}^d$ denote constant vectors. We define

$$\mathcal{K}_h := \{z_h : X_h \to \mathbb{R}^d \mid z_{h|T} = z_T \text{ on } T \text{ and } |z_{h|T}| \le \delta,\, T \in \mathcal{T}_h\}.$$

We approximate (3.90) by the following control problem depending on the mesh parameter h:

$$\begin{aligned} &\min_{u \in L^r(\Omega)} J_h(u) := \frac{1}{2}\int_\Omega |y_h - y_0|^2 + \frac{\alpha}{r}\int_\Omega |u|^r \\ &\text{subject to} \quad y_h = \mathcal{G}_h(u) \quad \text{and} \quad \nabla y_h \in \mathcal{K}_h. \end{aligned} \tag{3.97}$$

We first note that $\hat{y}_h := \mathcal{G}_h(\hat{u})$ satisfies a Slater condition similar to (3.92), since for $x_T \in T \in \mathcal{T}_h$ by (3.96)

$$|\nabla \hat{y}_h(x_T)| \le |\nabla(\hat{y}_h(x_T) - \hat{y}(x_T))| + |\nabla \hat{y}(x_T)| \le \|\nabla(\hat{y}_h - \hat{y})\|_{L^\infty} + \max_{x \in \bar{\Omega}} |\nabla \hat{y}(x)|$$

$$\le C h^{1-\frac{d}{r}} + (1-2\epsilon)\delta \le (1-\epsilon)\delta \quad \text{for all } T \in \mathcal{T}_h,$$

for some $\epsilon > 0$ and $0 < h \le h_0$, so that $(\nabla \hat{y}_h)_{T \in \mathcal{T}_h} \in \mathcal{K}_h$ satisfies the Slater condition

$$|\nabla \hat{y}_h(x)| < \delta \quad \text{for all } x \in \bar{\Omega}. \tag{3.98}$$

Therefore, as for problem (3.90) the setting of (1.138) with $\mathcal{K}$ replace by $\mathcal{K}_h$ applies to problem (3.97) and we have

Lemma 3.6 *Problem* (3.59) *has a unique solution* $u_h \in L^r(\Omega)$. *There exist* $\mu_T \in \mathbb{R}^d$, $T \in \mathcal{T}_h$ *and* $p_h \in X_{h0}$ *such that with* $y_h = \mathcal{G}_h(u_h)$ *we have*

$$a(v_h, p_h) = \int_\Omega (y_h - y_0)v_h + \sum_{T \in \mathcal{T}_h} |T| \nabla v_{h|T} \cdot \mu_T \quad \forall v_h \in X_{h0}, \tag{3.99}$$

$$p_h + \alpha |u_h|^{r-2} u_h = 0 \quad \text{in } \Omega, \tag{3.100}$$

$$\sum_{T \in \mathcal{T}_h} |T| \big(z_T - \nabla y_{h|T}\big) \cdot \mu_T \le 0 \quad \forall z_h \in \mathcal{K}_h. \tag{3.101}$$

In problem (3.97) we again apply variational discretization from [71]. From (3.100) we infer for the discrete optimal control

$$u_h = -\alpha^{-\frac{1}{r-1}} |p_h|^{\frac{2-r}{r-1}} p_h. \tag{3.102}$$

Further, according to Remark 3.7 we have the following representation of the discrete multipliers.

Lemma 3.7 *Let* u_h *denote the unique solution of* (3.97) *with corresponding state* $y_h = \mathcal{G}_h(u_h)$ *and multiplier* $(\boldsymbol{\mu}_T)_{T \in \mathcal{T}_h}$. *Then there holds*

$$\boldsymbol{\mu}_T = |\boldsymbol{\mu}_T| \frac{1}{\delta} \nabla y_{h|T} \quad \textit{for all } T \in \mathcal{T}_h. \tag{3.103}$$

Proof Fix $T \in \mathcal{T}_h$. The assertion is clear if $\boldsymbol{\mu}_T = 0$. Suppose that $\boldsymbol{\mu}_T \neq 0$ and define $z_h : \bar{\Omega} \to \mathbb{R}^d$ by

$$z_{h|\tilde{T}} := \begin{cases} \nabla y_{h|T}, & \tilde{T} \neq T, \\ \delta \frac{\boldsymbol{\mu}_T}{|\boldsymbol{\mu}_T|}, & \tilde{T} = T. \end{cases}$$

Clearly, $z_h \in \mathcal{K}_h$ so that (3.101) implies

$$\boldsymbol{\mu}_T \cdot \left(\delta \frac{\boldsymbol{\mu}_T}{|\boldsymbol{\mu}_T|} - \nabla y_{h|T} \right) \leq 0,$$

and therefore, since $(\nabla y_{h|T})_{T \in \mathcal{T}_h} \in \mathcal{K}_h$,

$$\delta |\boldsymbol{\mu}_T| \leq \boldsymbol{\mu}_T \cdot \nabla y_{h|T} \leq \delta |\boldsymbol{\mu}_T|.$$

Hence we obtain $\frac{\boldsymbol{\mu}_T}{|\boldsymbol{\mu}_T|} = \frac{1}{\delta} \nabla y_{h|T}$ and the lemma is proved.

As a consequence of Lemma 3.7 we immediately infer that

$$|\boldsymbol{\mu}_T| = \boldsymbol{\mu}_T \cdot \frac{1}{\delta} \nabla y_{h|T} \quad \text{for all } T \in \mathcal{T}_h. \tag{3.104}$$

We now use (3.104) in order to derive an important a priori estimate.

Lemma 3.8 *Let $u_h \in L^r(\Omega)$ be the optimal solution of* (3.97) *with corresponding state $y_h \in X_{h0}$ and adjoint variables $p_h \in X_{h0}$, $\boldsymbol{\mu}_T$, $T \in \mathcal{T}_h$. Then there exists $h_0 > 0$ such that*

$$\|y_h\|, \|u_h\|_{L^r}, \|p_h\|_{L^{\frac{r}{r-1}}}, \sum_{T \in \mathcal{T}_h} |T| |\boldsymbol{\mu}_T| \leq C \quad \textit{for all } 0 < h \leq h_0.$$

Proof Combining (3.104) with (3.98) we deduce

$$\boldsymbol{\mu}_T \cdot (\nabla y_{h|T} - \nabla \hat{y}_{h|T}) \geq \delta |\boldsymbol{\mu}_T| - (1 - \epsilon) \delta |\boldsymbol{\mu}_T| = \epsilon \delta |\boldsymbol{\mu}_T|.$$

Choosing $w_h = y_h - \hat{y}_h$ in (3.99) and using the definition of $\mathcal{G}_h$ together with (3.100) we hence obtain

$$\begin{aligned} \epsilon \delta \sum_{T \in \mathcal{T}_h} |T| |\boldsymbol{\mu}_T| &\leq \sum_{T \in \mathcal{T}_h} |T| \boldsymbol{\mu}_T \cdot (\nabla y_{h|T} - \nabla \hat{y}_{h|T}) \\ &= a(y_h - \hat{y}_h, p_h) - \int_\Omega (y_h - \hat{y}_0)(y_h - \hat{y}_h) \\ &= \int_\Omega (u_h - \hat{u}) p_h - \int_\Omega (y_h - \hat{y}_0)(y_h - \hat{y}_h) \\ &\leq -\frac{\alpha}{2} \int_\Omega |u_h|^r - \frac{1}{2} \int_\Omega |y_h|^2 + C(1 + \|y_0\|^2 + \|\hat{u}\|_{L^r}^r). \end{aligned}$$

This implies the bounds on y_h, u_h and $\boldsymbol{\mu}_T$. The bound on p_h follows from (3.100).

Remark 3.9 For the measure $\boldsymbol{\mu}_h \in \mathcal{M}(\bar{\Omega})^d$ defined by

$$\int_{\bar{\Omega}} f \cdot d\boldsymbol{\mu}_h := \sum_{T \in \mathcal{T}_h} \int_T f dx \cdot \boldsymbol{\mu}_T \quad \text{for all } f \in C^0(\bar{\Omega})^d,$$

it follows immediately that

$$\|\boldsymbol{\mu}_h\|_{\mathcal{M}(\bar{\Omega})^d} \leq C.$$

Now we are in the position to prove the following error estimates.

Theorem 3.17 *Let u and u_h be the solutions of* (3.90) *and* (3.97) *respectively. Then there exists $h_1 \leq h_0$ such that*

$$\|y - y_h\| \leq C h^{\frac{1}{2}(1-\frac{d}{r})}, \quad \textit{and} \quad \|u - u_h\|_{L^r} \leq C h^{\frac{1}{r}(1-\frac{d}{r})}$$

for all $0 < h \leq h_1$.

Proof Let us introduce $y^h := \mathcal{G}(u_h) \in W^{2,r}(\Omega) \cap W_0^{1,r}(\Omega)$, and $\tilde{y}_h := \mathcal{G}_h(u)$. In view of Lemma 3.8 and (3.96) we have

$$\|y^h - y_h\|_{W^{1,\infty}} \leq C h^{1-\frac{d}{r}} \|u_h\|_{L^r} \leq C h^{1-\frac{d}{r}}. \tag{3.105}$$

Let us now turn to the actual error estimate. To begin, we recall that for $r \geq 2$ there exists $\theta_r > 0$ such that

$$(|a|^{r-2}a - |b|^{r-2}b)(a - b) \geq \theta_r |a - b|^r \quad \forall a, b \in \mathbb{R}.$$

Hence, using (3.94) and (3.100),

$$\begin{aligned} \alpha\theta_r \int_\Omega |u - u_h|^r &\leq \alpha \int_\Omega (|u|^{r-2}u - |u_h|^{r-2}u_h)(u - u_h) \\ &= \int_\Omega (-p + p_h)(u - u_h) =: (1) + (2). \end{aligned}$$

Recalling (3.93) we have

$$\begin{aligned} (1) &= \int_\Omega p\big(\mathcal{A}y^h - \mathcal{A}y\big) \\ &= \int_\Omega (y - y_0)(y^h - y) + \int_{\bar{\Omega}} \big(\nabla y^h - \nabla y\big) \cdot d\boldsymbol{\mu} \\ &= \int_\Omega (y - y_0)(y^h - y) + \int_{\bar{\Omega}} \big(P_\delta(\nabla y^h) - \nabla y\big) \cdot d\boldsymbol{\mu} \\ &\quad + \int_{\bar{\Omega}} \big(\nabla y^h - P_\delta(\nabla y^h)\big) \cdot d\boldsymbol{\mu} \end{aligned}$$

where P_δ denotes the orthogonal projection onto $\bar{B}_\delta(0) = \{x \in \mathbb{R}^d \mid |x| \le \delta\}$. Note that

$$|P_\delta(x) - P_\delta(\tilde{x})| \le |x - \tilde{x}| \quad \forall x, \tilde{x} \in \mathbb{R}^d. \tag{3.106}$$

Since $x \mapsto P_\delta(\nabla y^h(x)) \in \mathcal{K}$ we infer from (3.95)

$$(1) \le \int_\Omega (y - y_0)(y^h - y) + \max_{x \in \bar{\Omega}} |\nabla y^h(x) - P_\delta(\nabla y^h(x))| \|\boldsymbol{\mu}\|_{\mathcal{M}(\bar{\Omega})^d}. \tag{3.107}$$

Let $x \in \bar{\Omega}$, say $x \in T$ for some $T \in \mathcal{T}_h$. Since u_h is feasible for (3.97) we have that $\nabla y_{h|T} \in \bar{B}_\delta(0)$ so that (3.106) together with (3.105) implies

$$\begin{aligned} \left|\nabla y^h(x) - P_\delta(\nabla y^h(x))\right| &\le \left|\nabla y^h(x) - \nabla y_{h|T}\right| + \left|P_\delta(\nabla y^h(x)) - P_\delta\big(\nabla y_{h|T}\big)\right| \\ &\le 2\left|\nabla y^h(x) - \nabla y_{h|T}\right| \le C h^{1-\frac{d}{r}} \|u_h\|_{L^r}. \end{aligned} \tag{3.108}$$

Thus

$$(1) \le \int_\Omega (y - y_0)(y^h - y) + C h^{1-\frac{d}{r}}. \tag{3.109}$$

Similarly,

$$\begin{aligned} (2) = a(\tilde{y}_h - y_h, p_h) &= \int_\Omega (y_h - y_0)(\tilde{y}_h - y_h) + \sum_{T \in \mathcal{T}_h} |T|\big(\nabla \tilde{y}_{h|T} - \nabla y_{h|T}\big) \cdot \boldsymbol{\mu}_T \\ &= \int_\Omega (y_h - y_0)(\tilde{y}_h - y) + \sum_{T \in \mathcal{T}_h} |T|\big(\nabla \tilde{y}_{h|T} - P_\delta(\nabla \tilde{y}_{h|T})\big) \cdot \boldsymbol{\mu}_T \\ &\quad + \sum_{T \in \mathcal{T}_h} |T|\big(P_\delta(\nabla \tilde{y}_{h|T} - \nabla y_{h|T}\big) \cdot \boldsymbol{\mu}_T \\ &\le \int_\Omega (y_h - y_0)(\tilde{y}_h - y) + \sum_{T \in \mathcal{T}_h} |T|\big(\nabla \tilde{y}_{h|T} - \nabla y(x_T)\big) \cdot \boldsymbol{\mu}_T \\ &\quad + \sum_{T \in \mathcal{T}_h} |T|\big(P_\delta(\nabla y(x_T) - P_\delta(\nabla y_{h|T})\big) \cdot \boldsymbol{\mu}_T, \end{aligned}$$

where $x_T \in T$, so that $(\nabla y(x_T))_{T \in \mathcal{T}_h} \in \mathcal{K}_h$. We infer from Lemma 3.8 and (3.1)

$$\begin{aligned} (2) &\le \int_\Omega (y_h - y_0)(\tilde{y}_h - y) + 2 \max_{T \in \mathcal{T}_h} |\nabla \tilde{y}_{h|T} - \nabla y(x_T)| \sum_{T \in \mathcal{T}_h} |T| |\boldsymbol{\mu}_T| \\ &\le \int_\Omega (y_h - y_0)(\tilde{y}_h - y) + C h^{1-\frac{d}{r}} \|u\|_{L^r}. \end{aligned} \tag{3.110}$$

Combining (1) and (2) we finally obtain

$$\alpha \theta_r \int_\Omega |u - u_h|^r \le \int_\Omega (y - y_0)\big(y^h - y\big) + \int_\Omega (y_h - y_0)\big(\tilde{y}_h - y_h\big) + C h^{1-\frac{d}{r}}$$

$$= -\int_\Omega |y - y_h|^2 + \int_\Omega ((y_0 - y_h)(y - \tilde{y}_h) + (y - y_0)(y^h - y_h))$$
$$+ Ch^{1-\frac{d}{r}}$$
$$\leq -\int_\Omega |y - y_h|^2 + C(\|y - \tilde{y}_h\| + \|y^h - y_h\|) + Ch^{1-\frac{d}{r}}$$
$$\leq -\int_\Omega |y - y_h|^2 + Ch(\|u\| + \|u_h\|) + Ch^{1-\frac{d}{r}}$$

in view of (3.55), and the result follows.

3.3.2.2 A Numerical Experiment with Pointwise Constraints on the Gradient

We now consider the finite element approximation of problem (3.90) with the data of Remark 3.8. Instead of variational discretization we also use piecewise linear, continuous Ansatz functions for the control u_h. For the numerical solution we use the routine `fmincon` contained in the MATLAB Optimization Toolbox. The state equation is approximated with piecewise linear, continuous finite elements on quasi-uniform triangulations $\mathcal{T}_h$ of $B_2(0)$. The gradient constraints are required element-wise. The resulting discretized optimization problem then reads

$$\min_{u_h \in X_h} J_h(u_h, y_h) = \frac{1}{2}\|y_h - y_0\|^2_{L^2(\Omega_h)} + \frac{\alpha}{r}\|u_h\|^r_{L^r(\Omega_h)}$$
$$\text{subject to} \quad y_h = \mathcal{G}_h(u_h)$$
$$|\nabla y_h|_T| \leq \delta = \tfrac{1}{2} \quad \forall T \in \mathcal{T}_h$$

In Figs. 3.15, 3.16 we present the numerical approximations y_h, u_h, and μ_h on a grid containing $m = 1089$ gridpoints. Figure 3.15 shows that a piecewise Ansatz for the control only delivers poor approximations to the piecewise constant continuous control, and we observe overshooting at the jump discontinuity. Figure 3.16 clearly shows that the support of μ_h is concentrated around $|x| = 1$ where $\mu_h = |\boldsymbol{\mu}_h|$ according to relation (3.103).

In Table 3.15 we investigate the experimental order of convergence defined in (3.25) for the error functionals

$$E_u(h) := \|u - u_h\|, \|u - u_h\|_{L^4}, \quad \text{and} \quad E_y(h) := \|y - y_h\|.$$

It turns out that the controls show an approximation behaviour which is slightly better than that predicted for the variational controls by Theorem 3.17. The L^2-norm of the state seems to converge linearly. This seems to be caused by the special structure of the example. In general we should not expect an approximation order for piecewise linear polynomial approximations to the control, since the exact solution does not admit weak derivatives.

In Table 3.16 we display the values of $\sum_{T \in \mathcal{T}_h} \mu_T$. These values are expected to converge to 2π as $h \to 0$, since this gives the value of μ applied to the function which is identically equal to 1 on $\bar{\Omega}$.

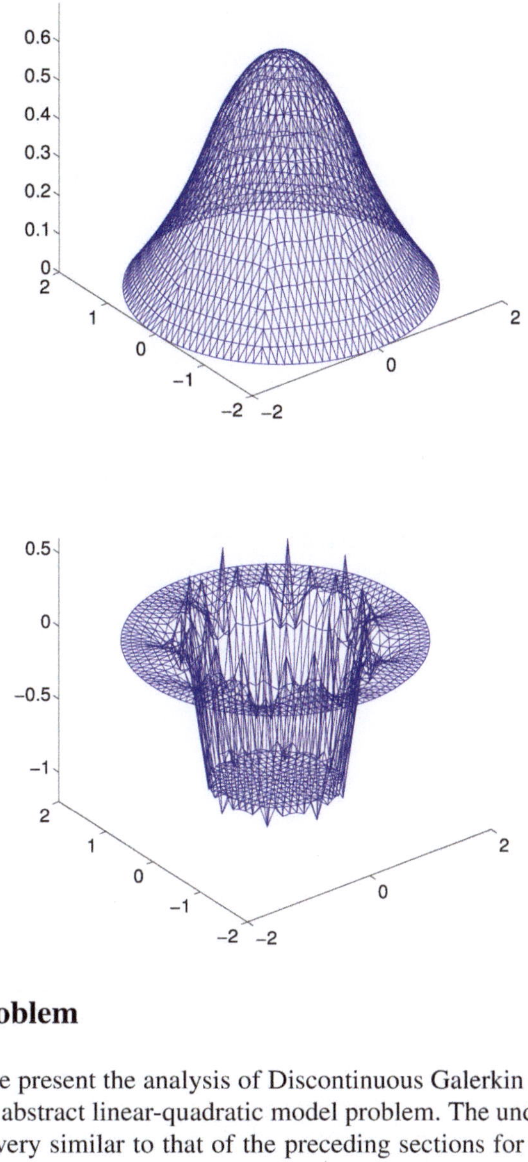

Fig. 3.15 State (*top*), and control (*bottom*)

3.4 Time Dependent Problem

For the time-dependent case we present the analysis of Discontinuous Galerkin approximations w.r.t. time for an abstract linear-quadratic model problem. The underlying analysis turns out to be very similar to that of the preceding sections for the stationary model problem.

3.4.1 Mathematical Model, State Equation

Let V, H denote separable Hilbert spaces, so that $(V, H = H^*, V^*)$ forms a Gelfand triple. We denote by $a : V \times V \to \mathbb{R}$ a bounded, coercive (and symmetric) bilinear form, by U the Hilbert space of controls, and by $B : U \to L^2(V^*)$ the linear

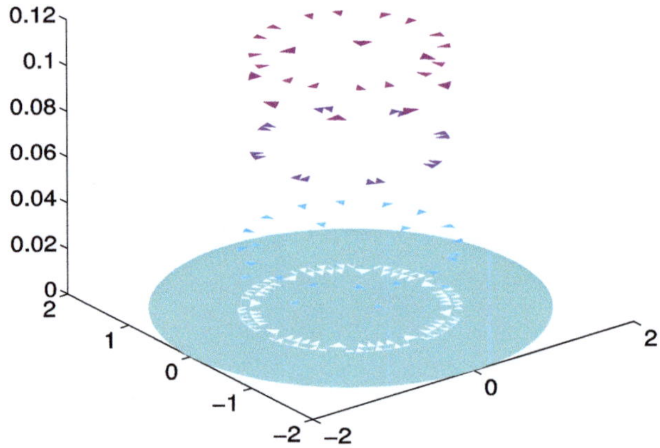

Fig. 3.16 Discrete multiplier

Table 3.15 Errors (top) and EOC for the numerical example

NT	$\|u-u_h\|_{L^4(\Omega_h)}$	$\|u-u_h\|_{L^2(\Omega_h)}$	$\|y-y_h\|_{L^2(\Omega_h)}$
32	8.26×10^{-1}	1.36	2.41×10^{-1}
128	6.18×10^{-1}	8.98×10^{-1}	8.66×10^{-2}
512	5.01×10^{-1}	6.33×10^{-1}	3.33×10^{-2}
2048	4.15×10^{-1}	4.36×10^{-1}	1.36×10^{-2}
	0.45487	0.65285	1.60659
	0.31573	0.52485	1.43072
	0.27582	0.54677	1.31906

Table 3.16 Behaviour of the discrete multiplier

NT	$\sum_{i=1}^{NT}\mu_i$
32	0
128	2.26
512	4.09
2048	5.14

bounded control operator. Here we recall $L^p(S)\equiv L^p((0,T);S)$ where S denotes a Banach space and $T>0$. For $y_0\in H$ we consider the state equation

$$\left.\begin{array}{ll}\int_0^T\langle y_t,v\rangle_{V^*,V}+a(y,v)dt=\int_0^T\langle (Bu)(t),v\rangle_{V^*,V}dt & \forall v\in L^2(V),\\ (y(0),v)_H=(y_0,v)_H & \forall v\in V,\end{array}\right\}\quad :\Longleftrightarrow$$
$$y=\mathcal{T}Bu,$$

which for every $u \in U$ admits a unique solution $y = y(u) \in W := \{w \in L^2(V), w_t \in L^2(V^*)\}$, see e.g. [146] and Theorem 1.37.

3.4.2 Optimization Problem

We consider the optimization problem

$$(TP) \quad \begin{cases} \min_{(y,u)\in W\times U_{\mathrm{ad}}} J(y,u) := \frac{1}{2}\|y-z\|^2_{L^2(H)} + \frac{\alpha}{2}\|u\|^2_U \\ \text{s.t. } y = \mathcal{T}Bu, \end{cases} \tag{3.111}$$

where $U_{\mathrm{ad}} \subseteq U$ denotes a closed, convex subset. Introducing the reduced cost functional

$$\hat{J}(u) := J(y(u), u),$$

the necessary (and in the present case also sufficient) optimality conditions take the form

$$\langle \hat{J}'(u), v - u\rangle_{U^*,U} \geq 0 \quad \text{for all } v \in U_{\mathrm{ad}}.$$

Here

$$\hat{J}'(u) = \alpha u + B^* p(y(u)),$$

where the adjoint state p solves the adjoint equation

$$\int_0^T \langle -p_t, w\rangle_{V^*,V} + a(w,p)dt = \int_0^T (y - z, w)_H \, dt \quad \forall w \in W,$$

$$(p(T), v)_H = 0, \quad v \in V.$$

This variational inequality is equivalent to the semi–smooth operator equation

$$u = P_{U_{\mathrm{ad}}}\left(-\frac{1}{\alpha} R B^* p(y(u))\right)$$

with $P_{U_{\mathrm{ad}}}$ denoting the orthogonal projection in U onto U_{ad}, and $R : U^* \to U$ the inverse of the Riesz isomorphism.

3.4.3 Discretization

Let $V_h \subset V$ denote a finite dimensional subspace, and let $0 = t_0 < t_1 < \cdots < t_m = T$ denote a time grid with grid width δt. We set $I_n := (t_{n-1}, t_n]$ for $n = 1, \ldots, m$ and seek discrete states in the space

$$V_{h,\delta t} := \{\phi : [0,T] \times \Omega \to \mathbb{R}, \phi(t,\cdot)_{|\bar{\Omega}} \in V_h, \phi(\cdot,x)_{|I_n} \in \mathbb{P}_r \text{ for } n = 1,\ldots,m\}.$$

i.e. $y_{h,\delta t}$ is a polynomial of degree $r \in \mathbb{N}$ w.r.t. time. Possible choices of V_h in applications include polynomial finite element spaces, and also wavelet spaces, say. We define the discontinuous Galerkin w.r.t. time approximation (dG(r)-approximation) $\tilde{y} = y_{h,\delta t}(u) \equiv \mathcal{T}_{h,\delta t} Bu \in V_{h,\delta t}$ of the state y as unique solution of

$$\begin{aligned} A(\tilde{y}, v) &:= \sum_{n=1}^{m} \int_{I_n} (\tilde{y}_t, v)_H + a(\tilde{y}, v) dt + \sum_{n=2}^{m} ([\tilde{y}]^{n-1}, v^{n-1+})_H + (\tilde{y}^{0+}, v^{0+})_H \\ &= (y_0, v^{0+})_H + \int_0^T \langle (Bu)(t), v \rangle_{V^*,V} dt \quad \text{for all } v \in V_{h,\delta t}. \end{aligned} \tag{3.112}$$

Here,

$$v^{n+} := \lim_{t \searrow t_n} v(t, \cdot), \;\; v^{n-} := \lim_{t \nearrow t_n} v(t, \cdot), \quad \text{and} \quad [v]^n := v^{n+} - v^{n-}.$$

The discrete counterpart of the optimal control problem for the variational approach of [71] reads

$$(P_{h,\delta t}) \qquad \min_{u \in U_{\text{ad}}} \hat{J}_{h,\delta t}(u) := J(y_{h,\delta t}(u), u)$$

and it admits a unique solution $u_{h,\delta t} \in U_{\text{ad}}$. We further have

$$\hat{J}'_{h,\delta t}(v) = \alpha v + B^* p_{h,\delta t}(y_{h,\delta t}(v)),$$

where $p_{h,\delta t}(y_{h,\delta t}(v)) \in V_{h,\delta t}$ denotes the unique solution of

$$A(v, p_{h,\delta t}) = \int_0^T (y_{h,\delta t} - z, v)_H dt \quad \text{for all } v \in V_{h,\delta t}.$$

Further, the unique discrete solution $u_{h,\delta t}$ satisfies

$$\langle u_{h,\delta t} + B^* p_{h,\delta t}, v - u_{h,\delta t} \rangle_{U^*,U} \geq 0 \quad \text{for all } v \in U_{\text{ad}}.$$

As in the continuous case this variational inequality is equivalent to a semi–smooth operator equation, namely

$$u_{h,\delta t} = P_{U_{\text{ad}}} \left(-\frac{1}{\alpha} R B^* p_{h,\delta t}(y_{h,\delta t}(u_{h,\delta t})) \right).$$

For this discrete approach the proof of the following theorem follows the lines of the proof of Theorem 3.4.

Theorem 3.18 *Let $u, u_{h,\delta t}$ denote the unique solutions of (P) and $(P_{h,\delta t})$, respectively. Then*

$$\alpha \|u - u_{h,\delta t}\|_U^2 + \|y_{h,\delta t}(u_{h,\delta t}) - y_{h,\delta t}(u)\|_{L^2(H)}^2$$

$$\leq \langle B^*(p(u) - \tilde{p}_{h,\delta t}(u)), u_{h,\delta t} - u\rangle_{U^*,U} + \|y(u) - y_{h,\delta t}(u)\|^2_{L^2(H)}, \quad (3.113)$$

where $\tilde{p}_{h,\delta t}(u) := \mathcal{T}^*_{h,\delta t}(\mathcal{T}Bu - z)$, $y_{h,\delta t}(u) := \mathcal{T}_{h,\delta t}Bu$, *and* $y(u) := \mathcal{T}Bu$.

As a result of estimate (3.113) we have that error estimates for the variational discretization are available if error estimates for the $dg(r)$-approximation to the state and the adjoint state are available. Using the setting for the heat equation investigated by Meidner and Vexler we recover with the help of [98, Prop. 4.3,4.4] their result of [98, Corollary 5.9] for variational discretization obtained with $dG(0)$ in time and piecewise linear and continuous finite elements in space, namely

$$\alpha\|u - u_{h,\delta t}\|^2_U + \|y_{h,\delta t}(u_{h,\delta t}) - y_{h,\delta t}(u))\|^2_{L^2(H)} \leq C\{\delta t + h^2\}.$$

Remark 3.10 If we choose a more specific setting in problem (3.111), say that for the heat equation with Neumann boundary conditions and $H = L^2(\Omega)$, $V = H^1(\Omega)$, and impose additional constraints on the state to be satisfied in the space-time domain $Q := (0,T) \times \Omega$ the results of Theorem 3.12 and Lemma 3.2 hold accordingly, if we require uniformly continuous states and a Slater condition similar to that of Assumption 3.11 and replace the operator $\mathcal{G}$ by $\mathcal{T}$. Moreover, a result of the following form can be obtained along the lines of the proof of Corollary 3.3, where the norms have to be taken w.r.t. the domain Q;

$$\begin{aligned}
&\alpha\|u - u_{h,\delta t}\|^2 + \|y - y_{h,\delta t}\|^2 \\
&\quad\leq C(\|u\|, \|u_{h,\delta t}\|)\left\{\|y - y_{h,\delta t}(u)\| + \|y^{h,\delta t}(u_{h,\delta t}) - y_{h,\delta t}\|\right\} \\
&\qquad + C(\|\mu\|_{\mathcal{M}(\bar{Q})}, \|\mu_{h,\delta t}\|_{\mathcal{M}(\bar{Q})})\left\{\|y - y_{h,\delta t}(u)\|_{L^\infty} + \|y^{h,\delta t}(u_{h,\delta t}) - y_{h,\delta t}\|_{L^\infty}\right\} \\
&\qquad + \alpha\|u_0 - u_{0,(h,\delta t)}\|^2.
\end{aligned}$$

Here $y^{h,\delta t}(u_{h,\delta t}) := \mathcal{T}Bu_{h,\delta t}$, $y_{h,\delta t}(u) := \mathcal{T}_{h,\delta t}Bu$.

This means that error estimates for the controls are available if we uniform estimates for the discrete states at hand. For the latter let us refer to the work of Ericsson and Johnson in [46].

3.4.4 Further Literature on Control of Time-Dependent Problems

In the literature only few contributions to numerical analysis for control problems with time dependent PDEs can be found. For unconstrained linear quadratic control problems with the time dependent Stokes equation in two- and three-dimensional domains Deckelnick and Hinze in [37] prove the error bound

$$\|u - u_{h,\sigma}\|_{L^2((0,T)\times\Omega)} = \mathcal{O}(\sigma + h^2).$$

Here and below σ denotes the discretization parameter in time, and h that in space. They use a fully implicit variant of Eulers method for the time discretization which is equivalent to the $dG(0)$ approximation introduced in Sect. 3.4.3. In space the use Taylor-Hood finite elements. Using [37, (3.1), (3.6)] combined with (3.113) this estimate directly extends also to the control constrained case.

Boundary control for the heat equation in one spatial dimension is considered by Malanowski in [93] with piecewise constant, and by Rösch in [118] with piecewise linear, continuous control approximations. Requiring strict complementarity for the continuous solution Rösch is able to prove the estimate

$$\|u - u_\sigma\| = \mathcal{O}(\sigma^{\frac{3}{2}}).$$

Malanowski proves the estimate

$$\|u - u_{h,\sigma}\|_{L^2((0,T)\times\Omega)} = \mathcal{O}(\sigma + h).$$

In a recent work [97, 98] Meidner and Vexler present extensive research for control problem (3.111) and its discrete approximation based on $dG(0)$ in time and finite element in space, where they consider the heat equation as mathematical model on a two- or three-dimensional convex polygonal domain. For variational discretization of [71] they prove the estimate

$$\|u - u_{h,\sigma}\|_{L^2((0,T)\times\Omega)} = \mathcal{O}(\sigma + h^2),$$

which under the assumption of strict complementarity of the continuous solution and further regularity requirements also holds for post-processing [100].

For control problems with nonlinear time dependent equations one only finds few contributions in the literature. In [59, 60] Gunzburger and Manservisi present a numerical approach to control of the instationary Navier-Stokes equations (1.145) using the first discretize then optimize approach discussed in Sect. 3.2.2. The first optimize then discretize approach of Sect. 3.2.3 applied to the same problem class is discussed by Hinze in [70]. Deckelnick and Hinze provide numerical analysis for a general class of control problems with the instationary Navier Stokes system (1.145) in [38]. Among other things they prove existence and local uniqueness of variational discrete controls in neighborhoods of nonsingular continuous solutions, and for semi-discretization in space with Taylor-Hood finite elements provide the error estimate

$$\int_0^T \|u - u_h\|_U^2\, dt \le Ch^4.$$

Here, u, u_h denote the continuous and variational discrete optimal control, respectively. For further references we refer to the papers [97, 98] of Meidner and Vexler.

Chapter 4
Applications

René Pinnau

Abstract The following chapter is devoted to the study of two industrial applications, in which optimization with partial differential equations plays a crucial role. It shall provide a survey of the different mathematical settings which can be handled with the general optimal control calculus presented in the previous chapters. We focus on large scale optimal control problems involving two well-known types of partial differential equations, namely elliptic and parabolic ones. Since real world applications lead generally to mathematically quite involved problems, we study in particular nonlinear systems of equations. The examples are chosen in such a way that they are up-to-date and modern mathematical tools are used for their specific solution. The industrial fields we cover are modern semiconductor design and glass production. We start each section with a modeling part to introduce the underlying physics and mathematical models, which are then followed by the analytical and numerical study of the related optimal control problems.

4.1 Optimal Semiconductor Design

Each student learns in the first lecture on numerical mathematics that the enormous speed-up of numerical simulations during the last 30 years is rooted in two facts, namely the significant improvement of algorithms and the ongoing miniaturization in electronics which allows for faster computing times. In the previous chapters we already learned in which way fast numerical algorithms can be developed. Now we study the impact of mathematical optimization on advanced semiconductor design. There are several stages at which optimization and control are used in semiconductor industry, e.g., in circuit design, thermal control of the circuit board or, on a smaller level, the design of the semiconductor device itself. Even the control of the whole production process itself is under mathematical investigation. The most popular semiconductor device is indeed the so-called MOSFET (metal oxide silicium field effect transistor), which is employed in many applications (see Fig. 4.1) [129].

In the design cycle one changes the geometry of the device (miniaturization!) and the so-called doping profile, which describes the density of charged background ions. This doping profile defines the type of the semiconductor device under consideration. In the conventional design cycle simulation tools are employed to compute

R. Pinnau (✉)
Fachbereich Mathematik, Technische Universität Kaiserslautern, Kaiserslautern,
Germany
e-mail: pinnau@mathematik.uni-kl.de

M. Hinze et al., *Optimization with PDE Constraints*,
Mathematical Modelling: Theory and Applications 23,

DOI 10.1007/978-1-4020-8839-1_4, © Springer Science + Business Media B.V. 2009

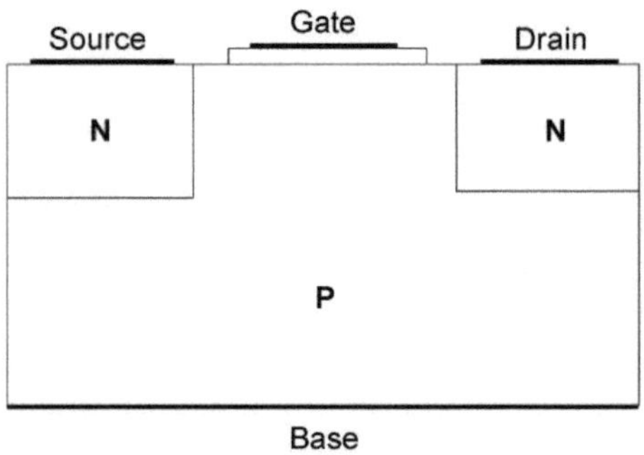

Fig. 4.1 MOSFET device

the so called current-voltage characteristics of the device, from which the electrical engineer can deduce many performance characteristics of the device. This is done for a certain set of design parameters, and then these parameters are adjusted empirically. Thus, the total design time depends crucially on the knowledge and experience of the engineer.

In standard applications a working point, i.e., a certain voltage-current pair, for the device is fixed. In particular for MOSFET devices in portable systems it is most important to have on the one hand a low leakage current (in the off-state), which maximizes the battery lifetime, and on the other hand one wants to maximize the drive current (in the on-state) [128]. Now, we want to study how one can apply the previously introduced techniques to optimize such a device and we pose the following design question [77]: *Is it possible to gain an amplified current at the working point only by a slight change of the doping profile*?

We proceed in several steps. First, we motivate the system of nonlinear equations, which is describing the electronic behavior of the semiconductor device. There are many semiconductor models at hand, but we will concentrate in the next section on the well established drift diffusion model [96]. Then, we state the optimization problem in mathematical terms, provide some analysis and study its numerical solution.

4.1.1 Semiconductor Device Physics

In this section we give a brief introduction into the physics (here we follow [141]) and numerical simulation of semiconductor devices, which is far from being comprehensive. If the reader wants to get deeper into the topic we suggest to study the excellent books by *Sze* [129] or *Selberherr* [124].

Clearly, the most important features of semiconductor devices are due to electromagnetic effects, i.e., such a device reacts on applied voltages. Here, we only consider electrostatic effects ignoring electrodynamics and magnetic phenomena. Further, we ignore quantum effects, which are getting increasingly important due to the shrinking device size [17, 96].

In general, a semiconductor is just a specifically modified crystal. The modification of the underlying crystal (consisting, e.g., of Silicium atoms) are due to a preparation of the surface (to build metallic or insulating contacts) and due to the implantation of impurities (e.g., Aluminum atoms). This has to be done since the electronic behavior of a homogeneous semiconductor is rather boring. Due to the replacement of atoms in the crystal, which is the so-called doping process, we get an inhomogeneous semiconductor exhibiting the desired electronic performance. There exist several sophisticated technologies to achieve the desired doping. And since these processes can be controlled on the nanometer scale, it is possible to fabricate nowadays devices with a gate length of less than 45 nanometers. Nevertheless, there is still a strong need for the (automated) design of the semiconductor device, i.e., how the doping profile has to be adjusted such that the device shows the desired behavior.

4.1.1.1 Charge Transport

Normally, one implants atoms into the semiconductor crystal which have more (donator atoms) or less (acceptor atoms) electrons participating at binding interactions. While Silicium atoms have four binding electrons, Phosphor atoms have five and Aluminum atoms have three. If a Silicium atom is replaced by a Phosphor atom we have one additional electron, which is not necessary for the binding and which can therefor move freely in the crystal. Hence, the Phosphor atom donates one electron to the conductivity band. But if the Silicium atom is replaced by an Aluminum atom, then the additional electron which is needed for the binding is taken from the surrounding atoms and a *hole* is generated.

In fact, also these holes contribute to the charge transport in the semiconductor crystal, since the Silicium atom (which is then positively charged) attracts an electron from one surrounding atom. This process repeats and charge transport takes place by the missing electrons, i.e., the holes. Experiments suggest that the charge transport by holes can be considered as charge transport by real particles which have a positive charge q.

Now, that we know how charge transport takes place in the semiconductor, let us assume that the semiconductor occupies a bounded domain $\Omega \subset \mathbb{R}^3$. So far, our assumptions imply that there is an instantaneous *electric field* $\mathbf{E}(x)$, $x \in \Omega$, which is only determined by the position of the charged particles.

Hence, we could describe the charge transport in the semiconductor just by considering an ensemble of charged particles interacting via the electric field. If we put an electron with velocity $\mathbf{v}_0$ in the point $x_0 \in \Omega$, then there will be an interaction of the electron with the electric field, which can be described by *Coulomb's law* and *Newton's second law*:

$$m_e \frac{d^2 x(t)}{dt^2} = -q\mathbf{E}(x(t)),$$

where m_e is the *electron mass* and q is the *elementary charge*. Further, we would have the initial conditions

$$x(t=0)=x_0 \quad \text{and} \quad \frac{dx}{dt}(t=0)=\mathbf{v}_0.$$

Clearly, this description is computationally not efficient since there will be billions of particles even in a very tiny piece of a semiconductor crystal. For this reason we introduce the *electron density* $n(x)$ (with unit m^{-3}, i.e., number of particles per cubic meter). This function can be interpreted as follows: Consider again that the semiconductor device occupies the domain $\Omega \subset \mathbb{R}^3$ and assume that this domain contains a large number of electrons. Now let there be a subdomain $\omega \subset \Omega$ which is large compared to the size of one electron. Then, the total number of electrons in this subdomain is given by

$$\int_\omega n(x)\,dx.$$

Since the number of particles in a domain is always nonnegative, we directly have $n \geq 0$. Note that at this stage we cannot exclude the occurrence of vacuum regions.

Further, we introduce the *mean electron velocity* $\mathbf{v}_n$. Assuming again that there is a subdomain $\omega \subset \Omega$ which is large compared to the size of one electron. Then the average velocity of electrons in this subdomain is given by

$$\int_\omega \mathbf{v}_n(x)\,dx.$$

Finally, we introduce the *electron current density* by

$$\mathbf{J}_n = qn\mathbf{v}_n.$$

Remark 4.1 The density of holes $p(x)$, the mean hole velocity v_p and the hole current density $\mathbf{J}_p$ are defined in analogy. Note that

$$\mathbf{J}_p = -qp\mathbf{v}_p.$$

Next, we motivate the set of partial differential equations connecting those quantities.

4.1.1.2 The Potential Equation

First, we give a mathematically tractable relation between the charge densities and the electric field. This can be done by introducing the *electrostatic potential* V which is defined as a solution of the *Poisson equation*

$$-\epsilon\Delta V = q(n-p+N_A-N_D),$$

where ϵ is the dielectric constant of the semiconductor material and N_A, N_D are the densities of acceptor and donator atoms, respectively. Here, we assumed that each donator atom contributes just one electron as well as each acceptor atom contributes just one hole. Then, the electric field can be expressed as

$$\mathbf{E} = -\nabla V.$$

Note that the potential is not uniquely defined by this equation, since one might add an arbitrary constant and still gets the same electric field. In particular, if the equation is posed on a bounded domain the prescription of boundary data will be essential.

Introducing the *doping profile*

$$C(x) := N_D(x) - N_A(x)$$

we finally get the equation

$$-\epsilon \Delta V = q(n - p - C), \tag{4.1}$$

where the function $q(n - p - C)$ is called the *total space charge*.

Remark 4.2 For the sake of simplicity and notational convenience we assume in the following that all physical parameters in our model are constant.

4.1.1.3 The Continuity Equations

The current density $\mathbf{J}(x)$ in the semiconductor consists of the sum of the electron and the hole current density, i.e.,

$$\mathbf{J} = \mathbf{J}_n + \mathbf{J}_p.$$

Clearly, only the combined current density can be measured. If we assume that we have conservation of charged particles and no generation and recombination processes are present, then it holds for each subdomain $\omega \subset \Omega$ with smooth boundary $\partial\omega$ that

$$I_{\partial\omega} = \int_{\partial\omega} \mathbf{J} \cdot \nu \, ds = 0.$$

Hence, Gauß' theorem implies directly

$$\int_{\omega} \operatorname{div} \mathbf{J} \, dx = 0.$$

This holds for any subdomain $\omega \subset \Omega$. Thus, the variational lemma yields the differential form of the continuity equation

$$\operatorname{div} \mathbf{J} = 0.$$

Remark 4.3 Taking into account $\mathbf{J} = \mathbf{J}_n + \mathbf{J}_p$ we get

$$\operatorname{div} \mathbf{J}_n = \operatorname{div} \mathbf{J}_p = 0.$$

4.1.1.4 The Current Densities

The above equations are by far not sufficient to describe charge transport in the semiconductor device. In particular, we need additional relations for the current densities. In many applications one can successfully assume that the current densities are entirely determined by the particle densities and by the electrostatic potential [124, 129]. Here, we only consider two contributions, namely the convective current density and the diffusive current density.

The *convective current density* describes the acceleration of charged particles in an electric field. It is assumed to be proportional to the electric field, i.e.,

$$\mathbf{J}_n^{\text{conv}} = q\mu_n n \nabla V, \qquad \mathbf{J}_p^{\text{conv}} = -q\mu_p n \nabla V,$$

where μ_n and μ_p are the mobilities of electrons and holes, respectively.

Remark 4.4 In general, the mobilities might depend on the electric field or even on the background doping profile. Here, we assume that they are constants.

The *diffusion current density* accounts for the compensation of density fluctuations for an ensemble of charged particles. Hence, this causes an additional movement of the particles, the so-called diffusion. We assume that these diffusion current densities are given by

$$\mathbf{J}_n^{\text{diff}} = qD_n \nabla n, \qquad \mathbf{J}_p^{\text{diff}} = qD_p \nabla p,$$

where the diffusion coefficients D_n and D_p are assumed to be positive constants.

Finally, we get the current density relations

$$\mathbf{J}_n = \mathbf{J}_n^{\text{diff}} + \mathbf{J}_n^{\text{conv}} = qD_n \nabla n + q\mu_n n \nabla V,$$

$$\mathbf{J}_p = -\left(\mathbf{J}_p^{\text{diff}} + \mathbf{J}_p^{\text{conv}}\right) = -qD_p \nabla p + q\mu_p p \nabla V.$$

These can be further simplified assuming the *Einstein relations* [129]

$$\frac{D_n}{\mu_n} = \frac{D_p}{\mu_p} = \frac{k_B T}{q} =: V_T,$$

where T is the (constant) temperature of electrons and holes and k_B is the Boltzmann constant. Here, V_T is called the *thermal voltage*.

Summarizing, we get the well-known *drift diffusion model* which was first introduced by Van Rosbroeck (cf. [95, 129] and the references therein):

$$\mathbf{J}_n = q(D_n \nabla n + \mu_n n \nabla V), \tag{4.2a}$$

$$\mathbf{J}_p = -q(D_p \nabla p - \mu_p p \nabla V), \tag{4.2b}$$

$$\operatorname{div} \mathbf{J}_n = 0, \tag{4.2c}$$

$$\operatorname{div} \mathbf{J}_p = 0, \tag{4.2d}$$

$$-\epsilon \Delta V = q(n - p - C). \tag{4.2e}$$

Hence, the drift diffusion model consists of a coupled system of nonlinear elliptic partial differential equations. This makes its mathematical analysis quite involved (see, e.g., [95, 96]).

To get a well posed problem we have to supplement (4.2) with additional boundary data. We assume that the boundary $\partial\Omega$ of the domain Ω splits into two disjoint parts Γ_D and Γ_N, where Γ_D models the Ohmic contacts of the device and Γ_N represents the insulating parts of the boundary. Let ν denote the unit outward normal vector along the boundary. First, assuming charge neutrality

$$n - p - C = 0$$

and thermal equilibrium

$$np = n_i^2$$

at the Ohmic contacts Γ_D and, secondly, zero current flow and vanishing electric field at the insulating part Γ_N yields the following set of boundary data

$$n = n_D, \qquad p = p_D, \qquad V = V_D \quad \text{on } \Gamma_D, \tag{4.2f}$$

$$\mathbf{J}_n \cdot \nu = \mathbf{J}_p \cdot \nu = \nabla V \cdot \nu = 0 \quad \text{on } \Gamma_N, \tag{4.2g}$$

where n_D, p_D, V_D are given on Γ_D by

$$n_D = \frac{C + \sqrt{C^2 + 4n_i^2}}{2}, \qquad p_D = \frac{-C + \sqrt{C^2 + 4n_i^2}}{2},$$

$$V_D = -V_T \log\left(\frac{n_D}{n_i}\right) + V_{\mathrm{bi}}.$$

Here, V_{bi} denotes the applied biasing voltage, which is, e.g., applied between the source and the drain contact of the MOSFET device, and n_i is the intrinsic carrier density of the semiconductor. Note, that the main unknowns in the above model are the densities n and p as well as the potential V.

4.1.1.5 Scaling

This model is not only challenging from the analytical point view, but also due to the severe numerical problems one has to encounter [95]. To understand this it is most convenient to rewrite the equations in nondimensional form using following diffusion scaling [96]

$$n \to C_m \tilde{n}, \qquad p \to C_m \tilde{p}, \qquad x \to L\tilde{x},$$
$$C \to C_m \tilde{C}, \qquad V \to V_T \tilde{V}, \qquad \mathbf{J}_{n,p} \to \frac{qU_T C_m \mu_{n,p}}{L} \tilde{\mathbf{J}}_{n,p},$$

where L denotes a characteristic device length, C_m the maximal absolute value of the background doping profile and $\mu_{n,p}$ a characteristic values for the respective mobilities. Defining the dimensionless *Debye length*

$$\lambda^2 = \frac{\epsilon V_T}{q C_m L^2}$$

the scaled equations read

$$\operatorname{div} \mathbf{J}_n = 0, \qquad \mathbf{J}_n = \nabla n + n \nabla V, \tag{4.3a}$$
$$\operatorname{div} \mathbf{J}_p = 0, \qquad \mathbf{J}_p = -(\nabla p - p \nabla V), \tag{4.3b}$$
$$-\lambda^2 \Delta V = n - p - C, \tag{4.3c}$$

where we omitted the tilde for notational convenience.

The Dirichlet boundary conditions on Γ_D transform to

$$n_D = \frac{C + \sqrt{C^2 + 4\delta^4}}{2}, \qquad p_D = \frac{-C + \sqrt{C^2 + 4\delta^4}}{2},$$
$$V_D = -\log\left(\frac{n_D}{\delta^2}\right) + V_{\mathrm{bi}}, \tag{4.3d}$$

where $\delta^2 = n_i / C_m$ denotes the scaled intrinsic density.

For typical device parameters [129] we get for the Debye length $\lambda^2 = 10^{-3}$ and for the scaled intrinsic density $\delta^2 = 10^{-4}$. Hence, the drift diffusion model is singular perturbed, which has to be taken into consideration in the numerical treatment.

Remark 4.5 There will be large gradients in the potential and thus also in the particle densities near to rapid changes in the doping profile, the so called junctions. In general, one employs the Scharfetter–Gummel discretization [95, 120] for the discretization, which can be interpreted as an exponentially fitted scheme. For a detailed discussion of numerical methods for semiconductor equations we refer to [16].

4.1.2 The Optimization Problem

After setting up the underlying model equations we now turn our attention to the design question. Remember that the main objective in optimal semiconductor design is to get an improved current flow at a specific contact of the device, e.g., focusing on the reduction of the leakage current (in the off-state) in MOSFET devices or maximizing the drive current (in the on-state) [76, 128]. In both cases a certain working point is fixed and one tries to achieve this objective by a change of the doping profile C. Hence, the objective of the optimization, the current flow over a contact Γ, is given by

$$I = \int_\Gamma \mathbf{J} \cdot \nu \, ds = \int_\Gamma (\mathbf{J}_n + \mathbf{J}_p) \cdot \nu \, ds, \tag{4.4}$$

where the current density $\mathbf{J}$ for a specific doping profile C is given by the solution of the drift diffusion model (4.3).

Next, we want to embed this design question into the optimal control context presented in the previous chapters. We intend to minimize a cost functional of tracking-type

$$J(n, p, V, C) := \frac{1}{2} \left| \int_\Gamma \mathbf{J} \cdot \nu \, ds - I^* \right|^2 + \frac{\gamma}{2} \int_\Omega \left| \nabla (C - \bar{C}) \right|^2 dx, \tag{4.5}$$

where $\bar{C}$ is a given reference doping profile, I^* is a desired current flow, and the parameter $\gamma > 0$ allows to adjust the deviations from $\bar{C}$. Clearly, C is acting here as the control parameter. The introduction of $\bar{C}$ is necessary to ensure that we change not the type of the semiconductor device during the optimization.

Remark 4.6 The definition of the cost functional already implies that we need $C \in H^1(\Omega)$, which is also related to the upcoming regularity theory for the drift diffusion model. Often, doping profiles are described as a superposition of Gaussian functions. This suggest the introduction of a control operator $B : U \to H^1(\Omega)$ with $B(u) = C$. In the following, we assume that $U = H^1(\Omega)$ and B is just the identity operator.

Remark 4.7 This problem can be clearly tackled by an optimization approach, but only recently efforts were made to solve the design problem using mathematical sound optimization techniques [17–19, 49, 50, 75–77]. In [89] *Lee et al.* present a finite-dimensional least-squares approach for adjusting the parameters of a semiconductor to fit a given, ideal current-voltage characteristics. Their work is purely numerical and has its focus on testing different approaches to solve the discrete least-squares problem.

Since the current density $\mathbf{J}$ is given by a solution of the drift diffusion model this yields altogether a constrained optimization problem in the spirit of (1.78). We describe in the following how one can use the adjoint approach (compare Sect. 1.6.2) to this problem. For this purpose we introduce the state $y \stackrel{\text{def}}{=} (n, p, V)$ and an admissible set of controls $U_{\text{ad}} \subset H^1(\Omega)$ and rewrite the state system (4.3) shortly as $e(y, u) = 0$. Due to the nonlinear structure of the equations we define the state space

$Y \stackrel{\text{def}}{=} y_D + Y_0$, where $y_D \stackrel{\text{def}}{=} (n_D, p_D, V_D)$ denotes the boundary data introduced in (4.3) and $Y_0 \stackrel{\text{def}}{=} (H^1_{0,\Gamma_D}(\Omega) \cap L^\infty(\Omega))^3$, where we define the spaces (for the definition of the trace operator see Theorem 1.12)

$$H^1_{0,\Gamma_D}(\Omega) \stackrel{\text{def}}{=} \left\{\phi \in H^1(\Omega) : T(\phi) = 0 \text{ on } \Gamma_D\right\},$$

as well as $Z \stackrel{\text{def}}{=} [H^{-1}(\Omega)]^3$. Then, one can show that $e : Y \times H^1(\Omega) \to Z$ is well-defined and infinitely often differentiable [76].

Now, the mathematically precise optimization problem reads

$$\min_{Y \times U_{\text{ad}}} J(y,u) \quad \text{such that} \quad e(y,u) = 0. \tag{4.6}$$

We restrict the set of admissible controls to

$$U_{\text{ad}} \stackrel{\text{def}}{=} \{u \in H^1(\Omega) : u = \bar{C} \text{ on } \Gamma_D\}. \tag{4.7}$$

This is necessary for the solvability of the state system and for the continuous dependence of the state on the control u, since the boundary data in (4.3) does depend on $C = B(u)$. In fact, there are various results on the solvability of the state system (c.f. [95, 96, 106] and the references therein). For completeness we state the following existence results, which is proved in [109].

Proposition 4.1 *Assume sufficient regularity of the boundary and the data. Then for each $C = B(u) \in H^1(\Omega)$ and all boundary data (n_D, p_D, V_D) with*

$$\frac{1}{K} \le n_D(x), p_D(x) \le K, \quad x \in \Omega, \quad \textit{and} \quad \|V_D\|_{L^\infty(\Omega)} \le K$$

for some $K \ge 1$, there exists a solution $(\mathbf{J}_n, \mathbf{J}_p, n, p, V) \in [L^2(\Omega)]^2 \times (H^1(\Omega) \cap L^\infty(\Omega))^3$ of system (4.3) fulfilling

$$\frac{1}{L} \le n(x), p(x) \le L, \quad x \in \Omega, \quad \textit{and} \quad \|V\|_{L^\infty(\Omega)} \le L$$

for some constant $L = L(\Omega, K, \|C\|_{L^p(\Omega)}) \ge 1$, where the embedding $H^1(\Omega) \hookrightarrow L^p(\Omega)$ holds.

Remark 4.8 The idea of the proof is to write down a fixed point mapping decoupling the equations and to use Schauder's fixed point theorem to get the existence of a fixed point. The compactness of the mapping is derived by energy estimates and Stampacchia's truncation method, which ensures the uniform bounds on the solution [109].

For the analysis of the optimization problem it is crucial to observe that, in general, there exists no unique solution of the state system (4.3). This is physically even reasonable, since there are devices, like the *thyristor*, whose performance relies on the multiplicity of solutions [96]. Nevertheless, one can ensure uniqueness near to the thermal equilibrium state, i.e., for small applied biasing voltages V_{bi}. This has also impact on the optimization problem. In particular, we cannot consider the reduced cost functional in each regime and also the linearized state operator e_y is in general not boundedly invertible. But one can still prove the existence of a minimizer [76].

Theorem 4.1 *The constrained minimization problem* (4.6) *admits at least one solution*

$$(n^*, p^*, V^*, C^*) \in Y \times U_{\text{ad}}.$$

Proof The proof follows the ideas presented in Sect. 1.5.2. Since J is bounded from below, $J_0 \stackrel{\text{def}}{=} \inf_{(y,u)\in Y\times U_{\text{ad}}} J(y,u)$ is finite. Consider a minimizing sequence $\{(y^k, u^k)\}_{k\in\mathbb{N}} \subset Y \times U_{\text{ad}}$. From the radial unboundedness of J we infer that $\{u^k\}_{k\in\mathbb{N}}$ is bounded in $H^1(\Omega)$. Hence, there exists a weakly convergent subsequence, again denoted by $\{u^k\}_{k\in\mathbb{N}}$, such that

$$u^k \rightharpoonup u^*, \quad \text{weakly in } H^1(\Omega).$$

Since U_{ad} is weakly closed with respect to the $H^1(\Omega)$-norm, we have $u^* \in U_{\text{ad}}$. By the continuous embedding $H^1(\Omega) \hookrightarrow L^p(\Omega)$ ($p \in [1,6)$) the sequence $\{u^k\}_{k\in\mathbb{N}}$ is also bounded in $L^p(\Omega)$. Now, one can employ Stampaccia's method [126] to derive the following estimates [109]

$$\|n^k\|_{H^1(\Omega)} + \|n^k\|_{L^\infty(\Omega)} \le K\left(\|n_D\|_{L^\infty(\Gamma_D)} + \|u^k\|_{L^p(\Omega)}\right), \tag{4.8a}$$

$$\|p^k\|_{H^1(\Omega)} + \|p^k\|_{L^\infty(\Omega)} \le K\left(\|p_D\|_{L^\infty(\Gamma_D)} + \|u^k\|_{L^p(\Omega)}\right), \tag{4.8b}$$

$$\|V^k\|_{H^1(\Omega)} + \|V^k\|_{L^\infty(\Omega)} \le K\left(\|V_D\|_{L^\infty(\Gamma_D)} + \|u^k\|_{L^p(\Omega)}\right), \tag{4.8c}$$

for some constant $K = K(\Omega) > 0$. These are by far sufficient to pass to the limit in the state equations (4.3), which can be seen as follows. First note that $\|(n_D, p_D, V_D)\|_{(L^\infty(\Gamma_D))^3}$ is bounded independently of k due to the definition of the admissible set (4.7). Every solution (n^k, p^k, V^k) of (4.3) associated to u^k satisfies the a priori estimates (4.8). Hence, there exists a subsequence, again denoted by $\{(n^k, p^k, V^k)\}_{k\in\mathbb{N}}$, such that

$$(n^k, p^k, V^k) \rightharpoonup (n^*, p^*, V^*) \quad \text{weakly in } (H^1(\Omega))^3,$$

which by Rellich's Theorem [152] implies strong convergence of $\{(n^k, p^k, V^k)\}_{k\in\mathbb{N}}$ in $(L^2(\Omega))^3$. Further, the uniform $L^\infty(\Omega)$-bounds imply

$$(n^k, p^k, V^k) \rightharpoonup (n^*, p^*, V^*) \quad \text{weak-* in } L^\infty(\Omega).$$

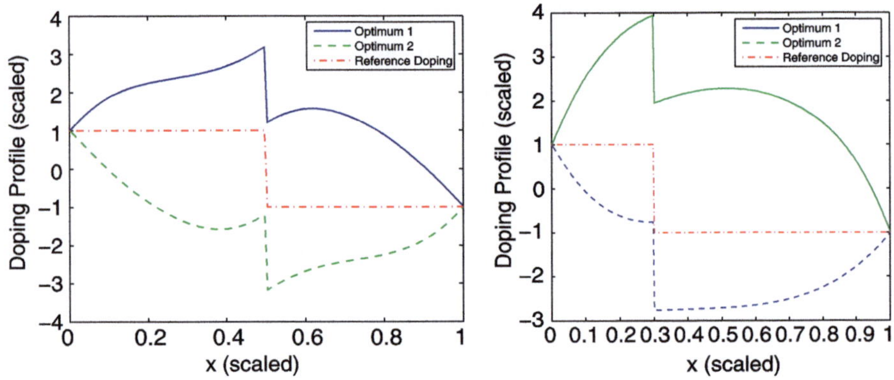

Fig. 4.2 Optimized doping profiles for a symmetric n–p–diode (*left*) and an unsymmetric n–p–diode (*right*)

Utilizing these convergences one can pass to the limit in the weak formulation of (4.3), which satisfies

$$
\begin{aligned}
\Delta n^* + \operatorname{div}(n^* \nabla V^*) &= 0, \\
\Delta p^* - \operatorname{div}(p^* \nabla V^*) &= 0, \\
-\lambda^2 \Delta V^* &= n^* - p^* - u^*
\end{aligned}
$$

together with the boundary conditions in (4.3). This completes the proof of the existence of a minimizer.

Since the set given by the constraint is not convex, we can in general not expect the uniqueness of the minimizer. In fact, one can show analytically that for special choices of the reference doping $\bar{C}$ there exist at least two solutions [77]. For other choices there is at least a numerical evidence of nonuniqueness. This is due to the fact that the minimizer has the possibility to interchange the roles of the electron and the hole current densities (see Fig. 4.2). Clearly, this has also some impact on the construction and convergence of numerical schemes. In particular, the choice of an appropriate starting point for iterative algorithms is then crucial.

4.1.2.1 The First-Order Optimality System

In this section we discuss the first-order optimality system which is, as we already know, the basis for all optimization methods seeking at least a stationary point. Since we have a constrained optimization problem, we write the first-order optimality system using the Lagrangian $L : Y \times U \times Z^* \to \mathbb{R}$ associated to problem (4.6) defined by

$$
L(y, u, \xi) \stackrel{\text{def}}{=} J(y, u) + \langle e(y, u), \xi \rangle_{Z, Z^*},
$$

where $\xi \stackrel{\text{def}}{=} (\xi^n, \xi^p, \xi^V) \in Z^* = [H^1(\Omega)]^3$ denotes the adjoint variable. For the existence of a Lagrange multiplier associated to an optimal solution (y^*, u^*) of (4.6) it is sufficient that the operator $e'(y^*, u^*)$ is surjective (compare Sect. 2.6.1).

For the drift diffusion model this does in general not hold, but one can ensure the bounded invertibility of $e'(y^*, u^*)$ for small current densities [96]. This idea can be used to prove the unique existence of an adjoint state [76].

Theorem 4.2 *There exists a constant $j = j(\Omega, \lambda, V_{\text{bi}}) > 0$ such that for each state $y \in Y$ with*

$$\left\| \frac{\mathbf{J}_n^2}{n} \right\|_{L^\infty(\Omega)} + \left\| \frac{\mathbf{J}_p^2}{p} \right\|_{L^\infty(\Omega)} \leq j$$

there exists an adjoint state $\xi \in Z^*$ fulfilling $e_y^*(y, u)\xi = -J_y(y, u)$.

Hence, at least for small current densities there exists a unique Lagrange multiplier $\xi^* \in Z^*$ such that together with an optimal solution (y^*, u^*) it fulfills the first-order optimality system

$$L'(y^*, u^*, \xi^*) = 0. \tag{4.9}$$

We can rewrite this equations in a more concise form:

$$\begin{aligned} e(y^*, u^*) &= 0 \quad \text{in } Z, \\ e_y^*(y^*, u^*)\xi^* + J_y(y^*, u^*) &= 0 \quad \text{in } Y^*, \\ e_u(y^*, u^*)\xi^* + J_u(y^*, y^*) &= 0 \quad \text{in } U^*. \end{aligned}$$

I.e., a critical point of the Lagrangian has to satisfy the state system (4.3), as well as the adjoint system and the optimality condition. The derivation of this system is an easy exercise just using the techniques presented in Sect. 1.6.4. Finally, one gets the coupled linear system

$$\Delta \xi^n - \nabla V \nabla \xi^n = \xi^V, \tag{4.10a}$$

$$\Delta \xi^p + \nabla V \nabla \xi^p = -\xi^V, \tag{4.10b}$$

$$-\lambda^2 \Delta \xi^V + \operatorname{div}\left(n \nabla \xi^n\right) - \operatorname{div}\left(p \nabla \xi^p\right) = 0, \tag{4.10c}$$

supplemented with the boundary data

$$\xi^{J_n} = \begin{cases} \int_\Gamma J_n \cdot \nu \, ds - I_n^*, & \text{on } \Gamma, \\ 0, & \text{on } \Gamma_D \setminus \Gamma, \end{cases} \tag{4.10d}$$

$$\xi^{J_p} = \begin{cases} \int_\Gamma J_p \cdot \nu \, ds - I_p^*, & \text{on } \Gamma, \\ 0, & \text{on } \Gamma_D \setminus \Gamma, \end{cases} \tag{4.10e}$$

$$\xi^V = 0, \quad \text{on } \Gamma_D, \tag{4.10f}$$

as well as

$$\nabla \xi^n \cdot \nu = \nabla \xi^p \cdot \nu = \nabla \xi^V \cdot \nu = 0 \quad \text{on } \Gamma_N. \tag{4.10g}$$

Further, we have the optimality condition

$$\gamma \Delta \left(u - \bar{C}\right) = \xi^V \quad \text{in } \Omega, \tag{4.11a}$$

$$u = \bar{C} \quad \text{on } \Gamma_D, \qquad \nabla u \cdot \nu = \nabla \bar{C} \cdot \nu \quad \text{on } \Gamma_N. \tag{4.11b}$$

4.1.3 Numerical Results

Next, we want to discuss the behavior of two numerical methods applied to this optimization problem.

4.1.3.1 Steepest Descent

The first adequate and easy to implement numerical method for the solution of (4.6) is the following gradient algorithm, which is a special variant of the general descent method Algorithm 2.1.

Algorithm 4.3

1. Choose $u_0 \in U_{\text{ad}}$.
2. For $k = 1, 2, \ldots$ compute $u_k = u_{k-1} - \sigma_k \nabla \hat{J}(u_{k-1})$.

Here, $\hat{J}(u) \overset{\text{def}}{=} J(y(u), u)$ denotes the reduced cost functional, which can be introduced near to the thermal equilibrium state. The evaluation of

$$\nabla \hat{J}(u) = J_u(y, u) + e_u^* \xi$$

requires the solution of the nonlinear state system (4.3) for y as well as a solution of the linear adjoint system (4.10) for ξ and finally a linear solve of a Poisson problem to get the correct Riesz representative.

Remark 4.9 We have seen that there exist various choices for the step sizes σ_k ensuring the convergence of this algorithm to a critical point, like the Armijo or the Goldstein rule (compare Sect. 2.2). The overall numerical performance of this algorithm relies on an appropriate choice of the step size rule for σ_k, since these methods require in general consecutive evaluations of the cost functional requiring additional solves of the nonlinear state system [92].

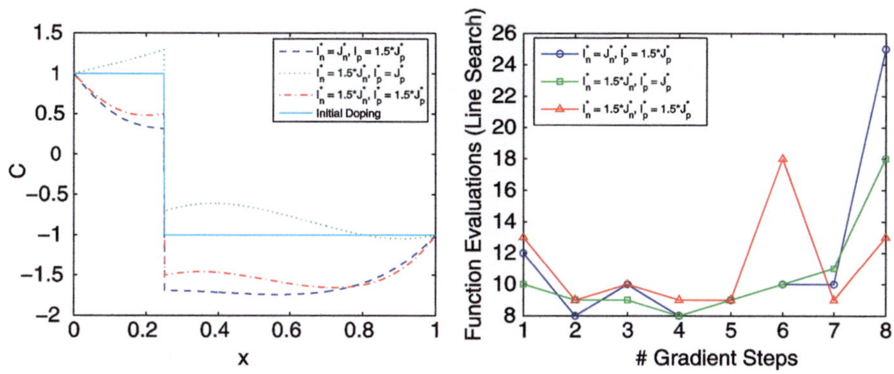

Fig. 4.3 Optimized doping profiles (*left*) and Function Evaluations for the Line Search (*right*)

We apply Algorithm 4.3 for the optimal design of an unsymmetric n–p–diode (for the reference doping profile see Fig. 4.3). We already learned that the cost functional employed so far might admit multiple minimizers. For this reason we study here a slightly different functional of the form

$$J(n,p,V,u) = \frac{1}{2}\left|\int_\Gamma \mathbf{J}_n \cdot \nu \, ds - I_n^*\right|^2 + \frac{1}{2}\left|\int_\Gamma \mathbf{J}_p \cdot \nu \, ds - I_p^*\right|^2 + \frac{\gamma}{2}\int_\Omega \left|\nabla(B(u) - \bar{C})\right|^2 dx.$$

This allows to adjust separately the contact electron and hole current. The computations were performed on a uniform grid with 1000 points and the scaled parameters were set to $\lambda^2 = 10^{-3}$, $\delta^2 = 10^{-2}$ and $V_{\mathrm{bi}} = 10$. For the parameter γ we chose 2×10^{-2}. The step-size σ_k is computed by an exact one dimensional line search

$$\sigma_k = \operatorname{argmin}_\sigma \hat{J}\left(u_{k-1} - \sigma \nabla \hat{J}(u_{k-1})\right),$$

which is performed using the matlab optimization toolbox. The iteration terminates when the relative error $\|\nabla \hat{J}(u_k)\|_{H^1} / \|\nabla \hat{J}(u_0)\|_{H^1}$ is less than 5×10^{-4}.

In Fig. 4.3 (left) we present the optimized doping profiles for different choices of I_n^*, I_p^*, i.e., we are seeking an amplification of either the hole current ($I_n^* = J_n^*$, $I_p^* = 1.5 \cdot J_p^*$) or of the electron current ($I_n^* = 1.5 \cdot J_n^*$, $I_p^* = J_p^*$) or of both of them $I_n^* = 1.5 \cdot J_n^*$, $I_p^* = 1.5 \cdot J_p^*$) by 50%.

To get an impression of the overall performance of the method we also have to consider the nonlinear solves needed for the exact one dimensional line search. These are presented in Fig. 4.3 (right). Indeed, this is the most expensive numerical part.

4.1.3.2 The Reduced Newton Method

Finally, we want to discuss the performance of the reduced Newton algorithm (compare also Sect. 2.4) which reads

Algorithm 4.4

1. Choose u_0 in a neighborhood of u^*.
2. For $k = 0, 1, 2, \dots$
   a. Solve $\hat{J}''(u_k)\delta u_k = -\hat{J}'(u_k)$,
   b. Set $u_{k+1} = u_k + \delta u_k$.

The solution of the system in step (ii.a) is done iteratively by using a conjugate gradient algorithm embedded inside the Newton algorithm, as the computation of a discretization of the Hessean would require a significant numerical effort, while a conjugate gradient based approach leads to the same result with a fraction of the demands on memory and computation time. The conjugate gradient algorithm only requires the applications of the Hessean on a sequence of direction vectors δu to be computed, so that no (direct) solution of the large system in (ii.a) is required.

Algorithm 4.5

1. Choose u_0 in a neighborhood of u^*.
2. For $k = 0, 1, 2, \dots$
   a. Evaluate $\hat{J}'(u_k)$ and set $\delta u_k^j = 0$
   b. For $j = 0, 1, 2, \dots$ do until convergence
      i. Evaluate $q_k^j = \hat{J}''(u_k)\delta u_k^j$
      ii. Compute an approximation δu_k^{j+1} for δu_k, e.g. by a cg–step
   c. Set $u_{k+1} = u_k + \delta u_k$.

Remark 4.10 Each application of the reduced Hessian $\hat{J}''(u_k)$ during the j-th cg–step requires two linear solves, in detail

$$v_k^j = e_y^{-1}(y_k, u_k) e_u(y_k, u_k) \delta u_k^j$$

and

$$w_k^j = e_y^{-*}(y_k, u_k) \left\{ J_{yy}(y_k, u_k)(v_k^j, \cdot) + \left\langle e_{yy}(y_k, u_k)(v_k^j, \cdot), \xi_k \right\rangle_{Z,Z^*} \right\}.$$

For the precise statement of these subproblems we refer to [78].

Again, we tried to achieve an increase of the electron and hole current by 50% and studied the influence of the regularization parameter γ. The different resulting optimal doping profiles can be found in Fig. 4.4 (left). As expected we get larger deviations from $\bar{C}$ for decreasing γ, which on the other hand also allows for a better

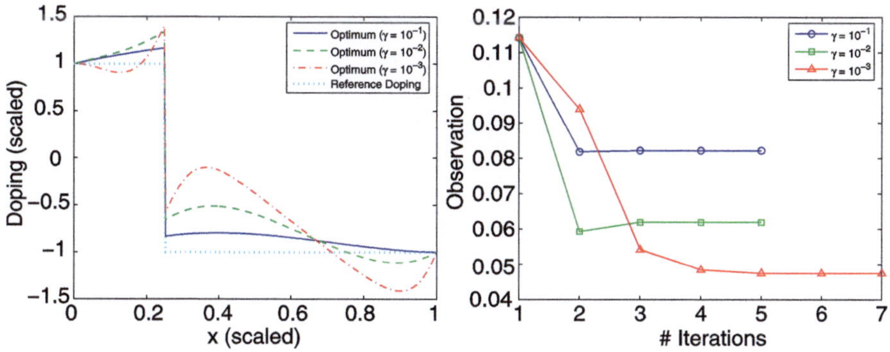

Fig. 4.4 Dependence of the optimum on γ (*left*) and dependence of the observation on γ (*right*)

reduction of the observation as can be seen in Fig. 4.4 (right). For all three cases we already get a significant reduction after two steps and the algorithm terminates rather quickly. Only for the smallest value of γ we need two more iterations to meet the stopping criterion, which can be explained by a loss of convexity or, equivalently, a weaker definiteness of the Hessean.

The conjugate gradient algorithm in the inner loop was terminated when the norm of the gradient became sufficiently small; to be more precise, in the j-th conjugate gradient step for the computation of the update in Newton step k we stop if the residual r_k^j satisfies

$$\frac{\|r_k^j\|}{\|\nabla\hat{J}(u^0)\|} \leq \min\left\{\left(\frac{\|\nabla\hat{J}(u_k)\|}{\|\nabla\hat{J}(u^0)\|}\right)^q, p\frac{\|\nabla\hat{J}(u_k)\|}{\|\nabla\hat{J}(u^0)\|}\right\} \quad \text{or} \quad j \geq 100. \tag{4.12}$$

Note, that q determines the order of the outer Newton algorithm, such that p should be chosen in the open interval $(1,2)$. The value of p is important for the first step of Newton's method, as for $k=0$ the norm quotients are all 1; for later steps, the influence of q becomes increasingly dominant.

To get deeper insight into the convergence behavior of the algorithm, we present in Fig. 4.5 (left) the norm of the residual during the iteration for different values of γ. Here, we used $q=2$ to get the desired quadratic convergence behavior. Again, one realizes that the convergence deteriorates with decreasing γ. Since the overall numerical effort is spend in the inner loop, we show the number of conjugate gradient steps in Fig. 4.5 (right). Here, one realizes the drastic influence of the regularization parameter.

The next numerical test was devoted to the stopping criterion of the inner iteration and the influence of the exponent q. In Fig. 4.6 (left) the decrease of the residual is depicted for different values of $q = 1$, 1.5, or 2. As predicted by the general theory one gets linear, superlinear and quadratic convergence. Note, that for all three cases we have a linear convergence behavior at the beginning of the iteration due to the globalization of the Newton algorithm. Clearly, the parameter q strongly influences the number of conjugate gradient steps, which can be seen from Fig. 4.6 (right).

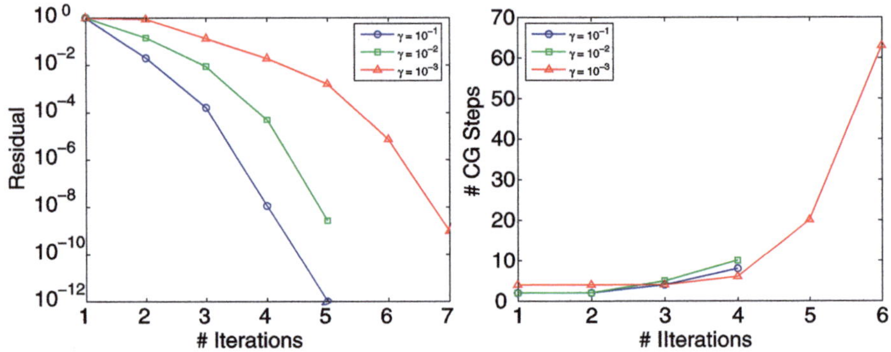

Fig. 4.5 Dependence of the residual on γ (*left*) and dependence of the CG iteration on γ (*right*)

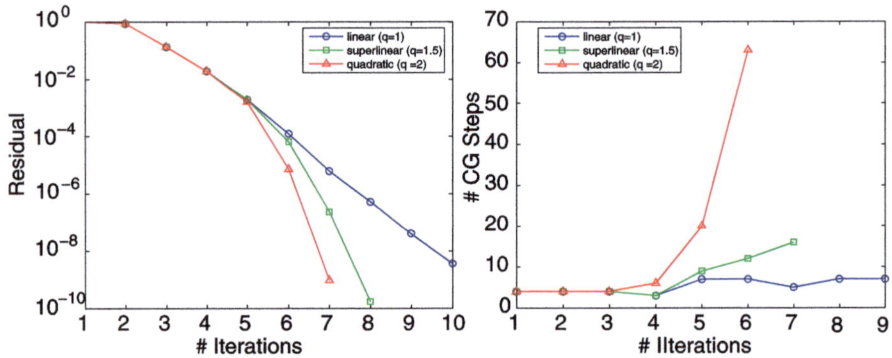

Fig. 4.6 Dependence of the residual on q (*left*) and dependence of the CG iteration on q (*right*)

While in the linear case ($q = 1$) we have an almost constant amount of CG steps in each iteration, we get, as expected, a drastic increase toward the end of the iteration for the quadratic case ($q = 2$). Hence, the overall numerical effort in terms of CG steps is despite of the quadratic convergence much larger compared to the relaxed stopping criterion, which only yields linear convergence!

4.2 Optimal Control of Glass Cooling

Glass manufacturing is a very old industry, but one has to be aware that nowadays it is technically rather advanced. There is a strong need for high quality glass products, like lenses for laser optics or mirrors for space telescopes. Further, one also wants to influence the production process of other products, like monitors or car windows. There are many stages in the production process where optimal control techniques can be used [32]. We focus here on the stage where a hot melt of glass is cooled in a controlled environment, e.g., a furnace. During cooling large temperature differences, i.e., large temperature gradients, have to be avoided since they lead to thermal

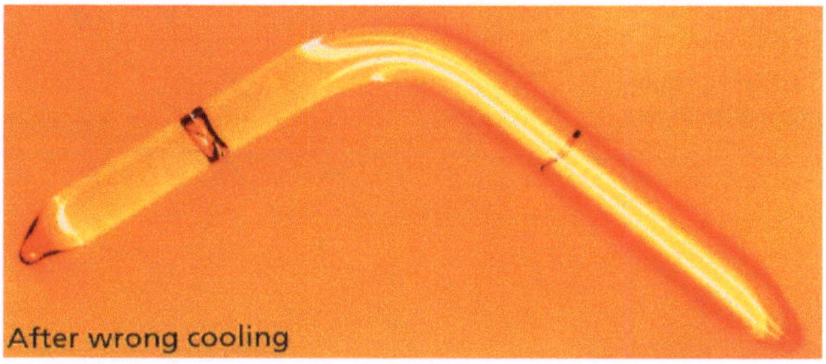

Fig. 4.7 This happens after wrong cooling! (Photo by courtesy of N. Siedow)

stress in the material. This may cause cracks or affect the optical quality of the resulting product (see Fig. 4.7). Hence, the process has to be managed in such a way that temperature gradients are sufficiently small. A related question concerns chemical reactions during the cooling process, which have to be activated and triggered. Again, one wants to avoid spatial temperature gradients since these reactions have to take place homogeneously in the glass melt. We embed these two different questions into the same mathematical optimal control context. The following discussion of this optimal control problem is done in three steps. First, we introduce the equations which are used for the simulation of the cooling process. Then, we state and discuss the optimal control problem and, finally, we present numerical results.

4.2.1 Modeling

The modeling of glass cooling has to take into account that this process involves very high temperatures up to 1500 K. In this temperature range heat transfer will be dominated by radiation and not by diffusion anymore. Hence, we have first to understand how radiation can be modeled [107, 131].

4.2.1.1 Radiation

Thermal radiation can be viewed as electro-magnetic waves or, alternatively, as photons. It is characterized by its speed of propagation c, wavelength λ and frequency ν, which are related by $c = \lambda \cdot \nu$. The most important difference to heat conduction and convection is that it is a long-range, non-local phenomenon in contrast to the local, microscopic diffusion effect. Note, that the magnitude of conduction and convection is linear in the temperature T, whereas radiation depends essentially on the fourth power of T, which shows that this effect gets increasingly important for higher temperatures.

In general, engineers are only interested in the energy of the radiative field and they describe it using the radiative intensity $I = I(x, t, \omega, \nu)$, which depends on the position x, the time t, the angular direction ω and on the frequency ν. To derive an

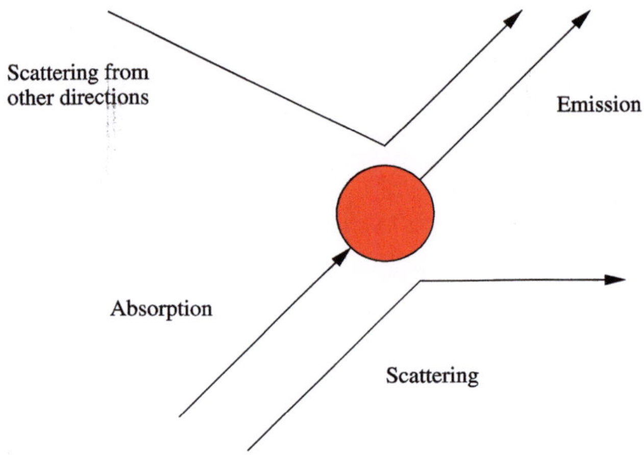

Fig. 4.8 Radiative effects [131]

equation for the intensity I, we consider a small portion Δx of a ray in direction ω (compare Fig. 4.8).

In this region, one looses energy due to absorption $-\kappa I \Delta x$, where κ is the absorption coefficient of the material. Note, that the absorption coefficient might also depend on the temperature T and the frequency ν as well as on the spatial position x. Further, one gains energy due to emission $+\kappa B \Delta x$, where

$$B(T,\nu) = n_G^2 \frac{2h\nu^3}{c^2} \left(e^{\frac{h\nu}{kT}} - 1\right)^{-1}$$

is Planck's function for black body radiation [107]. Another source for energy loss is scattering $-\sigma I \Delta x$, where σ is the scattering constant of the material. Again, this constant might depend on the temperature T and the frequency ν as well as on the spatial position x. But one can also gain energy due to back scattering, i.e., one has to collect the distributions from all incoming directions $+\frac{\sigma}{4\pi}\int_{S^2} I(\omega')d\omega'\Delta x$.

Now, we can write down the balance equation for the radiative intensity

$$I(x + c\omega\Delta t, \omega, t + \Delta t) - I(x,\omega,t) = \left(-\kappa I + \kappa B - \sigma I + \frac{\sigma}{4\pi}\int_{S^2} I(\omega')\,d\omega'\right)\Delta x.$$

Going to the limit $\Delta t \to 0$, $\Delta x = c\Delta t$ yields

$$\frac{1}{c}\partial_t I + \omega \cdot \nabla I + (\kappa + \sigma) I = \frac{\sigma}{4\pi}\int_{S^2} I(\omega')\,d\omega' + \kappa B. \qquad (4.13)$$

This equation holds for all times $t \in \mathbb{R}^+$, all spatial points $x \in \Omega$, all angles $\omega \in S^2$ (where S^2 denotes the unit sphere) and all frequencies $\nu \in \mathbb{R}^+$! To get an impression of the computational effort let us assume that we use a discretization with 60 angles $\times$ 10 frequency bands $\times$ 8000 spatial points $\times$ 100 time steps. This yields altogether 500 millions discrete variables! Indeed, this leads to a large scale optimization problem. To be honest, we will not even dare to use this equation directly, but instead we use techniques from asymptotic analysis to derive a numerically tractable model. Nevertheless, there are recent papers on optimal control problems for the full stationary radiative transfer equation (4.13), see e.g., [63, 64].

Finally, we pose some physically reasonable assumptions which will significantly simplify the upcoming presentation. Note that the speed of propagation c is large and hence we will drop the time derivative, i.e., we assume that radiation takes place in a quasi-static manner compared to the diffusion time scale. Further, we assume that no scattering occurs in the glass, i.e., $\sigma \equiv 0$ and that we have a gray medium, in which the intensity is independent of the frequency ν. In this case we can average the equations with respect to ν and can use the fact that

$$\int_0^\infty B(T,\nu)\,d\nu = aT^4,$$

where a is related to the Stefan-Boltzmann constant.

4.2.1.2 SP_N-approximations

The are several ways to perform an asymptotic analysis of the radiative transfer equation (4.13) (c.f. [107] and the references therein). Here, we follow a new approach first presented in [88]. Using a diffusion scaling one can introduce the optical thickness of the material as a small parameter

$$\varepsilon = \frac{1}{\kappa_{\text{ref}} x_{\text{ref}}} \approx \frac{\text{mean free path}}{\text{reference length}}.$$

Then, the remaining scaled radiative transfer equation reads

$$\varepsilon\omega \cdot \nabla I = \kappa(B - I).$$

Now, the idea is to invert the transport operator

$$\left(1 + \frac{\varepsilon}{\kappa}\omega \cdot \nabla\right) I = B$$

formally using the Neumann series.

Remark 4.11 Clearly, this inversion holds only on a formal level due to the unboundedness of the derivative operator. After a discretization of the model equation this approach can be made mathematically sound [64].

Then it holds for the mean radiative intensity $\rho := \int_{S^2} I\,d\omega$ in the limit $\varepsilon \to 0$ the asymptotic expansion

$$4\pi B = \left[1 - \frac{\varepsilon^2}{3\kappa^2}\Delta - \frac{4\varepsilon^4}{45\kappa^4}\Delta^2 - \frac{44^\varepsilon 6}{945\kappa^6}\Delta^3\right]\rho + O(\varepsilon^8).$$

This yields the SP_N-approximations of order $\mathcal{O}(\varepsilon^{2N})$ [88]. In the following we only employ the SP_1-approximation.

The radiative intensity depends crucially on the temperature, which enters via the Planck function. To resolve also the temperature changes, we need to couple our approximate equation with the heat equation for the temperature which yields the following system of partial differential equations

$$\partial_t T = k\Delta T + \frac{1}{3\kappa}\Delta\rho,$$

$$0 = -\varepsilon^2 \frac{1}{3\kappa}\Delta\rho + \kappa\rho - 4\pi\kappa a T^4.$$

This system has to be supplemented with appropriate initial conditions $T(x,0) = T_0(x)$ and boundary data

$$\frac{h}{\varepsilon k}T + n\cdot\nabla T = \frac{h}{\varepsilon k}u,$$

$$\frac{3\kappa}{2\varepsilon}\rho + n\cdot\nabla\rho = \frac{3\kappa}{2\varepsilon}4\pi a u^4.$$

Here, we assume that we have heat loss over the boundary only due to Newton's cooling law, where h is the heat transfer coefficient, and that we have semi-transparent boundary data for the mean radiative intensity ρ. Further, u denotes the ambient temperature which will act in the following as the control variable.

This leads altogether to an optimal boundary control problem for a parabolic/elliptic system, which can be treated numerically with standard finite element techniques.

4.2.2 Optimal Boundary Control

We intend to minimize cost functionals of tracking-type having the form

$$J(T,u) = \frac{1}{2}\|T - T_d\|^2_{L^2(0,1;L^2(\Omega))} + \frac{\delta}{2}\|u - u_d\|^2_{H^1(0,1;\mathbb{R})}. \tag{4.15}$$

Here, $T_d = T_d(t,x)$ is a specified temperature profile, which is typically given by engineers. In glass manufacturing processes, T_d is used to control chemical reactions in the glass, in particular their activation energy and the reaction time. For the quality of the glass it is essential that these reactions happen spatially homogeneous, such that we will later on require that T_d is constant in space. The control variable u, which is considered to be space-independent, enters the cost functional as regularizing term, where additionally a known cooling curve u_d can be prescribed. The parameter δ allows to adjust the effective heating costs of the cooling process.

The main subject is now the study of the following boundary control problem

$$\begin{aligned} &\min J(T,\rho,u) \text{ w.r.t. } (T,\rho,u),\\ &\text{subject to the } SP_1\text{-system (4.14).} \end{aligned} \tag{4.16}$$

For notational convenience we introduce the following notations and spaces:

$$\begin{aligned} Q &\stackrel{\text{def}}{=} (0,1)\times\Omega, & \Sigma &\stackrel{\text{def}}{=} (0,1)\times\partial\Omega,\\ V &\stackrel{\text{def}}{=} L^2(0,1;H^1(\Omega)), & W &\stackrel{\text{def}}{=} \left\{\phi\in V:\phi_t\in V^*\right\}. \end{aligned}$$

Based on these we set $Y \stackrel{\text{def}}{=} W\times V$ and as the space of controls we choose $U \stackrel{\text{def}}{=} H^1(0,1;\mathbb{R})$. Further, we define $Z \stackrel{\text{def}}{=} V\times V\times L^2(\Omega)$ and $Y_\infty \stackrel{\text{def}}{=} Y\cap[L^\infty(Q)]^2$ as the space of states $y \stackrel{\text{def}}{=} (T,\rho)$. Finally, we set $\alpha=\frac{h}{\varepsilon k}$ and $\gamma=\frac{3\kappa}{2\varepsilon}$.

We define the state/control pair $(y,u)\in Y_\infty\times U$ and the nonlinear operator $e \stackrel{\text{def}}{=} (e_1,e_2,e_3): Y_\infty\times U\to Z^*$ via

$$\begin{aligned} \langle e_1(y,u),\phi\rangle_{V^*,V} \stackrel{\text{def}}{=} {}& \langle\partial_t T,\phi\rangle_{V^*,V} + k\,(\nabla T,\nabla\phi)_{L^2(Q)} + \frac{1}{3\kappa}(\nabla\rho,\nabla\phi)_{L^2(Q)}\\ &+ k\alpha(T-u,\phi)_{L^2(\Sigma)} + \frac{1}{3\kappa}\gamma(\rho-4\pi a u^4,\phi)_{L^2(\Sigma)} \end{aligned} \tag{4.17a}$$

and

$$\begin{aligned} \langle e_2(y,u),\phi\rangle_{V^*,V} \stackrel{\text{def}}{=} {}& \frac{\varepsilon^2}{3\kappa}(\nabla\rho,\nabla\phi)_{L^2(Q)} + \kappa(\rho-4\pi\kappa a T^4,\phi)_{L^2(Q)}\\ &+ \frac{\varepsilon^2}{3\kappa}\gamma(\rho-4\pi a u^4,\phi)_{L^2(\Sigma)} \end{aligned} \tag{4.17b}$$

for all $\phi\in V$. Further, we define $e_3(y,u) \stackrel{\text{def}}{=} T(0)-T_0$.

Remark 4.12 Note, that for $d\le 2$ it is in fact possible to use Y itself as the state space, but for $d=3$ we cannot guarantee that e_2 is well defined due to the fourth-order nonlinearity in T. For the special case $d=1$ the spaces Y and Y_∞ coincide due to Sobolev's embedding theorem (see Sect. 1.14).

Reasonable regularity assumptions on the data ensure the existence of a unique solution to system (4.14) [111].

Theorem 4.6 *Assume that the domain Ω is sufficiently regular and let $u\in U$ and $T_0\in L^\infty(\Omega)$ be given. Then, the SP_1 system $e(y,u)=0$, where e is defined by* (4.17) *has a unique solution $(T,\rho)\in Y_\infty$ and there exists a constant $c>0$ such that the following energy estimate holds*

$$\|T\|_W + \|\rho\|_V \le c\left\{\|T_0\|^4_{L^\infty(\Omega)} + \|u\|^4_U\right\}. \tag{4.18}$$

Further, the solution is uniformly bounded, i.e. $(T,\rho)\in[L^\infty(Q)]^2$ *and we have*

$$\underline{T}\le T\le\overline{T},\qquad \underline{\rho}\le\rho\le\overline{\rho},\tag{4.19}$$

where $\underline{T}=\min(\inf_{t\in(0,1)}u(t),\inf_{x\in\Omega}T_0(x))$ *and* $\overline{T}=\max(\sup_{t\in(0,1)}u(t),\sup_{x\in\Omega}T_0(x))$ *as well as* $\underline{\rho}=4\pi a|\underline{T}|^3\underline{T}$ *and* $\overline{\rho}=4\pi a|\overline{T}|^3\overline{T}$.

Then the minimization problem (4.16) can be shortly written as

$$\begin{aligned}&\min J(y,u)\text{ over }(y,u)\in Y_\infty\times U,\\&\text{subject to }e(y,u)=0\text{ in }Z^*.\end{aligned}\tag{4.20}$$

In fact, one can show the existence of a minimizer.

Theorem 4.7 *There exists a minimizer* $(y^*,u^*)\in Y_\infty\times U$ *of the constrained minimization problem* (4.20).

Proof We have $J_0\stackrel{\text{def}}{=}\inf_{Y_\infty\times U}J(y,u)>-\infty$. We can choose a minimizing sequence $(y_k,u_k)_{k\in\mathbb{N}}\in Y_\infty\times U$. Then, the radial unboundedness of J with respect to u implies that $(u_k)_{k\in\mathbb{N}}$ is bounded in U. Hence, there exists a weakly convergent subsequence, again denoted by $(u_k)_{k\in\mathbb{N}}$ such that

$$u_k\rightharpoonup u^*,\qquad\text{weakly in }U$$

for $k\to\infty$. From Sobolev's embedding theorem (see Sect. 1.14) we deduce that up to a subsequence we also have $u_k\to u^*$ strongly in $C^0(0,1;\mathbb{R})$ for $k\to\infty$. Now, the bounds stated in Theorem 4.6 imply the boundedness of $(\|y_k\|_Y)_{k\in\mathbb{N}}$. Hence, there exist subsequences such that

$$\begin{aligned}T_k&\rightharpoonup T^*,&&\text{weakly in }V,\\\partial_tT_k&\rightharpoonup\partial_tT^*,&&\text{weakly in }V^*,\\\rho_k&\rightharpoonup\rho^*,&&\text{weakly in }V,\end{aligned}$$

for $k\to\infty$, i.e., $y_k=(T_k,\rho_k)\rightharpoonup(T^*,\rho^*)=y^*$ weakly in $W\times V$. The weak lower semi-continuity of J implies

$$J(y^*,u^*)=J_0.$$

Finally, we have to show the constraint $e(y^*,u^*)=0$. Aubin's Lemma [125] implies the strong convergence of $(T_k)_{k\in\mathbb{N}}$ in $L^2(0,1;L^2(\Omega))$. Further, note the uniform boundedness of the solution, which yields

$$(T_k,\rho_k)\rightharpoonup(T^*,\rho^*),\qquad\text{weak-*-ly in }L^\infty(Q),$$

for $k\to\infty$. These convergences are by far sufficient to pass to the limit in (4.17), yielding

$$e(y^*,u^*)=0\quad\text{in }Z^*,$$

which finally proves the assertion.

4.2.2.1 Derivatives

In the following we provide the derivative information, which is necessary for the application of the Newton method for the reduced problem. Owing to the fact that the system (4.14) is uniquely solvable, we may reformulate the minimization problem (4.20) introducing the *reduced cost functional* $\hat{J}$ as

$$\begin{aligned} &\text{minimize} \quad \hat{J}(u) \stackrel{\text{def}}{=} J(y(u), u) \quad \text{over} \quad u \in U \\ &\text{where } y(u) \in Y \text{ satisfies } e(y(u), u) = 0. \end{aligned} \tag{4.21}$$

The numerical realization of Newton's method relies on derivative information on J and e, or $\hat{J}$, respectively. Following the discussion in Sect. 1.6.2, these can derived as follows: First, the implicit function theorem leads to the following derivative of y at u in a direction δu:

$$y'(u)\delta u = -e_y^{-1}(y(u), u)e_u(y(u), u)\delta u.$$

Using the chain rule one obtains

$$\left\langle \hat{J}'(u), \delta u \right\rangle_{U^*,U} = \left\langle J_u(y(u), u) - e_u^*(y(u), u)e_y^{-*}(y(u), u)J_y(y(u), u), \delta u \right\rangle_{U^*,U}.$$

Here, $e_y^*(y, u)\xi$ denotes the adjoint of the linearization of e at (y, u) in the direction ξ. We define the adjoint variable $\xi = (\xi_T, \xi_\rho, \xi_{T0}) \in Z$ by

$$\xi = -e_y^{-*}(y(u), u)J_y(y(u), u) \in Z.$$

Assuming enough regularity of the solution one gets the derivative

$$\hat{J}'(u) = J_u(y(u), u) + e_u^*(y(u), u)\xi. \tag{4.22}$$

In case of the cost functional (4.15), the adjoint variable can be characterized [111] as the variational solution of

$$-\partial_t \xi_T = k\Delta\xi_T + 16\pi a\kappa T^3 \xi_\rho - (T - T_d), \tag{4.23a}$$

$$-\frac{\varepsilon^2}{3\kappa}\Delta\xi_\rho + \kappa\xi_\rho = \frac{1}{3\kappa}\Delta\xi_T, \quad \text{in } Q \tag{4.23b}$$

with boundary conditions

$$k(n \cdot \nabla\xi_T + \alpha\xi_T) = 0, \tag{4.23c}$$

$$n \cdot \nabla\xi_T + \gamma\xi_T + \varepsilon^2(n \cdot \nabla\xi_\rho + \gamma\xi_\rho) = 0, \quad \text{on } \Sigma \tag{4.23d}$$

and terminal condition

$$\xi_T(1) = 0 \quad \text{in } \Omega. \tag{4.23e}$$

Introducing the Lagrangian $L : Y_\infty \times U \times Z \to \mathbb{R}$ associated to (4.20) defined by

$$L(y,u,\xi) \stackrel{\text{def}}{=} J(y,u) + \langle e(y,u), \xi \rangle_{Z^*,Z} \cdot$$

we know that there exists a critical point of the Lagrangian. In fact, for an optimal solution there exists a unique Lagrange multiplier [111].

Theorem 4.8 *Let $(y^*, u^*) \in Y_\infty \times U$ denote an optimal solution. Then there exists a unique Lagrange multiplier $\xi^* \in Z^*$ such that the triple (y^*, u^*, ξ^*) satisfies*

$$L'(y^*, u^*, \xi^*) = 0.$$

Let $(y^*, u^*) \in Y_\infty \times U$ denote an optimal solution. Following the discussion in Sect. 1.6.5, the second derivative of the Lagrangian is given by

$$\begin{aligned} &L''(y^*, u^*, \xi^*) \\ &= \begin{pmatrix} J_{yy}(y^*,u^*) + \langle e_{yy}(y^*,u^*)(\cdot,\cdot), \xi^* \rangle & 0 & e_y^*(y^*,u^*) \\ 0 & J_{uu}(y^*,u^*) + \langle e_{uu}(y^*,u^*)(\cdot,\cdot), \xi^* \rangle & e_u^*(y^*,u^*) \\ e_y(y^*,u^*) & e_u(y^*,u^*) & 0 \end{pmatrix}. \end{aligned}$$

Defining the operator

$$T(x,u) \stackrel{\text{def}}{=} \begin{pmatrix} -e_y^{-1}(y,u) e_u(y,u) \\ id_U \end{pmatrix}$$

we can write the reduced Hessean as

$$\hat{J}''(u) \stackrel{\text{def}}{=} T^*(y,u) L_{ww}(w,\xi) T(y,u), \tag{4.24}$$

where $w \stackrel{\text{def}}{=} (y,u)$, i.e., L_{ww} is the upper left 2×2-block of L''.

4.2.2.2 Newton's Method

In this section we describe the second order optimization algorithm, i.e., we apply Newton's method for the computation of an optimal control for the reduced cost functional. The algorithm reads formally

Algorithm 4.9

1. Choose u_0 in a neighborhood of u^*.
2. For $k = 0, 1, 2, \ldots$
   a. Solve $\hat{J}''(u_k)\delta u_k = -\hat{J}'(u_k)$,

b. Set $u_{k+1} = u_k + \delta u_k$.

Remark 4.13 The solution of the system in step (ii.a) is done again iteratively using a conjugate gradient algorithm embedded inside the Newton algorithm (compare Algorithm 4.5).

In particular, for the cost functional (4.15) one has to apply successively the following steps [112]

1. Solve the linearized state system (see system 4.14)

$$\partial_t v_T = k\Delta v_T + \frac{1}{3\kappa}\Delta v_\rho \tag{4.25a}$$

$$-\frac{\varepsilon^2}{3\kappa}\Delta v_\rho + \kappa v_\rho = 16\pi\kappa a T_k^3 v_T \tag{4.25b}$$

with boundary conditions

$$n \cdot \nabla v_T + \alpha v_T = \alpha \delta u_k^j \tag{4.25c}$$

$$n \cdot \nabla v_\rho + \gamma v_\rho = \gamma 16\pi a u_k^3 \delta u_k^j \tag{4.25d}$$

and initial condition

$$v_T(0) = 0 \tag{4.25e}$$

for $v_k^j \stackrel{\text{def}}{=} (v_T, v_\rho) \in Y$, where $y_k = (T_k, \rho_k)$ is given.

2. Evaluate

$$J_{yy}(y_k, u_k)(v_k^j, \cdot) + \left\langle e_{yy}(y_k, u_k)(v_k^j, \cdot), \xi_k \right\rangle = v_T + 48\pi\kappa a T_k^2 v_T \xi_{T,k}.$$

3. Solve the linearized adjoint system (see system 4.23)

$$-\partial_t w_T = k\Delta w_T + 16\pi\kappa a T_k^3 w_\rho + v_T - 48\pi\kappa a T_k^2 v_T \xi_{T,k} \tag{4.26a}$$

$$-\frac{\varepsilon^2}{3\kappa}\Delta w_\rho + \kappa w_\rho = \frac{1}{3\kappa}\Delta w_T \tag{4.26b}$$

with boundary conditions

$$k(n \cdot \nabla w_T + \alpha w_T) = 0 \tag{4.26c}$$

$$\varepsilon^2(n \cdot \nabla w_\rho + \gamma w_\rho) + n \cdot \nabla w_T + \gamma w_T = 0 \tag{4.26d}$$

and terminal condition

$$w_T(1) = 0 \tag{4.26e}$$

for $w_k^j \stackrel{\text{def}}{=} (w_T, w_\rho) \in Y$.

4. Set

$$q_k^j(t) = \frac{1}{|\partial\Omega|}\int_{\partial\Omega} k\alpha w_T + \frac{\gamma 16\pi a}{3\kappa} u^2(u(w_T + \varepsilon^2 w_\rho) - 3\delta u_k^j(\xi_T + \varepsilon^2\xi_\rho))\,ds$$
$$+ \frac{\delta}{|\partial\Omega|}\int_{\partial\Omega} \delta u_k^j + \partial_{tt}\delta u_k^j\,ds.$$

4.2.3 Numerical Results

The spatial discretization of the PDEs is based on linear finite elements. We use a non-uniform grid with an increasing point density toward the boundary of the medium, consisting of 109 points. The temporal discretization uses a uniform grid consisting of 180 points for the temperature-tracking problem. We employ the implicit backward Euler method to compute the state (T, ρ). The adjoint systems are discretized using a modified implicit Euler backward method taking into account the symmetry of the discrete reduced Hessean [61].

The conjugate gradient algorithm was terminated when the norm of the gradient became sufficiently small; to be more precise, in the j-th conjugate gradient step for the computation of the update in Newton step k we stop if the residual r_k^j satisfies

$$\frac{\|r_k^j\|}{\|J'(u^0)\|} \le \min\left\{\left(\frac{\|\hat{J}'(u_k)\|}{\|\hat{J}'(u^0)\|}\right)^p, q\frac{\|\hat{J}'(u_k)\|}{\|\hat{J}'(u^0)\|}\right\} \quad \text{or} \quad j \ge 100. \tag{4.27}$$

Note, that p determines the order of the outer Newton algorithm, such that p should be chosen in the open interval $(1, 2)$. The value of q is important for the first step of Newton's method, as for $k = 0$ the norm quotients are all 1; for later steps, the

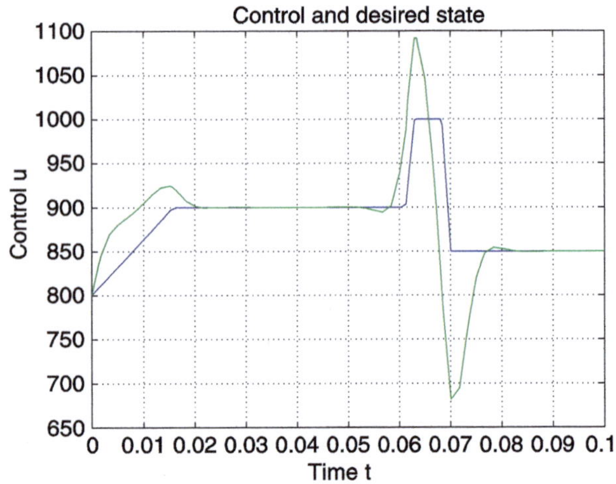

Fig. 4.9 Unoptimized (*dark*) and optimized (*light*) cooling profile

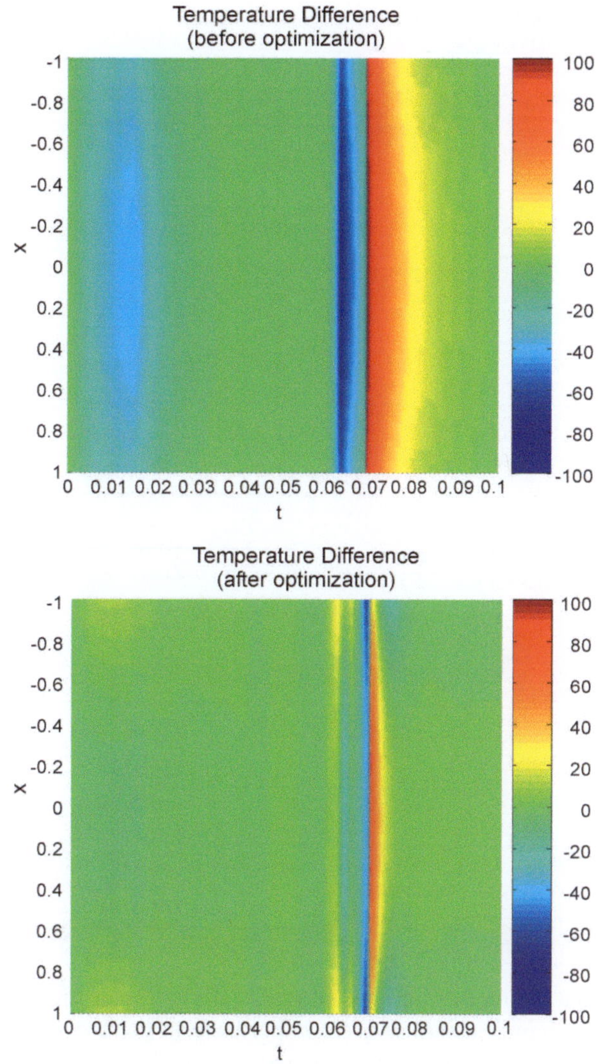

Fig. 4.10 Temperature differences for the uncontrolled (*top*) and controlled (*bottom*) state

influence of p becomes increasingly dominant. In our numerical experiments, $p = 1.5$ and $q = 0.1$ proved to be a suitable choice.

Remark 4.14 In the Newton algorithm, one might use

$$J_{uu}(u) = \delta(I - \partial_{tt})$$

as a preconditioning operator for the Newton system (ii.a).

Now we present numerical results underlining the feasibility of our approach. For a given (time dependent) temperature profile T_d we compute an optimal u such

Table 4.1 Convergence statistics for $\delta = 3.5 \times 10^{-7}$

k	$J(u_k)$	$\|\hat{J}'(u_{k+1})\|_2$	#cg
1	224.7359	$1.605777 \times 10^{+01}$	26
2	184.5375	$1.306437 \times 10^{+01}$	16
3	142.9351	$1.038065 \times 10^{+01}$	14
4	112.5493	$7.985859 \times 10^{+00}$	13
5	90.52294	$5.861017 \times 10^{+00}$	13
6	74.95062	$3.989118 \times 10^{+00}$	14
7	64.45030	$2.357674 \times 10^{+00}$	14
8	57.97925	9.541926×10^{-01}	16
9	54.70802	4.762934×10^{-02}	17
10	53.96191	5.101231×10^{-04}	17
11	53.96017	2.086531×10^{-06}	17
12	53.96017	1.590937×10^{-09}	25

Table 4.2 Convergence statistics for $\delta = 3.5 \times 10^{-6}$

k	$J(u_k)$	$\|\hat{J}'(u_{k+1})\|_2$	#cg
1	337.5395	$3.912697 \times 10^{+01}$	29
2	254.1703	$2.918320 \times 10^{+01}$	27
3	193.4364	$1.978074 \times 10^{+01}$	27
4	151.9171	$1.094969 \times 10^{+01}$	28
5	126.9592	$2.751367 \times 10^{+00}$	29
6	116.2621	3.163388×10^{-02}	29
7	115.4742	2.184202×10^{-04}	31
8	115.4741	4.352735×10^{-07}	37
9	115.4741	4.542256×10^{-09}	28

that the temperature of the glass follows the desired profile T_d as good as possible. Such profiles are of great importance in glass manufacturing in order to control at which time, at which place and for how long certain chemical reactions take place, which is essential for the quality of the glass. Intervals of constant temperature allow for lengthy reactions in a controlled manner; short peaks of high temperature trigger reactions that have a high activation energy. In particular, it can be desirable to attain a spatially constant temperature, which is in contradiction to the boundary layers of the temperature due the radiative heat loss over the boundary.

The dark line in Fig. 4.9 describes the desired temperature profile $T_d(t)$ which shall be attained homogeneously in space. From the engineering point of view it is an educated guess to use the same profile for the boundary control. Clearly, this leads to deviations which can be seen in the left graphic of Fig. 4.10. Our optimal control approach results now in the light line in Fig. 4.9, which yields in turn the improved temperature differences on the right in Fig. 4.10. One realizes a significant improvement although we have still a large peak. But note that we want to resolve a very sharp jump in the temperature. Due to diffusive part of the equations it is almost impossible to resolve such fast change in the cooling.

Finally, let us discuss the influence of the penalizing parameter δ on the convergence of the iterative Newton method. In Tables 4.1 and 4.2 we compare the number of Newton iterations, the evolution of the cost functional and the residual as well as number of cg iteration in each Newton step. As expected we get a better performance for the "more convex" problem. Note that we used a globalization strategy based on the trust-region Newton-CG iteration in the Steiaug variant [127] to ensure the global convergence of the Newton iteration. This can be also seen in Table 4.1 where we reach the region of quadratic convergence after seven iterations.

References

1. Adams, R.A.: Sobolev Spaces. Academic, San Diego (1975)
2. Agmon, S., Douglis, A., Nirenberg, L.: Estimates near the boundary for solutions of elliptic partial differential equations satisfying general boundary conditions. Commun. Pure Appl. Math. **12**, 623–727 (1959)
3. Allgower, E.L., Böhmer, K., Potra, F.A., Rheinboldt, W.C.: A mesh-independence principle for operator equations and their discretizations. SIAM J. Numer. Anal. **23**, 160–169 (1986)
4. Alt, H.W.: Lineare Funktionalanalysis. Springer, Berlin (1999)
5. Alt, W.: The Lagrange-Newton method for infinite-dimensional optimization problems. Numer. Funct. Anal. Optim. **11**, 201–224 (1990)
6. Alt, W.: Discretization and mesh-independence of Newton's method for generalized equations, in Mathematical programming with data perturbations, pp. 1–30. Dekker, New York (1998)
7. Apel, T., Rösch, A., Winkler, G.: Optimal control in non-convex domains: a priori discretization error estimates. Calcolo **44**(3), 137–158 (2007)
8. Arada, N., Casas, E., Tröltzsch, F.: Error estimates for the numerical approximation of a semilinear elliptic control problem. Comput. Optim. Appl. **23**, 201–229 (2002)
9. Becker, R., Rannacher, R.: An optimal control approach to a posteriori error estimation in finite element methods. Acta Numer. **37**, 1–225 (2001)
10. Benedix, O., Vexler, B.: A posteriori error estimation and adaptivity for elliptic optimal control problems with state constraints. Priority Programme 1253, Preprint SPP1253-02-02 (2007)
11. Berggren, M.: Approximation of very weak solutions to boundary value problems. SIAM J. Numer. Anal. **42**, 860–877 (2004)
12. Bergounioux, M., Kunisch, K.: Primal-dual strategy for state-constrained optimal control problems. Comput. Optim. Appl. **22**, 193–224 (2002)
13. Bergounioux, M., Ito, K., Kunisch, K.: Primal-dual strategy for constrained optimal control problems. SIAM J. Control Optim. **37**, 1176–1194 (1999)
14. Bertsekas, D.P.: Nonlinear Programming, 2nd edn. Athena Scientific (1999)
15. Bonnans, J.F., Shapiro, A.: Optimization problems with perturbations: A guided tour. SIAM Rev. **40**, 228–264 (1998)
16. Brezzi, F., Marini, L.D., Micheletti, S., Pietra, P., Sacco, R., Wang, S.: Discretization of semiconductor device problems. I. In: Schilders, W.H.A., et al. (eds.) Special Volume: Numerical Methods in Electromagnetics. Handbook of Numerical Analysis, vol. XIII, pp. 317–441. Elsevier/North-Holland, Amsterdam (2005)
17. Burger, M., Pinnau, R.: Optimization Models for Semiconductor Devices. Birkhäuser (2009, to be published)
18. Burger, M., Engl, H.W., Markowich, P., Pietra, P.: Identification of doping profiles in semiconductor devices. Inverse Probl. **17**, 1765–1795 (2001)
19. Burger, M., Engl, H.W., Markowich, P.: Inverse doping problems for semiconductor devices. In: Tang, T., Xu, J.A., Ying, L.A., Chan, T.F., Huang, Y. (eds.) Recent Progress in Computational and Applied PDEs, pp. 39–54. Kluwer Academic, Dordrecht (2002)
20. Casas, E.: L^2 estimates for the finite element method for the Dirichlet problem with singular data. Numer. Math. **47**, 627–632 (1985)
21. Casas, E.: Control of an elliptic problem with pointwise state constraints. SIAM J. Cont. Optim. **4**, 1309–1322 (1986)
22. Casas, E.: Boundary control of semilinear elliptic equations with pointwise state constraints. SIAM J. Cont. Optim. **31**, 993–1006 (1993)

23. Casas, E.: Pontryagin's principle for state-constrained boundary control problems of semilinear parabolic equations. SIAM J. Cont. Optim. **35**, 1297–1327 (1997)
24. Casas, E.: Error estimates for the numerical approximation of semilinear elliptic control problems with finitely many state constraints. ESAIM, Control Optim. Calc. Var. **8**, 345–374 (2002)
25. Casas, E., Fernández, L.: Optimal control of semilinear elliptic equations with pointwise constraints on the gradient of the state. Appl. Math. Optimization **27**, 35–56 (1993)
26. Casas, E., Mateos, M.: Uniform convergence of the FEM. Applications to state constrained control problems. Comput. Appl. Math. **21** (2002)
27. Casas, E., Mateos, M.: Error estimates for the numerical approximation of Neumann control problems. Comput. Appl. Math. (to appear)
28. Casas, E., Raymond, J.P.: Error estimates for the numerical approximation of Dirichlet Boundary control for semilinear elliptic equations. SIAM J. Control Optim. **45**, 1586–1611 (2006)
29. Casas, E., Tröltzsch, F.: Error estimates for the finite element approximation of a semilinear elliptic control problems. Contr. Cybern. **31**, 695–712 (2005)
30. Casas, E., Mateos, M., Tröltzsch, F.: Error estimates for the numerical approximation of boundary semilinear elliptic control problems. Comput. Optim. Appl. **31**, 193–219 (2005)
31. Casas, E., de Los Reyes, J.C., Tröltzsch, F.: Sufficient second-order optimality conditions for semilinear control problems with pointwise state constraints. SIAM J. Optim. (to appear)
32. Choudhary, M.K., Huff, N.T.: Mathematical modeling in the glass industry: An overview of status and needs. Glastech. Ber. Glass Sci. Technol. **70**, 363–370 (1997)
33. Ciarlet, P.G.: The Finite Element Method for Elliptic Problems. North-Holland, Amsterdam (1978)
34. Clarke, F.H.: Optimization and Nonsmooth Analysis. Wiley, New York (1983)
35. Conway, J.B.: A Course in Functional Analysis. Springer, Berlin (1990)
36. Dennis, J.E., Schnabel, R.B.: Numerical Methods for Unconstrained Optimization and Nonlinear Equations. Prentice Hall, Englewood Cliffs (1983)
37. Deckelnick, K., Hinze, M.: Error estimates in space and time for tracking-type control of the instationary Stokes system. ISNM **143**, 87–103 (2002)
38. Deckelnick, K., Hinze, M.: Semidiscretization and error estimates for distributed control of the instationary Navier-Stokes equations. Numer. Math. **97**, 297–320 (2004)
39. Deckelnick, K., Hinze, M.: Convergence of a finite element approximation to a state constrained elliptic control problem. SIAM J. Numer. Anal. **45**, 1937–1953 (2007)
40. Deckelnick, K., Hinze, M.: A finite element approximation to elliptic control problems in the presence of control and state constraints. Hamburger Beiträge zur Angewandten Mathematik HBAM2007-01 (2007)
41. Deckelnick, K., Hinze, M.: Numerical analysis of a control and state constrained elliptic control problem with piecewise constant control approximations. In: Proceedings of the ENUMATH (2007)
42. Deckelnick, K., Günther, A., Hinze, M.: Finite element approximations of elliptic control problems with constraints on the gradient. Numer. Math. DOI:10.1007/s00211-008-0185-3 (2008)
43. Deckelnick, K., Günther, A., Hinze, M.: Finite element approximation of Dirichlet boundary control for elliptic PDEs on two- and three-dimensional curved domains. Priority Programme 1253, Preprint-Number SPP1253-08-05 (2008)
44. Deuflhard, P., Potra, F.A.: Asymptotic mesh independence of Newton-Galerkin methods via a refined Mysovskiuı theorem. SIAM J. Numer. Anal. **29**, 1395–1412 (1992)
45. Dontchev, A.L., Hager, W.W., Veliov, V.: Uniform convergence and mesh independence of Newton's method for discretized variational problems. SIAM J. Control Optim. **39**, 961–980 (2000)
46. Eriksson, K., Johnson, C.: Adaptive finite element methods for parabolic problems II: Optimal error estimates in $L_\infty L_2$ and $L_\infty L_\infty$. SIAM J. Numer. Anal. **32**, 706–740 (1995)
47. Evans, L.C.: Partial Differential Equations. American Mathematical Society, Providence (1998)

48. Falk, R.: Approximation of a class of optimal control problems with order of convergence estimates. J. Math. Anal. Appl. **44**, 28–47 (19973)
49. Fang, W., Cumberbatch, E.: Inverse problems for metal oxide semiconductor field-effect transistor contact resistivity. SIAM J. Appl. Math. **52**, 699–709 (1992)
50. Fang, W., Ito, K.: Reconstruction of semiconductor doping profile from laser-beam-induced current image. SIAM J. Appl. Math. **54**, 1067–1082 (1994)
51. Gaevskaya, A., Hoppe, R.H.W., Repin, S.: Functional approach to a posteriori error estimation for elliptic optimal control problems with distributed control. J. Math. Sci. **144**, 4535–4547 (2007)
52. Gajewski, H., Gröger, K., Zacharias, K.: Nichtlineare Operatorgleichungen und Operatordifferentialgleichungen. Akademie-Verlag, Berlin (1974)
53. Gastaldi, L., Nochetto, R.H.: On L^∞-accuracy of mixed finite element methods for second order elliptic problems. Math. Appl. Comput. **7**, 13–39 (1988)
54. Geveci, T.: On the approximation of the solution of an optimal control problem governed by an elliptic equation. Math. Model. Numer. Anal. **13**, 313–328 (1979)
55. Gilbarg, D., Trudinger, N.S.: Elliptic Partial Differential Equations of Second Order, 2nd edn. Springer, Berlin (1983)
56. Goldberg, H., Tröltzsch, F.: On a Lagrange-Newton method for a nonlinear parabolic boundary control problem. Optim. Methods Softw. **8**, 225–247 (1998)
57. Griesse, R., Metla, N., Rsch, A.: Local quadratic convergence of SQP for elliptic optimal control problems with nonlinear mixed control-state constraints. Preprint, RICAM (2007)
58. Günther, A., Hinze, M.: A posteriori error control of a state constrained elliptic control problem. J. Numer. Math. DOI:10.1515/JNUM.2008.00 (2008)
59. Gunzburger, M.D., Manservisi, S.: Analysis and approximation of the velocity tracking problem for Navier-Stokes flows with distributed control. SIAM J. Numer. Anal. **37**, 1481–1512 (2000)
60. Gunzburger, M.D., Manservisi, S.: The velocity tracking problem for Navier-Stokes flows with boundary controls. SIAM J. Control Optim. **39**, 594–634 (2000)
61. Hager, W.W.: Runge-Kutta methods in optimal control and the transformed adjoint system. Numer. Math. **87**(2), 247–282 (2000)
62. Hackbusch, W.: Multi-Grid Methods and Applications. Springer, New York (1985)
63. Herty, M., Pinnau, R., Seaid, M.: On optimal control problems in radiative transfer. OMS **22**, 917–936 (2007)
64. Herty, M., Pinnau, R., Thömmes, G.: Asymptotic and discrete concepts for optimal control in radiative transfer. ZAMM **87**, 333–347 (2007)
65. Hintermüller, M., Hinze, M.: A note on an optimal parameter adjustment in a Moreau-Yosida-based approach to state constrained elliptic control problems. DFG Priority Programme 1253, Preprint SPP1253-08-04 (2008)
66. Hintermüller, M., Kunisch, K.: Feasible and noninterior path-following in constrained minimization with low multiplier regularity. SIAM J. Control Optim. **45**, 1198–1221 (2006)
67. Hintermüller, M., Kunisch, K.: Path-following methods for a class of constrained minimization problems in function space. SIAM J. Optim. **17**, 159–187 (2006)
68. Hintermüller, M., Ulbrich, M.: A mesh-independence result for semismooth Newton methods. Math. Program. **101**, 151–184 (2004)
69. Hintermüller, M., Ito, K., Kunisch, K.: The primal-dual active set strategy as a semi-smooth Newton method. SIAM J. Optim. **13**, 865–888 (2003)
70. Hinze, M.: Optimal and instantaneous control of the instationary Navier-Stokes equations. Habilitationsschrift, Fachbereich Mathematik, Technische Universität, Berlin (2000)
71. Hinze, M.: A variational discretization concept in control constrained optimization: The linear-quadratic case. Comput. Optim. Appl. **30**, 45–63 (2005)
72. Hinze, M., Kunisch, K.: Second order methods for optimal control of time-dependent fluid flow. SIAM J. Control Optim. **40**, 925–946 (2001)
73. Hinze, M., Matthes, U.: A note on variational discretization of elliptic Neumann boundary control. Hamburger Beiträge zur Angewandten Mathematik 2008-01 (2008)

74. Hinze, M., Meyer, C.: Variational discretization of Lavrentiev-regularized state constrained elliptic control problems. Comput. Optim. Appl. DOI:10.1007/s10589-008-9198-1 (2008)
75. Hinze, M., Pinnau, R.: Optimal control of the drift diffusion model for semiconductor devices. In: Hoffmann, K.-H., Lasiecka, I., Leugering, G., Sprekels, J. (eds.) Optimal Control of Complex Structures. ISNM, vol. 139, pp. 95–106. Birkhäuser, Basel (2001)
76. Hinze, M., Pinnau, R.: An optimal control approach to semiconductor design. Math. Models Methods Appl. Sci. **12**(1), 89–107 (2002)
77. Hinze, M., Pinnau, R.: Mathematical tools in optimal semiconductor design. Bull. Inst. Math. Acad. Sin. (N.S.) **2**, 569–586 (2007)
78. Hinze, M., Pinnau, R.: A second order approach to optimal semiconductor design. J. Optim. Theory Appl. **133**, 179–199 (2007)
79. Hinze, M., Schiela, A.: Discretization of interior point methods for state constrained elliptic optimal control problems: optimal error estimates and parameter adjustment. Priority Program 1253, Preprint SPP1253-08-03 (2007)
80. Hinze, M., Vierling, M.: Semi-discretization and semi-smooth Newton methods; implementation, convergence and globalization in pde constrained optimization with control constraints. Preprint (2008, to appear)
81. Hoppe, R.H.W., Kieweg, M.: A posteriori error estimation of finite element approximations of pointwise state constrained distributed control problems. SIAM J. Control Optim. (2007, to appear)
82. Jäger, H., Sachs, E.W.: Global convergence of inexact reduced SQP methods. Optim. Methods Softw. **7**, 83–110 (1997)
83. Jost, J.: Postmodern Analysis. Springer, Berlin (1998)
84. Kelley, C.T.: Iterative Methods for Optimization. SIAM, Philadelphia (1999)
85. Kelley, C.T., Sachs, E.W.: Multilevel algorithms for constrained compact fixed point problems. SIAM J. Sci. Comput. **15**, 645–667 (1994)
86. Kinderlehrer, D., Stampacchia, G.: Introduction to Variational Inequalities and their Applications. Academic, San Diego (1980)
87. Kummer, B.: Newton's method for nondifferentiable functions. In: Advances in Mathematical Optimization, pp. 114–125. Akademie-Verlag, Berlin (1988)
88. Larsen, E.W., Thömmes, G., Seaid, M., Götz, Th., Klar, A.: Simplified P_N approximations to the equations of radiative heat transfer and applications to glass manufacturing. J. Comput. Phys. **183**(2), 652–675 (2002)
89. Lee, W.R., Wang, S., Teo, K.L.: An optimization approach to a finite dimensional parameter estimation problem in semiconductor device design. J. Comput. Phys. **156**, 241–256 (1999)
90. Lions, J.-L.: Optimal Control of Systems Governed by Partial Differential Equations. Springer, Berlin (1971)
91. Liu, W., Yan, N.: A posteriori error estimates for nonlinear elliptic control problems. Appl. Numer. Anal. **47**, 173–187 (2003)
92. Luenberger, D.G.: Linear and Nonlinear Programming, 2nd edn. Addison–Wesley, Reading (1989)
93. Malanowski, K.: Convergence of approximations vs. regularity of solutions for convex, control constrained optimal-control problems. Appl. Math. Optim. **8**, 69–95 (1981)
94. May, S., Rannacher, R., Vexler, B.: A priori error analysis for the finite element approximation of elliptic Dirichlet boundary control problems. In: Proceedings of ENUMATH 2007, Graz (2008)
95. Markowich, P.A.: The Stationary Semiconductor Device Equations, 1st edn. Springer, Wien (1986)
96. Markowich, P.A., Ringhofer, Ch.A., Schmeiser, Ch.: Semiconductor Equations, 1st edn. Springer, Wien (1990)
97. Meidner, D., Vexler, B.: A priori error estimates for space-time finite element discretization of parabolic optimal control problems. Part I: Problems without control constraints. SIAM J. Control Optim. **47**, 1150–1177 (2008)

98. Meidner, D., Vexler, B.: A priori error estimates for space-time finite element discretization of parabolic optimal control problems. Part II: Problems with control constraints. SIAM J. Control Optim. **47**, 1301–1329 (2008)
99. Meyer, C.: Error estimates for the finite element approximation of an elliptic control problem with pointwise constraints on the state and the control. WIAS Preprint 1159 (2006)
100. Meyer, C., Rösch, A.: Superconvergence properties of optimal control problems. SIAM J. Control Optim. **43**, 970–985 (2004)
101. Meyer, C., Rösch, A.: L^{∞}-estimates for approximated optimal control problems. SIAM J. Control Optim. **44**, 1636–1649 (2005)
102. Meyer, C., Rösch, A., Tröltzsch, F.: Optimal control problems of PDEs with regularized pointwise state constraints. Preprint 14, Inst. f. Mathematik, TU Berlin. Comput. Optim. Appl. (2004, to appear)
103. Meyer, C., Rösch, A., Tröltzsch, F.: Optimal control of PDEs with regularized pointwise state constraints. Comput. Optim. Appl. **33**, 209–228 (2006)
104. Meyer, C., Prüfert, U., Tröltzsch, F.: On two numerical methods for state-constrained elliptic control problems. Optim. Methods Softw. **22**, 871–899 (2007)
105. Mifflin, R.: Semismooth and semiconvex functions in constrained optimization. SIAM J. Control Optim. **15**, 959–972 (1977)
106. Mock, M.S.: Analysis of Mathematical Models of Semiconductor Devices, 1st edn. Boole Press, Dublin (1983)
107. Modest, M.F.: Radiative Heat Transfer. McGraw–Hill, New York (1993)
108. Murthy, M.K.V., Stampacchia, G.: A variational inequality with mixed boundary conditions. Israel J. Math. **13**, 188–224 (1972)
109. Naumann, J., Wolff, N.: A uniqueness theorem for weak solutions of the stationary semiconductor equations. Appl. Math. Optim. **24**, 223–232 (1991)
110. Nielsen, O.A.: An Introduction to Integration and Measure Theory. Wiley, New York (1997)
111. Pinnau, R.: Analysis of optimal boundary control for radiative heat transfer modelled by the SP_1-system. Commun. Math. Sci. **5**, 951–969 (2007)
112. Pinnau, R., Schulze, A.: Newton's method for optimal temperature-tracking of glass cooling processes. IPSE **15**(4), 303–323 (2007)
113. Qi, L., Sun, J.: A nonsmooth version of Newton's method. Math. Program. **58**, 353–367 (1993)
114. Raymond, J.-P., Zidani, H.: Hamiltonian Pontryagin's principle for control problems governed by semilinear parabolic equations. Appl. Math. Optim. **39**, 143–177 (1999)
115. Renardy, M., Rogers, R.C.: An Introduction to Partial Differential Equations. Springer, Berlin (1993)
116. Robinson, S.M.: Stability theory for systems of inequalities in nonlinear programming, part II: differentiable nonlinear systems. SIAM J. Num. Anal. **13**, 497–513 (1976)
117. Robinson, S.M.: Strongly regular generalized equations. Math. Oper. Res. **5**, 43–62 (1980)
118. Rösch, A.: Error estimates for parabolic optimal control problems with control constraints. Z. Anal. Ihre Anwend. ZAA **23**, 353376 (2004)
119. Schatz, A.H.: Pointwise error estimates and asymptotic error expansion inequalities for the finite element method on irregular grids. I: Global estimates. Math. Comput. **67**(223), 877–899 (1998)
120. Scharfetter, D.L., Gummel, H.K.: Large signal analysis of a silicon read diode oscillator. IEEE Trans. Electr. Dev. **15**, 64–77 (1969)
121. Schiela, A.: Barrier methods for optimal control problems with state constraints. Konrad Zuse Zentrum Berlin, ZIB Report 07-07 (2007)
122. Schiela, A., Weiser, M.: Superlinear convergence of the control reduced interior point method for PDE constrained optimization. Comput. Optim. Appl. (2008, to appear)
123. Scholtes, S.: Introduction to piecewise differentiable equations. Technical report no. 53/1994, Universitat Karlsruhe, Institut fur Statistik und Mathematische Wirtschaftstheorie (1994)
124. Selberherr, S.: Analysis and Simulation of Semiconductor Devices. Springer, Wien (1984)
125. Simon, J.: Compact sets in the space $L^p(0, T; B)$. Ann. Math. Pura Appl. **146**, 65–96 (1987)

126. Stampaccia, G.: Contributi alla regolarizzazione delle soluzioni dei problemi al contorno per secondo ordine ellittiche. Ann. Sc. Norm. Super. Pisa **12**, 223–245 (1958)
127. Steihaug, T.: The conjugate gradient method and trust region in large scale optimization. SIAM J. Numer. Anal. **20**(3), 626–637 (1983)
128. Stockinger, M., Strasser, R., Plasun, R., Wild, A., Selberherr, S.: A qualitative study on optimized MOSFET doping profiles. In: Proceedings SISPAD 98 Conf., pp. 77–80 (1998)
129. Sze, S.M.: Physics of Semiconductor Devices, 2nd edn. Wiley, New York (1981)
130. Temam, R.: Navier-Stokes Equations, 3rd edn. North-Holland, Amsterdam (1984)
131. Thömmes, G.: Radiative heat transfer equations for glass cooling problems: Analysis and numerics. PhD Thesis (2002)
132. Thomée, V.: Galerkin Finite Element Methods for Parabolic Problems. Springer, Berlin (1997)
133. Tröltzsch, F.: Optimale Steuerung mit partiellen Differentialgleichungen. (2005)
134. Ulbrich, M.: Nonsmooth Newton-like methods for variational inequalities and constrained optimization problems in function spaces. Habilitationsschrift, Zentrum Mathematik, Technische Universität München, München, Germany (2001)
135. Ulbrich, M.: On a nonsmooth Newton method for nonlinear complementarity problems in function space with applications to optimal control. In: Complementarity: Applications, Algorithms and Extensions, Madison, WI, 1999, pp. 341–360. Kluwer Academic, Dordrecht (2001)
136. Ulbrich, M.: Semismooth Newton methods for operator equations in function spaces. SIAM J. Optim. **13**, 805–841 (2003)
137. Ulbrich, M.: Constrained optimal control of Navier-Stokes flow by semismooth Newton methods. Syst. Control Lett. **48**, 297–311 (2003)
138. Ulbrich, M., Ulbrich, S.: Superlinear convergence of affine-scaling interior-point Newton methods for infinite-dimensional nonlinear problems with pointwise bounds. SIAM J. Control Optim. **38**, 1938–1984 (2000)
139. Ulbrich, M., Ulbrich, S.: A multigrid semismooth Newton method for contact problems in linear elasticity. Preprint, TU München (2008)
140. Ulbrich, M., Ulbrich, S.: Primal-dual interior-point methods for PDE-constrained optimization. Math. Program. **117**, 435–485 (2009)
141. Unterreiter, A.: Halbleitergleichungen. Skript, TU Kaiserslautern
142. Vexler, B.: Finite element approximation of elliptic Dirichlet optimal control problems. Numer. Funct. Anal. Optim. **28**, 957–975 (2007)
143. Vexler, B., Wollner, W.: Adaptive finite elements for elliptic optimization problems with control constraints. SIAM J. Contr. Optim. **47**, 509–534 (2008)
144. Walter, W.: Gewöhnliche Differentialgleichungen. Springer, Berlin (1986)
145. Weiser, M., Gänzler, T., Schiela, A.: A control reduced primal interior point method for PDE constrained optimization. Comput. Optim. Appl. (2008, to appear)
146. Wloka, J.: Funktionalanalysis unf ihre Anwendungen. De Gruyter (1971)
147. Wollner, W.: Adaptive finite elements and interior point methods for an elliptic optimization problem with state constraints. Priority Programme 1253, Preprint SPP1253-23-02 (2008)
148. Xu, J., Zou, J.: Some nonoverlapping domain decomposition methods. Siam Rev. **40**, 857–914 (1998)
149. Yosida, K.: Functional Analysis. Springer, Berlin (1980)
150. Zowe, J., Kurcyusz, S.: Regularity and stability for the mathematical programming problem in Banach spaces. Appl. Math. Optim. **5**, 49–62 (1979)
151. Zeidler, E.: Nonlinear Functional Analysis and Its Applications. I, Fixed-Point Theorems. Springer, Berlin (1986)
152. Zeidler, E.: Nonlinear Functional Analysis and its Applications, vols. II/A, II/B, 1st edn. Springer, Berlin (1990)

MIX
Papier aus verantwortungsvollen Quellen
Paper from responsible sources
FSC® C105338

If you have any concerns about our products,
you can contact us on
ProductSafety@springernature.com

In case Publisher is established outside the EU,
the EU authorized representative is:
Springer Nature Customer Service Center GmbH
Europaplatz 3, 69115 Heidelberg, Germany

Printed by Libri Plureos GmbH
in Hamburg, Germany